GIScience and Geo-environmental Modelling

The "GIScience and Geo-environmental Modelling" book series seeks to publish a broad portfolio of scientific books addressing the interface between geography and the environment. The aim of the book series is to present geospatial technology approaches to data mining techniques, data analytics, modeling, risk assessment, visualization, and management strategies. The series includes peer-reviewed monographs, edited volumes, textbooks, and conference proceedings. The focus of Geo-environmental is on geospatial modelling in the frontier area of earth-environment related fields, such as urban and peri-urban environmental issues, ecology, hydrology, basin management, geohazards, estuarine-ecology, groundwater, agriculture, climate change, land-water, and forest resources, and related topics.

Geo-environmental modelling techniques have enjoyed an overwhelming interest in recent decades among the earth environmental and social sciences research communities for their powerful ability to solve and understand various complex problems and develop novel approaches toward sustainable earth and human society. Geo-environmental modelling using data mining, machine learning, and cloud computing technology is focused on spatiotemporal data analysis and modeling for sustainability in our environment.

Swapan Talukdar • Atiqur Rahman •
Somnath Bera • G. V. Ramana •
Atish Prashar
Editors

Environmental Risk and Resilience in the Changing World

Integrated Geospatial AI and Multidimensional Approach

Springer

Editors
Swapan Talukdar
Department of Geography
Asutosh College
University of Calcutta
Kolkata, West Bengal, India

Somnath Bera
Department of Geography
Central University of South Bihar
Gaya, Bihar, India

Atish Prashar
Department of Mass
Communication and Media
Central University of South Bihar
Gaya, India

Atiqur Rahman
Department of Geography
Jamia Millia Islamia
New Delhi, Delhi, India

G. V. Ramana
Department of Civil Engineering
Indian Institute of Technology Delhi
New Delhi, Delhi, India

ISSN 2730-7506 ISSN 2730-7514 (electronic)
GIScience and Geo-environmental Modelling
ISBN 978-3-031-62441-4 ISBN 978-3-031-62442-1 (eBook)
https://doi.org/10.1007/978-3-031-62442-1

This Springer imprint is published by the registered company Springer Nature Switzerland AG
The registered company address is: Gewerbestrasse 11, 6330 Cham, Switzerland

To our revered guru, the guiding light illuminating our paths of knowledge and wisdom, this book is lovingly dedicated.

Foreword

I am pleased and honoured to introduce the book *Environmental Risk and Resilience in the Changing World—Integrated Geospatial AI and Multidimensional Approach*. As I had the privilege of reviewing the contents of this remarkable publication, I am convinced of its significance and potential to contribute to the field of environmental sciences.

In today's rapidly changing world, our planet faces numerous risks and threats due to various factors such as land use changes, population growth, and pollution. These challenges significantly burden our ecosystems and their ability to sustain long-term service levels while supporting sustainable development across economic, social, and environmental aspects.

Risk assessment has emerged as a pivotal research topic and a powerful tool for environmental restoration and health improvement towards sustainable development. However, traditional risk assessment approaches alone are no longer sufficient. Scientists and researchers increasingly recognize the need for resilience strategies and advanced technologies such as artificial intelligence and geospatial analysis to address environmental risks effectively.

Environmental Risk and Resilience in the Changing World is a compilation of cutting-edge research and innovative methodologies highlighting the integration of risk assessment and resilience strategies. The book encompasses a multidisciplinary approach, incorporating disciplines such as environmental science, ecosystem services, environmental management, disaster management, remote sensing and GIS, and artificial intelligence.

The chapters presented in this book cover a wide range of topics, including spatiotemporal analysis of natural hazards, forest landscapes, floodplain wetlands, urban heat islands, micro-plastic pollution, water pollution, aerosol concentration, ecosystem health, eutrophication modelling, vulnerability assessment, and the impact of environmental risks on society. Each chapter is authored by leading experts in their respective fields, ensuring the highest quality of research and insights.

This book provides a comprehensive understanding of environmental risks and presents novel approaches to resilience and mitigation strategies. It showcases the power of geospatial technologies, artificial intelligence, and state-of-the-art frameworks in addressing complex environmental challenges. The knowledge and data shared within this book are valuable resources for academics, scientists, students, environmentalists, meteorologists, urban planners, remote sensing, and GIS experts.

I am confident that *Environmental Risk and Resilience in the Changing World* will significantly contribute to the geospatial, earth, environmental, humanities, and social sciences. Its findings and methodologies will inspire further research and foster collaboration among researchers, practitioners, and policymakers working towards a sustainable and resilient future.

I sincerely congratulate the editors and contributors for their outstanding efforts in compiling this valuable resource. I hope that readers find immense value in the content of this book and that it will stimulate innovative thinking and action to address the environmental challenges we face.

Prof. Thomas Oommen
Chair, Department of Geology
and Geological Engineering
University of Mississippi
Oxford, USA

Preface

Our world is undergoing rapid transformations, with the geosphere, biosphere, hydrosphere, atmosphere, and cryosphere experiencing immense pressure from various risks and threats. Global land use changes, population growth, and pollution are disrupting the delicate ecological balance, leading to vulnerabilities that jeopardize our environment and undermine sustainable development efforts.

Environmental Risk and Resilience in the Changing World takes a bold step in addressing these pressing issues by exploring innovative approaches to risk assessment and resilience strategies. The book brings together a distinguished group of contributors, each leading expert in their respective fields, covering environmental science, ecosystem services, environmental management, disaster management, remote sensing, GIS, artificial intelligence, and data science.

A notable strength of this book lies in its emphasis on the integration of geospatial technologies and artificial intelligence in the realm of risk assessment and mitigation. Leveraging cutting-edge frameworks like Pressure-State-Response (PSR), Vulnerability, Impact, and Option Assessment (VIOR), and AI-based solutions, the authors present compelling case studies and methodologies that offer invaluable insights into environmental risk management.

This book comprises four parts with twenty articles, each exploring critical aspects of the changing environmental landscape and its impact on society:

Part I: "A Changing Environmental Landscape: Threats, Risks, and Vulnerability" sets the foundation by presenting a comprehensive exploration of the challenges and strategies related to environmental risk and resilience. It further delves into specific issues such as land cover changes, forest degradation, groundwater arsenic contamination, and its impacts over the Indian subcontinent.

Part II: "Environmental Risk, Vulnerability, and Health Assessment: Methods, Innovations, and Technologies" sheds light on advanced methodologies for assessing environmental risk and vulnerability, including the use of artificial intelligence and machine learning to understand water pollution, evaluating factors influencing organizational sustainability, and examining landslide susceptibility in the Indian Himalayas.

Part III: "Impact of Environmental Risk on Society" examines the implications of environmental risks on various sectors, such as waste management practices in the apparel, textile, and leather industry, the intersection of COVID-19 and climate change, and the assessment of ecosystem health in specific regions.

Part IV: "Integrated Risk Mitigation and Resilience Measures" focuses on innovative strategies for mitigating environmental risks and enhancing resilience. Chapters in this section explore topics such as mangrove restoration, agricultural waste management, water quality assessment, urbanization impacts on land use and carbon dynamics, digital twin applications, and land resource impact assessment.

We extend our heartfelt gratitude to all the contributors who demonstrated unwavering dedication in completing their papers promptly. Their scholarly and diligent efforts have made a significant contribution to the creation of this enlightening and invaluable book. This multidisciplinary book is tailored to appeal to a wide range of readers, including academics, scientists, students, environmentalists, meteorologists, urban planners, remote sensing experts, and GIS specialists. With a plethora of cutting-edge research and practical insights, *Environmental Risk and Resilience in the Changing World* stands as a valuable resource for enhancing the understanding of environmental risk and resilience.

Our sincere thanks also go to the esteemed Springer Publishing House, Prof. Prabhat Kumar Shit, Springer Series Editor, and Doris Bleier, along with her highly efficient team. Their utmost professionalism and collaborative efforts have been instrumental in ensuring the timely publication of this book, and we are immensely grateful for their support and assistance.

Kolkata, India — Dr. Swapan Talukdar
New Delhi, India — Prof. Atiqur Rahman
Gaya, India — Dr. Somnath Bera
New Delhi, India — Prof. G. V. Ramana
Gaya, India — Prof. Atish Prashar

Acknowledgements

The preparation and successful completion of this volume would not have been possible without the unwavering support and contributions of numerous individuals and organizations, to whom we extend our heartfelt gratitude. First and foremost, we express our sincere appreciation to all the authors who wholeheartedly embraced and supported the original idea behind this book. Your dedication, expertise, and commitment to advancing the field of environmental management have been instrumental in shaping the contents of this volume. Your valuable contributions have enriched the discourse and knowledge in this area, and we are deeply grateful for your involvement. We also extend our gratitude to the anonymous reviewers whose diligent efforts and insightful comments played a pivotal role in elevating the quality and rigour of the work presented in this book. Your scholarly critiques and thoughtful suggestions have undoubtedly contributed to refining the ideas and findings contained within these pages. We acknowledge the time you devoted to reviewing the chapters, and we are truly grateful for your invaluable input.

Furthermore, we would like to express our heartfelt thanks to Mr. Shivam Priyadarshi, and Ms. Pritha Ghosh for their invaluable editorial assistance throughout the publishing process. Your attention to detail, meticulous editing, and dedication to ensuring the clarity and coherence of the content have been indispensable in bringing this book to fruition. Our sincere appreciation also goes to the entire Springer team for their continuous and unwavering support throughout this journey. Your timely provision of information, guidance, and encouragement at every stage of this project has been invaluable. We extend our gratitude to all the individuals who worked tirelessly behind the scenes to make this publication possible, from the editorial and production teams to the marketing and distribution personnel.

Finally, we wish to express our hope that this book's publication will serve as a catalyst for inspiring students and researchers to delve into more in-depth investigations and analyses of environment-human interaction. As global urbanization and environmental changes continue to accelerate, and with an ever-growing interest in remotely sensed big data, the field of environmental risk management has emerged as a dynamic and critical area of study. May this volume ignite curiosity and enthusiasm in contributing to global environment observation and sustainable development goals.

Once again, we extend our deepest gratitude to all those who played a part in making this endeavour a reality. Your support and encouragement have been a source of inspiration, and we are genuinely thankful for your invaluable contributions.

Contents

Editors and Contributors

About the Editors

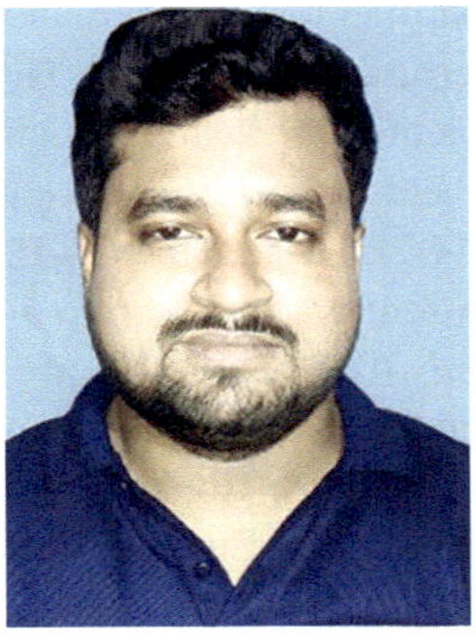

Dr. Swapan Talukdar is an assistant professor in the Department of Geography, Asutosh college, University of Calcutta, Kolkata, India. He has worked as Dr. D. S. Kothari postdoctoral fellow in the Department of Geography, Faculty of Natural Sciences, Jamia Millia Islamia, New Delhi. He completed his Ph.D. from the Department of Geography, University of Gour Banga, West Bengal, India, in 2020. He served as the junior research fellow (JRF) during 2015–17 and senior research fellow (SRF) during 2017–19 in the Department of Geography, University of Gour Banga. He has completed two international projects (as Co-PI) in collaboration with the King Khalid University, Saudi Arabia. He has published more than 105 scientific research articles in the journals of international repute such as *Journal of Cleaner Production*, *Environmental Pollution*, *Scientific Reports*, *Ecological Indicators*, *Sustainable Cities and Society*, and *Remote Sensing*. He has also published one edited book in Springer Nature. His main areas of research are urban environment, urban wetlands, natural hazards, remote sensing, and AI and their applications in environmental management. He has served as the reviewer for several reputed international journals.

Google Scholar: https://scholar.google.co.in/citations?user=e9MgpZkAAAAJ&hl=en

Prof. Dr. Atiqur Rahman is a professor of Geography in the Department of Geography, Faculty of Natural Sciences, Jamia Millia Islamia, New Delhi. He completed B.Sc. (Hons.), M.Sc., M.Phil., and Ph.D. from Aligarh Muslim University, Aligarh. His research interest is urban environmental management, water resources, and use of remote sensing satellite data, GIS and GPS. He was a post-doctoral fellow (PDF) at UFZ-Centre for Environmental Research, Leipzig, Germany (1999). He was the Co-PI of Indo-Germany DST-DAAD major project. He is the recipient of prestigious Young Scientist Project Grant Award (2001–2004) from the Department of Science and Technology (DST), Government of India. He worked as a member of Scientific Research Team of NASA funded UEM project on *Urban Ecology and Sustainability* (2004–2007). He was the Co-PI Indo-Canadian major research projects on *Economic Transformation and Childhood Obesity* funded by ICMR (India) and CHI (Canada) during 2010–2015. He was the Co-PI of a major research project funded by the Ministry of Environment and Forest, Government of India. He was the collaborating scientist of another NASA funded major project on *Impacts of Desert Urbanization on Climate using Remote Sensing and Numerical Modeling* (2012–2015). He has also done professional development initiative training on *Climate Change and Its Impacts: Resilience and Adaptation to Changes in Precipitation*, at Brown University USA (2010). He served as a Steering Committee (SC) member (2009–2012) of Population and Environment Research Network (PERN), CISIN, Columbia University USA. He was the advisor-technical cum consultant on Remote Sensing and GIS for the project (2013–18) in the PHFI and AIIMS, and Emory University, USA project. He also served as a member, Governing Body (GB), of some colleges of University of Delhi as a University Nominee. He was the faculty coordinator, Inspired Teachers In-Residence Programme organized by the Office of the President of India at Rashtrapati Bhavan, New Delhi, 6–12 June 2015. He was a honorary deputy director, (Academic) of Jamia Millia Islamia, New Delhi (2015–2018). He is the coordinator, UGC-DRS SAP-I, Special Assistance

Programme (SAP-I), Department of Geography, JMI. Ten students have been awarded Ph.D. degrees under his supervision. He has published 15 books and 150 peer-reviewed research papers in the journals of international repute such as applied geography, ecological indicators, urban climate, and nature's scientific reports. He has widely travelled across the globe. He is among the top 2% Global Elite Scientists in a study by the University of Stanford, USA, published in the journal *PLOS Biology in 2020 and 2021*.

Google Scholar: https://scholar.google.co.in/citations?user=CWcZvxMAAAAJ&hl=en

Research Gate: https://www.researchgate.net/profile/Atiqur-Rahman-36

Dr. Somnath Bera is currently an assistant professor in the Department of Geography, Central University of South Bihar, India. He did his M.Phil. and Ph.D. in Disaster Studies from the Tata Institute of Social Sciences, Mumbai. After his Ph.D., he worked as a research associate at the Department of Civil Engineering, Indian Institute of Technology Delhi. His research area focuses on disaster risk and resilience in the changing environment. Throughout his academic journey, he explored disaster affected region across India, Nepal, China, and Japan. He received several prestigious fellowships, grants, and awards. He published numerous research articles in journals of international repute, such as *International Journal of Disaster Risk Reduction*, *Natural Hazards*, *Bulletin of Engineering Geology and the Environment*, *Progress in Disaster Science*, *Remote Sensing Applications: Society and Environment*, *Human and Ecological Risk Assessment: An International Journal*, *Journal of Cleaner Production*. Additionally, he serves as a reviewer for prestigious international journals published by "Elsevier", "Springer", and "Taylor and Francis". Beyound academia, he finds joy in landscape painting.

Google Scholar: https://scholar.google.com/citations?user=g0fSpo0AAAAJ&hl=en

Prof. Dr. G. V. Ramana is a professor (Higher Administrative Grade) of geotechnical engineering and obtained masters and doctoral degrees from Rensselaer Polytechnic Institute, Troy, New York, and has been teaching at IIT Delhi for the last 25 years. His areas of interest are (i) geoenviornmental engineering, (ii) geoenvironmental risk, (iii) ground improvement, and (iv) geotechnical earthquake engineering. He was one of the two co-authors of the textbook *Principles of Soil Dynamics*, (published by CENGAGE in January 2010). He was deputed to the Asian Institute of Technology, Bangkok, to teach a course on *soil dynamics and earthquake engineering*. Has guided 19 Ph.D. thesis, one M.S. thesis, and 60 M.Tech. theses primarily in the area of geotechnical/geoenvironmental engineering and documented the research efforts in various reputed journals and conferences. He had been a principal investigator for several research projects. He is actively associated and contributed to the following dam projects: Salma Dam, Afghanistan; Polavaram Irrigation Project, Andhra Pradesh; TLDP Stage 3 Project, Tamanthi HEP, several ash ponds and tailings dams. He is been a member of several panels of Expert Committee Members. He is associated with over 100 consultancy projects dealing with specification of geotechnical investigations, recommendation of bearing capacity of shallow and deep foundations, soil-structure interaction of high-rise buildings, design of deep excavations, vibration assessment and isolation, liquefaction assessment and mitigation, testing of geosynthetics, design of ash ponds and landfills, and slope instability assessment and remedial measures. He also peer reviews design of RE walls and RE Slopes. He actively disseminates geotechnical expertise through invited lectures at short courses, training programmes for graduate engineers, and at professional organizations.

Google Scholar: https://scholar.google.com/citations?user=c_jMH2MAAAAJ&hl=en

Research Gate: https://www.researchgate.net/profile/R-Gunturi

Prof. Atish Prashar is a leading newspaper columnist and a well-known media educator with expertise in all media platforms. He was the dean of the School of Media, Arts, and Aesthetics and worked relentlessly as the head of the Department of Mass Communication at the Central University of South Bihar (CUSB). He made his entry in academia in the year 2002 and since then has been associated with Delhi University, IP University, and many other institutions of excellence in the country. He believes in maintaining high standards in media education which has led him to make significant contributions in various universities and institutes as a selection committee member also he is contributing in Bihar State University Service Commission as a subject expert. He is a well-known trainer and associated with Bihar Institute of Public Administration and Rural Development (BIPARD) and Bihar Police Academy. He has been training participants from Indian Administrative Services and Bihar Administrative Services. He is a prolific writer with many reputed publications to his credit with an authorship of five books, 12 chapters, and 30 research papers. He is currently leading a major project as a project director at CUSB, funded by ICSSR. He has been closely worked with NIDM and organized many workshops on disaster management. Quite often his editorials appear in reputed newspapers and magazines. He is a member of various Board of Studies like Central University of Jharkhand, Assam University, CUSB, Patna University, and Amity University. He is also an academic council member at CUSB.

Contributors

Isam Mohammed Abdel-Magid Dean Graduate College & Scientific Studies, Al-Razi University, Khartoum, Sudan

Shukla Acharjee Centre for Studies in Geography, Dibruragh University, Dibrugarh, Assam, India

Mansoor Ahmad Department of Geography, University of Jammu, Jammu, Jammu & Kashmir, India

Shakil Ahmad Directorate of Library Affairs, Imam Abdulrahman Bin Faisal University, Dammam, Saudi Arabia

Syed Naushad Ahmad Department of Geography, Faculty of Science, Aligarh Muslim University (Central University), Aligarh, Uttar Pradesh, India

Ishita Afreen Ahmed Department of Geography, Faculty of Natural Sciences, Jamia Millia Islamia, New Delhi, India;
Department of Civil Engineering, Indian Institute of Technology, New Delhi, India

Shafiq Ahmed Department of Geography, University of Jammu, Jammu, Jammu & Kashmir, India

Md Kaikubad Ali Department of Geography, Aligarh Muslim University, Aligarh, India

R. B. Archana Centre for Geoinformatics, Jamsetji Tata School of Disaster Studies, Tata Institute of Social Sciences, Deonar, Mumbai, India

Md. Sarfaraz Asgher Department of Geography, University of Jammu, Jammu, Jammu & Kashmir, India

Enayat Bano Department of Geography, Faculty of Science, Aligarh Muslim University (Central University), Aligarh, Uttar Pradesh, India

Somnath Bera Department of Geography, Central University of South Bihar, Gaya, India

Das Ambika Bharti Department of Psychological Sciences, Central University of South Bihar, Gaya, Bihar, India

Rajib Tarani Das Centre for Studies in Geography, Dibruragh University, Dibrugarh, Assam, India

Somen Das Department of Geography, Kazi Nazrul University, Asansol, West Bengal, India

Tanmoy Das Department of Geography, Faculty of Natural Sciences, Jamia Millia Islamia, New Delhi, India

Mrinalendra Narayan Dutta Department of Applied Geology, Dibrugarh University, Dibrugarh, Assam, India

Shahab Fazal Department of Geography, Aligarh Muslim University, Aligarh, India

Soumen Ghosh School of Management and Labour Studies, Tata Institute of Social Sciences (TISS), Mumbai, India

Balamurugan Guru Department of Geology, School of Earth Sciences, Central University of Tamil Nadu, Thiruvarur, India

Habibullah Department of Geography, University of Jammu, Jammu, Jammu & Kashmir, India

Aznarul Islam Department of Geography, Aliah University, Kolkata, West Bengal, India

Pasarul Islam Indian Council of Social Science Research (ICSSR), Ministry of Education, Government of India, New Delhi, India

Chetna Jaiswal Department of Psychological Sciences, Central University of South Bihar, Gaya, Bihar, India

Rumki Khatun Department of Geography, Kazi Nazrul University, Asansol, West Bengal, India

Narsingh Kumar Department of Psychology, University of Allahabad, Allahabad, India

Naveen Kumar Department of Geography, University of Jammu, Jammu, Jammu & Kashmir, India

Roshan Kumar Department of Commerce and Business Studies, Central University of South Bihar, Gaya, India

Shishu Kesh Kumar Department of Psychological Sciences, Central University of South Bihar, Gaya, Bihar, India

Unnikannan P. Santhosh Kumar Department of Psychological Sciences, Central University of South Bihar, Gaya, Bihar, India

Manisha Kumari Department of Geography, University of Jammu, Jammu, Jammu & Kashmir, India

K. N. Kusuma Department of Earth Sciences, Pondicherry University, Kalapet, Puducherry, India

Kaustubh Lahiri Department of Geography, Dr. Kanailal Bhattacharyya College, Howrah, West Bengal, India

Srija Mitra Centre for Economic and Social Studies (CESS), Hyderabad, India

Pradheeps Muthulingam Chennai Metropolitan Development Authority (CMDA), Egmore, Chennai, India

Thomas Oommen Department of Geology and Geological Engineering, University of Mississippi, Oxford, MS, USA

Atish Prashar Department of Mass Communication and Media, Central University of South Bihar, Gaya, India

Abdur Rahman Department of Geography, Dr. Kanailal Bhattacharyya College, Howrah, West Bengal, India

Atiqur Rahman Department of Geography, Faculty of Natural Sciences, Jamia Millia Islamia, New Delhi, India

Pradip Ram Department of Commerce and Business Studies, Central University of South Bihar, Gaya, India

G. V. Ramana Department of Civil Engineering, Indian Institute of Technology Delhi, Hauz Khas, New Delhi, India

Mohd. Rihan Department of Geography, Faculty of Natural Sciences, Jamia Millia Islamia, New Delhi, India

Yamini Sahu Department of Data Science and Engineering, IISER, Bhopal, India

Lohra Salew Centre for Geo-Informatics, Tata Institute of Social Sciences, Mumbai, India

Biplab Sarkar Department of Geography, Aliah University, Kolkata, West Bengal, India

Sekh Sayeed Department of Geography, Kazi Nazrul University, Asansol, West Bengal, India

Shahfahad Department of Geography, Faculty of Natural Sciences, Jamia Millia Islamia, New Delhi, India

Sunita Singh Department of Geography, Central University of South Bihar, Gaya, India

Rajani Srivastava Department of Environmental Science (Environmental Technology), RGSC, Institute for Environment & Sustainable Development, Banaras Hindu University, Varanasi, India

Swapan Talukdar Department of Geography, Asutosh College, University of Calcutta, Kolkata, India

Neha Thappa Department of Geography, University of Jammu, Jammu, Jammu & Kashmir, India

Vaibhav Kumar Upadhyay Department of Data Science and Engineering, IISER, Bhopal, India

Part I

A Changing Environmental Landscape: Threats, Risks, and Vulnerability

1 Environmental Risk and Resilience in a Changing World: A Comprehensive Exploration and Interplay of Challenges and Strategies

Swapan Talukdar, Atiqur Rahman, Somnath Bera, G. V. Ramana, and Atish Prashar

Abstract

Environmental risk is a fundamental concept in environmental economics, encompassing potential adverse impacts on the environment and human well-being due to various factors. The escalating challenges posed by climate change, biodiversity loss, pollution, and resource depletion highlight the urgency of understanding and managing these risks for sustainable development and decision-making. The intricate interconnections between human activities and the environment in the modern world lead to severe consequences, including hazardous chemicals, pollution, habitat destruction, climate change, and ecosystem disruption. As a result, environmental risks have significant implications for human health, economies, ecosystems, and biodiversity. Addressing threats like air pollution, climate change, and biodiversity loss requires effective strategies to mitigate negative consequences and foster sustainable development. Quantifying environmental risks is vital for comprehension and mitigation. Advanced methodologies like machine learning are often employed to analyze the complex relationships between environmental factors and risks. Additionally, resilience, the ability of systems and communities to withstand and recover from disturbances, plays a crucial role in responding to environmental risks. This chapter aims to explore the definition and types of environmental risk, governance, resilience pursuit, and recent trends in research on environmental risk. It investigates the interplay between risk and resilience to understand how systems and societies respond to environmental challenges. By gaining a comprehensive understanding of environmental risks and their implications, we can better inform strategies for sustainable development and safeguarding human well-being and the environment.

S. Talukdar
Department of Geography, Asutosh College, University of Calcutta, Kolkata, India

A. Rahman (✉)
Department of Geography, Faculty of Sciences, Jamia Millia Islamia, New Delhi 110025, India
e-mail: arahman2@jmi.ac.in

S. Bera
Department of Geography, Central University of South Bihar, Gaya 824236, India

G. V. Ramana
Department of Civil Engineering, Indian Institute of Technology Delhi, Hauz Khas, New Delhi 110016, India

A. Prashar
Department of Mass Communication and Media, Central University of South Bihar, Gaya 824236, India

S. Talukdar et al. (eds.), *Environmental Risk and Resilience in the Changing World*, GIScience and Geo-environmental Modelling, https://doi.org/10.1007/978-3-031-62442-1_1

Keywords

Environmental risk · Resilience · Climate change · Biodiversity loss · Pollution · Sustainable development

1.1 Introduction

Environmental risk is a fundamental concept in environmental economics and risk assessment, encompassing the potential for adverse impacts on the environment and human well-being due to various factors and activities. As our planet confronts escalating challenges related to climate change, biodiversity loss, pollution, and resource depletion, understanding and managing environmental risks have become paramount for sustainable development and effective decision-making.

The contemporary world is characterized by intricate interconnections between human activities and the environment. Economic development, industrialization, urbanization, and technological advancements have yielded significant benefits, but they have also given rise to environmental risks that can have severe consequences. These risks manifest as hazardous chemicals, air and water pollution, habitat destruction, climate change, and ecosystem disruption. The uncertainty and potential magnitude of these risks pose substantial challenges for policymakers, scientists, businesses, and society as a whole.

In this edited volume, titled 'Environmental Risk and Resilience in a Changing World,' we delve into a comprehensive exploration of environmental risk, with a specific focus on its quantification, governance, and the pursuit of resilience. Our aim is to provide a contemporary and robust understanding of environmental risks and their implications for present and future generations.

Quantifying environmental risks is a pivotal step in comprehending their potential impacts and developing effective mitigation strategies. Various methodologies and algorithms are employed to analyze and model the complex relationships between environmental factors and associated risks (Fig. 1.1). We have search on Google Scholar the word 'environmental risk using machine learning' and collected 52 articles. Then we analyzed the title of all literature and generated word cloud (Fig. 1.1), which shows different environmental issues have been researched with the help of recent machine learning algorithms.

Notably, regression models such as multiple regression, logistic regression, and geographically weighted regression have been utilized to predict continuous or binary outcomes concerning climate change (Jha et al. 2023; Passow and Donner 2020), biodiversity loss (Powell and Lenton 2013), water scarcity and quality (Madani et al. 2022), air pollution (Jiang 2021), natural hazards (Bera et al. 2021a), land degradation (Abdel-Kader 2019), and other environmental risks. Neural networks, including artificial neural networks (ANN) and deep learning models like convolutional neural networks (CNNs) and recurrent neural networks (RNN), offer significant promise in image analysis, time series forecasting, and predicting environmental changes. They have been applied to understand and address risks associated with climate change, biodiversity loss, water scarcity and quality (Wang et al. 2013; Talukdar et al. 2023; Mallik et al. 2023), air pollution (Chen and Zheng 2022), natural hazards (Bera et al. 2021b), and land degradation and soil erosion (Ahmed et al. 2023). Clustering algorithms, such as k-means and hierarchical clustering, are employed to identify spatial clusters of environmental risks or segment areas based on shared characteristics. These algorithms have contributed to the understanding of risk patterns related to climate change, biodiversity loss, water scarcity and quality (Du et al. 2017), air pollution, natural hazards (Tang et al. 2020; Wu et al. 2021), land degradation, and various other environmental risks.

To truly grasp the significance of environmental risks, it is essential to explore the concept of resilience. Resilience refers to the capacity of systems and communities to withstand, adapt to, and recover from disturbances or shocks. As environmental risks continue to intensify, building resilience becomes a critical aspect of maintaining ecological balance and human well-being. In this book, we seek to examine the interplay between environmental risk and resilience,

Fig. 1.1 Word cloud for showing priority research topic with recent techniques

understanding how resilient systems and societies respond to and recover from environmental challenges. Therefore, the primary objectives of this edited book are multifaceted and aimed at promoting sustainability and resilience in a rapidly changing world. They include:

Understanding the Complexity of Environmental Risks: This book endeavors to shed light on the complex and interconnected nature of environmental risks. By exploring various risk assessment models, algorithms, and case studies, we aim to provide a comprehensive understanding of the diverse challenges faced by our environment.

Advancing Proactive Risk Governance: Effective risk governance is crucial for managing environmental risks. Drawing from Renn's (2008) advocacy for a proactive approach to risk governance, we aim to highlight the importance of stakeholder engagement, transparent decision-making, and collaborative governance structures in addressing environmental challenges.

Exploring Risk Perception and Communication: Human perception of environmental risks plays a pivotal role in shaping individual and collective responses. We will draw on Slovic's (1999) insights to examine the intricate relationships between trust, emotions, politics, science, and risk assessment, and its impact on the decision-making process.

Promoting Sustainable Development: Integrating environmental risk considerations into development plans is essential for fostering sustainable practices. By showcasing successful examples of sustainable development, we seek to underscore the importance of harmonizing economic progress with environmental well-being.

Fostering Resilience: Resilience is a critical aspect of responding to environmental risks. Through case studies and research findings, we will explore how resilience can be cultivated at the individual, community, and ecosystem levels.

1.2 Types of Environmental Risks

Environmental risks encompass a wide array of potential hazards that can adversely impact the environment and human societies. These risks can be broadly classified into four main categories: pollution, biodiversity loss, climate change, and natural disasters.

1.2.1 Pollution

Pollution refers to the introduction of harmful substances or pollutants into the environment, leading to degradation and negative health effects. Air pollution, water pollution, and soil contamination are prominent examples of this category.

Air Pollution: The burning of fossil fuels, industrial activities, and transportation emit various pollutants into the atmosphere, such as particulate matter, nitrogen oxides, sulfur dioxide, and volatile organic compounds. These pollutants contribute to smog formation, respiratory diseases, and climate change (IPCC 2013). Sorek-Hamer et al. (2013) conducted a study on Delhi's air pollution using remote sensing techniques. They utilized satellite remotely sensed aerosol products to classify dust days, providing crucial information for monitoring air pollution levels and assessing its potential health impacts. The study highlighted the significance of remote sensing in tracking air pollution patterns and aiding policymakers in implementing effective air quality management measures (Somvanshi et al. 2019). Similarly, Ju et al. (2022) conducted a case study in Lanzhou City, monitoring air pollution using remote sensing data from 2010 to 2019. Their research further emphasized the importance of remote sensing in analyzing air pollution trends and their implications for public health and the environment (Ju et al. 2022).

Water Pollution: Discharge of untreated industrial waste, agricultural runoff, and improper waste disposal contaminate water bodies, threatening aquatic ecosystems and human health. The degradation of water quality can lead to eutrophication, algal blooms, and the presence of toxic substances (White 2018).

Soil Contamination: Improper use of pesticides, fertilizers, and industrial activities can contaminate soil, making it unsuitable for agriculture and posing risks to ecosystems and groundwater resources (Daily and Ehrlich 1992).

1.2.2 Biodiversity Loss

Biodiversity loss refers to the decline in the variety and abundance of living organisms in different ecosystems. Human activities, such as deforestation, habitat destruction, overexploitation of resources, and introduction of invasive species, significantly contribute to this environmental risk (IPCC 2013; UNEP 2020). Dewan and Yamaguchi (2009) conducted a case study in Dhaka Metropolitan, Bangladesh, using remote sensing and GIS technologies to detect and monitor land use and land cover changes. The study revealed the significant loss of natural habitats due to urban expansion, highlighting the urgency of conservation efforts to mitigate biodiversity loss (Dewan and Yamaguchi 2009). Another study by Abdi et al. (2021) utilized Google Earth Engine to conduct a spatio-temporal analysis of air pollutants over Turkey before and during the first wave of the COVID-19 outbreak. The integration of remote sensing data enabled a comprehensive assessment of air pollution levels, aiding in understanding the impacts of human activities on air quality and the environment (Ghasempour et al. 2021).

1.2.3 Climate Change

Climate change is a global environmental risk resulting from human-induced greenhouse gas emissions, primarily from burning fossil fuels and deforestation. Climate change can alter rainfall patterns, causing more frequent and intense droughts or floods, impacting agriculture and water availability. Hoegh-Guldberg et al. (2019) emphasized the urgent need to stabilize global climate change at 1.5 °C through their study on the impacts of climate change on coral reefs and marine ecosystems. The research demonstrated the critical role of remote sensing in monitoring and assessing climate-related changes in marine environments

(Hoegh-Guldberg et al. 2019). The warming climate increases the frequency and intensity of extreme weather events, such as hurricanes, cyclones, and wildfires, posing significant risks to human lives and infrastructure (IPCC 2014). Additionally, Liu et al. (2021) investigated the influences of climate change and human activities on normalized difference vegetation index (NDVI) changes in China using remote sensing data. Their study highlighted the value of remote sensing in understanding climate change's ecological impacts and its implications for terrestrial ecosystems (Liu et al. 2021).

1.2.4 Natural Hazards and Disasters

Disasters are the exposition of environmental risks caused by natural processes, such as earthquakes, tsunamis, volcanic eruptions, and hurricanes. Earthquakes result from tectonic plate movements, often triggering tsunamis that can devastate coastal regions (White 2018). Volcanic activity releases ash, gases, and lava, posing risks to local ecosystems, air quality, and human settlements (White 2018). These powerful storms can cause extensive damage to coastal regions through strong winds, storm surges, and heavy rainfall (IPCC 2014). Wildfires are often sparked by natural causes or human activities and can have severe impacts on ecosystems, air quality, and human health (White et al. 1996). Taubenböck et al. (2011) utilized multi-sensoral remote sensing approaches to assess flood risks in urbanized areas, enabling better disaster preparedness and risk assessment. The integration of remote sensing data provided valuable insights into flood-prone areas, supporting disaster management and response efforts (Taubenböck et al. 2011). Teodoro and Duarte (2022a, b) highlighted the synergy of remote sensing and GIS in the management of natural disasters. Their review emphasized the pivotal role of remote sensing technologies in disaster response, damage assessment, and continuous monitoring of affected areas (Teodoro and Duarte 2022a, b).

Therefore, these case studies demonstrate the significance of remote sensing and GIS technologies in monitoring and understanding environmental risks, ranging from air pollution and biodiversity loss to climate change and natural disasters. These technologies play a crucial role in providing accurate and timely information for informed decision-making, enabling effective risk management, and supporting sustainable development practices to protect the environment and promote resilience.

As researchers continue to explore the potential of remote sensing and GIS technologies, the application of these tools in environmental risk assessment and management will undoubtedly contribute to better environmental conservation and sustainable development efforts. By utilizing advanced remote sensing techniques, scientists can enhance their understanding of complex environmental processes and contribute to the protection of our planet's health and well-being (Rocchini et al. 2019). This interdisciplinary approach, integrating remote sensing with ecological studies and biodiversity conservation, holds great promise for advancing environmental science and ensuring a sustainable future for generations to come (Pettorelli et al. 2014). The combination of satellite imagery, aerial photography, and ground-based sensors allows for comprehensive and continuous monitoring of environmental changes at various spatial and temporal scales (Piao et al. 2019). For instance, remote sensing can be used to assess the impact of land use changes on biodiversity in the Western Ghats (Menon and Bawa 1997) and to monitor forest biodiversity and changes in spectral diversity (Kacic and Kuenzer 2022). Moreover, remote sensing technology has proved invaluable in monitoring climate change impacts on vegetation phenology and productivity (Piao et al. 2019).

With the growing threat of climate change and its far-reaching impacts on ecosystems and human societies, remote sensing has become a powerful tool for assessing the extent and severity of environmental risks related to climate change (Cui et al. 2019). For example, thermal remote sensing and spectral mixture analysis have been employed to examine the relations between urban heat island effects, land use, and air pollution (Feizizadeh and Blaschke 2013). These studies provide critical insights for urban planning and climate change mitigation strategies. Furthermore, remote sensing techniques have been applied to study the effects of climate change on water resources, such as changes in precipitation patterns and water availability (Cui et al. 2019). These findings can help inform sustainable water resource management and adaptation strategies.

In agriculture, remote sensing has proven to be instrumental in monitoring crop health, estimating crop yields, and assessing the impacts of climate change on agricultural productivity (Jung et al. 2021). By combining satellite imagery with artificial intelligence, researchers can improve the resilience of agricultural production systems and optimize resource use (Jung et al. 2021). Additionally, remote sensing technologies have been employed to assess soil carbon sequestration and monitor changes in soil properties (Mandal et al. 2022). These insights are crucial for promoting sustainable agricultural practices and mitigating greenhouse gas emissions.

Another critical application of remote sensing lies in disaster management and monitoring natural hazards (Teodoro and Duarte 2022a, b). Remote sensing technologies enable rapid and accurate assessment of damages caused by disasters, such as earthquakes, floods, and wildfires (Teodoro and Duarte 2022a, b). The timely information provided by remote sensing can facilitate effective disaster response and relief efforts, saving lives and minimizing damages.

Overall, the applications of remote sensing in environmental risk assessment are diverse and far-reaching. Remote sensing technologies provide valuable data and insights for understanding environmental processes, assessing risks, and developing strategies for sustainable resource management and conservation. As these technologies continue to advance, they will play an increasingly critical role in environmental science, policy-making, and efforts to safeguard the planet's health and biodiversity.

1.3 Impacts of Environmental Risks

Environmental risks have profound consequences on human health, economies, ecosystems, and biodiversity. Understanding these impacts is crucial for devising effective strategies to mitigate the negative consequences and foster sustainable development.

1.3.1 Impacts on Human Health

Environmental risks, such as air pollution, climate change, and emerging pathogens, have significant implications for human health. Air pollution, a major environmental risk, is associated with respiratory and cardiovascular diseases, leading to increased morbidity and mortality (Kim et al. 2014). Climate change exacerbates heat-related illnesses, vector-borne diseases, and food and waterborne infections, posing additional health challenges (Kim et al. 2014). Emerging pathogens, often associated with biodiversity loss, can lead to zoonotic diseases that threaten human populations (Schmeller et al. 2020). Addressing environmental risks is essential to safeguarding human health and well-being.

1.3.2 Impacts on Economies

Environmental risks can exert considerable economic burdens on societies. Marine plastic pollution, for instance, incurs substantial costs in terms of beach clean-up, waste management, and damage to marine ecosystems (Beaumont et al. 2019). Loss of biodiversity and ecosystems due to climate change can affect industries like agriculture, forestry, and fisheries, compromising food security and livelihoods (Pecl et al. 2017). The economic implications of environmental risks highlight the urgency of sustainable resource management and transitioning toward a green economy (Gasparatos et al. 2017).

1.3.3 Impacts on Ecosystems and Biodiversity

Environmental risks, particularly those associated with climate change and habitat loss, pose significant threats to ecosystems and biodiversity. Changes in temperature and precipitation patterns disrupt ecosystems, affecting species distributions and phenology (Pecl et al. 2017). Loss of wetlands, an essential ecosystem, negatively impacts water quality, flood regulation, and habitat for numerous species (Horwitz and Finlayson 2011). The impact of urban growth on biodiversity underscores the importance of integrating conservation efforts with urban planning (McDonald et al. 2020). Biodiversity is a vital component of ecosystem functioning, and its decline can lead to ecological imbalances with far-reaching consequences (Pereira et al. 2022).

1.3.4 Direct and Indirect Effects

The impacts of environmental risks can be both direct and indirect. For instance, the Russian–Ukrainian war has direct consequences on the environment, such as habitat destruction and pollution, affecting ecosystems and biodiversity (Pereira et al. 2022). Indirectly, the conflict can disrupt access to essential ecosystem services, leading to negative consequences for human well-being (Pereira et al. 2022). Similarly, the COVID-19 pandemic has had both direct socio-economic losses and indirect environmental gains. While the pandemic led to economic downturns, it also resulted in reduced greenhouse gas emissions and improved air quality (Lenzen et al. 2020). Understanding these direct and indirect effects is essential for comprehensive risk assessment and management.

Environmental risks have a cascading effect on various components of the Earth's systems, from the intrinsic value of biodiversity to the economic value of ecosystem services (Costanza et al. 2014). Biodiversity loss, driven by environmental risks, can lead to the loss of vital ecosystem services, such as pollination, nutrient cycling, and climate regulation (Costanza et al. 2014). These ecosystem services are fundamental for human well-being and sustainable development. Furthermore, the burden of disease attributable to environmental determinants highlights the significance of addressing environmental risks to reduce the global burden of disease (Prüss-Ustün et al. 2016).

Therefore, environmental risks have wide-ranging and profound impacts on human health, economies, ecosystems, and biodiversity. From the direct health effects of air pollution to the economic consequences of marine plastic pollution, understanding these impacts is essential for devising strategies to manage and mitigate environmental risks. Moreover, the interconnectedness of environmental risks with human societies and natural systems underscores the importance of adopting a holistic approach to address these challenges. By promoting sustainable practices and conservation efforts, we can safeguard human well-being, protect ecosystems, and ensure a resilient and thriving planet for future generations.

1.4 Current Strategies for Managing Environmental Risks

There are a multitude of strategies for managing environmental risks. This includes both systemic interventions and grassroots approaches, integrating both technological innovations and traditional methods. All these strategies aim to combat a variety of environmental issues, from pollution and biodiversity loss to climate change and natural disasters.

1.4.1 Policy Interventions

Policies and regulatory frameworks play a critical role in environmental risk management. These can range from international agreements to local ordinances that address specific environmental risks (Burby et al. 2013). Adger et al. (2005) discuss how policy interventions across scales can support successful adaptation to climate change, highlighting the importance of regulatory measures that promote resilience and adaptation.

1.4.2 Technology-Based Solutions

Technological advancements provide powerful tools for monitoring, predicting, and mitigating environmental risks (Abid et al. 2021; Munawar et al. 2022). Remote sensing technologies and Geographic Information Systems (GIS), for example, have revolutionized the way we assess and respond to various environmental risks, from forest fires (Xu et al. 2016; Naderpour et al. 2021) to coastal hazards (Youssef et al. 2021). More recently, the advent of disruptive technologies such as drones (Koutalakis et al. 2021) and artificial intelligence (Table 1.1) have expanded the capabilities in environmental risk management, offering greater efficiency and precision.

1.4.3 Conservation Efforts

Nature-based solutions are increasingly recognized for their potential to address environmental risks while providing multiple social, economic, and ecological co-benefits (Raymond et al. 2017; Ruckelshaus et al. 2020; Singh et al. 2020). Such solutions can include ecosystem restoration and conservation initiatives, as well as the integration of natural elements into urban environments (Pei et al. 2021; Branny et al. 2022; Li et al. 2022). Recent studies have emphasized the role of stakeholder engagement in implementing successful nature-based solutions, underscoring the importance of local involvement and co-management (Ferreira et al. 2020; Cundy et al. 2013).

1.4.4 Community-Based Initiatives

The active involvement of local communities in environmental risk management is increasingly recognized as a key component of successful adaptation and mitigation strategies (Reid et al. 2010). This involves empowering local communities to take action in response to the specific environmental risks they face, from adapting to climate change (Mercer et al. 2012; Sultana and Mallick 2015) to managing the aftermath of natural disasters (Dasan et al. 2022). As Amaru and Chhetri (2013) note, flexible and participatory approaches are needed to enhance institutional responses to environmental constraints.

1.4.5 Strengths and Weaknesses of Current Strategies

While current strategies for managing environmental risks offer a range of potential benefits, they also face significant challenges. For instance, the success of policy interventions often depends on their alignment with local contexts and needs, as well as their integration across different scales of governance (Adger et al. 2005; Burby et al. 2013). Technology-based solutions, while offering increasing precision and efficiency, require resources and technical capacities that may not be available in all regions or communities (Khan et al. 2020; Munawar et al. 2022). Conservation efforts and nature-based solutions can offer multiple co-benefits, but their success often depends on extensive stakeholder engagement and the integration of local and external knowledge (Raymond et al. 2017; Ferreira et al. 2020). Similarly, while community-based initiatives empower local populations and leverage local knowledge, they can be constrained by resource limitations and institutional barriers (Reid et al. 2010; Amaru and Chhetri 2013).

Therefore, the management of environmental risks requires a multifaceted and integrative approach, incorporating policy interventions, technology-based solutions, conservation efforts, and community-based initiatives. As the world continues to face escalating environmental challenges, these strategies will need to be continually assessed, adapted, and improved in response to emerging needs and contexts.

Table 1.1 Application of AI in environmental risk modeling

Environmental risk model	Algorithms	Utility	Application
Regression models	Multiple regression; logistic regression; stepwise regression; multivariate adaptive regression splines; geographically weighted regression	Predicting continuous or binary outcomes	Climate Change (Jha et al. 2023; Passow and Donner 2020) Biodiversity Loss (Powell and Lenton 2013) Water Scarcity and Quality (Madani et al. 2022) Air Pollution (Jiang 2021) Natural hazard (Bera et al. 2021a) Land Degradation (Abdel-Kader 2019) Environmental Risks Renewable energy (Tampubolon et al. 2023) Social and Economic Impacts (Yang and Geng 2022) Policy and Governance Risks (Zeng et al. 2020)
Decision trees	Classification and Regression Trees (CART); Iterative Dichotomiser 3 (ID3)	Handle both categorical and continuous data and are effective in identifying critical predictors and assessing variable interactions	Climate Change (Cai et al. 2022; Nourani et al. 2019) Biodiversity Loss (Branco et al. 2023) Water Scarcity and Quality (Birnbaum et al. 2022; Jeihouni et al. 2020) Air Pollution (Putra and Sitanggang 2020; Althuwaynee et al. 2020) Natural Disasters (Guo et al. 2021) Land Degradation (Vieira et al. 2021) Environmental Risks Renewable energy (Shorabeh et al. 2022)
Support vector machines (SVM)	Linear SVM; Nonlinear SVM; Polynomial Kernel SVM	Dealing with complex, high-dimensional datasets; distinguish between different risk categories	Climate Change (Khan et al. 2021) Biodiversity Loss (Branco et al. 2023) Water Scarcity and Quality (Liu and Lu 2014) Air Pollution (Leong et al. 2020) Natural hazards (Liu et al. 2022) Land Degradation (Torabi et al. 2021) Environmental Risks Renewable energy Social and Economic Vulnerability and Impacts (Singh et al. 2020)

(continued)

Table 1.1 (continued)

Environmental risk model	Algorithms	Utility	Application
Neural networks	Artificial neural networks (ANNs) and deep learning models, such as convolutional neural networks (CNNs) and recurrent neural networks (RNNs)	Image analysis, time series forecasting, and predicting environmental changes	Climate Change Biodiversity Water Scarcity and Quality (Wang et al. 2013; Talukdar et al. 2023; Mallik et al. 2023) Air Pollution (Chen and Zheng 2022) Natural hazards (Bera et al. 2021b) Land Degradation and soil erosion (Ahmed et al. 2023) Social and Economic Vulnerability (Abarca-Alvarez et al. 2019)
Clustering algorithms	k-means and hierarchical clustering	Identify spatial clusters of environmental risks or segment areas based on shared characteristics	Climate change Water Scarcity and Quality (Du et al. 2017) Natural hazards (Tang et al. 2020; Wu et al. 2021)
Gaussian process regression	Gaussian process regression; Gaussian process Kernel	Uncertainty estimates involving sparse or irregular data	Climate Change Biodiversity Water Scarcity and Quality Air Pollution Natural hazards
Anomaly detection	Isolation forest and one-class SVM, active learning	Detect irregular patterns	Climate Change (Liu et al. 2020) Water Scarcity and Quality (Yan and Tao 2022) Air Pollution (Rollo et al. 2023) Natural hazards (Luz et al. 2022) Environmental Risks (Russo et al. 2020)
Genetic algorithms (GA)	Generational GA Steady-state GA Steady-generational GA	Find optimal solutions for complex problems	Climate Change (Park and Park 2021; Demir 2022) Air Pollution (Demir 2022) Environmental Risks (Davis et al. 2019)
Ensemble methods	Bagging and boosting random forest	Increase the robustness and accuracy of model	Climate Change (Dawkins et al. 2023) Water Scarcity and Quality (Yadav et al. 2020; Bhagat et al. 2021) Natural hazards (Bera et al. 2023) Land Degradation (Yousefi et al. 2021) Social and Economic Vulnerability and Impacts

References

Abarca-Alvarez FJ, Reinoso-Bellido R, Campos-Sánchez FS (2019) Decision model for predicting social vulnerability using artificial intelligence. ISPRS Int J Geo-Information 8:10–12. https://doi.org/10.3390/ijgi8120575

Abdel-Kader FH (2019) Assessment and monitoring of land degradation in the northwest coast region, Egypt using Earth observations data. Egypt J Remote Sens Sp Sci 22:165–173. https://doi.org/10.1016/j.ejrs.2018.02.001

Abdi AM, Carrié R, Sidemo-Holm W, Cai Z, Boke-Olén N, Smith HG, Eklundh L, Ekroos J (2021) Biodiversity decline with increasing crop productivity in agricultural fields revealed by satellite remote sensing. Ecol Indicat 130:108098

Abid SK, Sulaiman N, Chan SW, Nazir U, Abid M, Han H, Ariza-Montes A, Vega-Muñoz A (2021) Toward an integrated disaster management approach: how artificial intelligence can boost disaster management. Sustainability 13(22):12560

Adger WN et al (2005) Successful adaptation to climate change across scales. Global Environ Change 15 (2):77–86

Ahmed IA, Talukdar S, Islam ARMT et al (2023) Contribution and behavioral assessment of physical and anthropogenic factors for soil erosion using integrated deep learning and game theory. J Clean Prod 416:137689. https://doi.org/10.1016/j.jclepro.2023.137689

Althuwaynee OF, Balogun A, Al Madhoun W (2020) Air pollution hazard assessment using decision tree algorithms and bivariate probability cluster polar function: evaluating inter-correlation clusters of PM10 and other air pollutants. Gisci Remote Sens 57:207–226. https://doi.org/10.1080/15481603.2020.1712064

Amaru S, Chhetri NB (2013) Climate adaptation: institutional response to environmental constraints, and the need for increased flexibility, participation, and integration of approaches. Appl Geogr 39:128–139

Beaumont NJ, Aanesen M, Austen MC, Börger T, Clark JR, Cole M, Hooper T, Lindeque PK, Pascoe C, Wyles KJ (2019) Global ecological, social and economic impacts of marine plastic. Mar Pollut Bull 142:189–195

Bera S, Melo R, Guru B (2021a) Assessment of exposed elements in a changing built environment by using an integrated model of debris flow initiation and runout (Kalimpong region, Himalaya). Bull Eng Geol Environ 80:7131–7152. https://doi.org/10.1007/s10064-021-02352-w

Bera S, Upadhyay VK, Guru B, Oommen T (2021b) Landslide inventory and susceptibility models considering the landslide typology using deep learning: Himalayas, India. Nat Hazards 108:1257–1289. https://doi.org/10.1007/s11069-021-04731-8

Bera S, Gnyawali K, Dahal K et al (2023) Assessment of shelter location-allocation for multi-hazard emergency evacuation. Int J Disaster Risk Reduct 84:103435. https://doi.org/10.1016/j.ijdrr.2022.103435

Bhagat SK, Tiyasha T, Awadh SM et al (2021) Prediction of sediment heavy metal at the Australian Bays using newly developed hybrid artificial intelligence models. Environ Pollut 268:115663. https://doi.org/10.1016/j.envpol.2020.115663

Birnbaum A, Lamontagne J, Wild T et al (2022) Drivers of future physical water scarcity and its economic impacts in Latin America and the Caribbean. Earth's Fut 10:e2022EF002764. https://doi.org/10.1029/2022EF002764

Branco VV, Correia L, Cardoso P (2023) The use of machine learning in species threats and conservation analysis. Biol Conserv 283:110091. https://doi.org/10.1016/j.biocon.2023.110091

Branny A, Møller MS, Korpilo S, McPhearson T, Gulsrud N, Olafsson AS, Raymond CM, Andersson E (2022) Smarter greener cities through a social-ecological-technological systems approach. Curr Opin Environ Sustain 55:101168

Burby R, Dixon J, Ericksen N, Handmer J, May P, Michaels S, Smith DI (2013) Environmental management and governance: intergovernmental approaches to hazards and sustainability. Routledge, New York

Cai W, Wei R, Xu L, Ding X (2022) A method for modelling greenhouse temperature using gradient boost decision tree. Inf Process Agric 9:343–354. https://doi.org/10.1016/j.inpa.2021.08.004

Chen S, Zheng L (2022) Complementary ensemble empirical mode decomposition and independent recurrent neural network model for predicting air quality index. Appl Soft Comput 131:109757. https://doi.org/10.1016/j.asoc.2022.109757

Costanza R et al (2014) Changes in the global value of ecosystem services. Global Environ Change 26:152–158

Cui X, Guo X, Wang Y, Wang X, Zhu W, Shi J, Lin C, Gao X (2019) Application of remote sensing to water environmental processes under a changing climate. J Hydrol 574:892–902

Cundy AB, Bardos RP, Church A, Puschenreiter M, Friesl-Hanl W, Müller I, Neu S, Mench M, Witters N, Vangronsveld J (2013) Developing principles of sustainability and stakeholder engagement for "gentle" remediation approaches: the European context. J Environ Manag 129:283–291

Daily GC, Ehrlich PR (1992) Population, sustainability, and Earth's carrying capacity

Dasan J, Ahmedy F, Shamsul S, Godoong E, Sipaut CS, Jeffree MS (2022) Post-COVID-19 challenges for a sustainable community-based ecotourism: a case study of rural community in Sabah. North of Borneo. SAGE Open 12(4):21582440221138816

Davis JK, Gebrehiwot T, Worku M et al (2019) A genetic algorithm for identifying spatially-varying environmental drivers in a malaria time series model. Environ Model Softw 119:275–284. https://doi.org/10.1016/j.envsoft.2019.06.010

Dawkins LC, Bernie DJ, Lowe JA, Economou T (2023) Assessing climate risk using ensembles: a novel framework for applying and extending open-source climate risk assessment platforms. Clim Risk Manag 40:100510. https://doi.org/10.1016/j.crm.2023.100510

Demir AS (2022) Modeling and forecasting of CO_2 emissions resulting from air transport with genetic algorithms: the United Kingdom case. Theor Appl Climatol 150:777–785. https://doi.org/10.1007/s00704-022-04203-4

Dewan AM, Yamaguchi Y (2009) Using remote sensing and GIS to detect and monitor land use and land cover change in Dhaka Metropolitan of Bangladesh during 1960–2005. Environ Monit Assess 150:237–249

Du X, Shao F, Wu S et al (2017) Water quality assessment with hierarchical cluster analysis based on Mahalanobis distance. Environ Monit Assess 189:1–12. https://doi.org/10.1007/s10661-017-6035-y

Feizizadeh B, Blaschke T (2013) Examining urban heat island relations to land use and air pollution: Multiple endmember spectral mixture analysis for thermal remote sensing. IEEE J Select Top Appl Earth Observ Remote Sens 6(3):1749–1756

Ferreira V, Barreira AP, Loures L, Antunes D, Panagopoulos T (2020) Stakeholders' engagement on nature-based solutions: a systematic literature review. Sustainability 12(2):640

Gasparatos A, Doll CNH, Esteban M, Ahmed A, Olang TA (2017) Renewable energy and biodiversity: implications for transitioning to a green economy. Renew Sustain Energy Rev 70:161–184

Ghasempour F, Sekertekin A, Kutoglu SH (2021) Google earth engine based spatio-temporal analysis of air pollutants before and during the first wave COVID-19 outbreak over Turkey via remote sensing. J Clean Product 319:128599

Guo Z, Shi Y, Huang F et al (2021) Landslide susceptibility zonation method based on C5.0 decision tree and K-means cluster algorithms to improve the efficiency of risk management. Geosci Front 12:101249. https://doi.org/10.1016/j.gsf.2021.101249

Hoegh-Guldberg O et al (2019) The human imperative of stabilizing global climate change at 1.5 °C

Horwitz P, Finlayson CM (2011) Wetlands as settings for human health: incorporating ecosystem services and health impact assessment into water resource management. Bioscience 61(9):678–688

IPCC (2013) Climate change 2013: the physical science basis

Jeihouni M, Toomanian A, Mansourian A (2020) Decision tree-based data mining and rule induction for identifying high quality groundwater zones to water supply management: a novel hybrid use of data mining and GIS. Water Resour Manag 34:139–154. https://doi.org/10.1007/s11269-019-02447-w

Jha P, Chinngaihlian S, Upreti P, Handa A (2023) A machine learning approach to assess implications of climate risk factors on agriculture: the Indian case. Clim Risk Manag 41:100523. https://doi.org/10.1016/j.crm.2023.100523

Jiang W (2021) The data analysis of shanghai air quality index based on linear regression analysis. J Phys Conf Ser 1813:012031

Ju T, Liang Z, Liu W, Li B, Huang R, Geng T (2022) Monitoring of air pollution by remote sensing in Lanzhou City from 2010 to 2019. Water Air Soil Pollut 233(9):359

Jung J, Maeda M, Chang A, Bhandari M, Ashapure A, Landivar-Bowles J (2021) The potential of remote sensing and artificial intelligence as tools to improve the resilience of agriculture production systems. Curr Opin Biotechnol 70:15–22

Kacic P, Kuenzer C (2022) Forest biodiversity monitoring based on remotely sensed spectral diversity—a review. Remote Sens 14(21):5363

Khan A, Gupta S, Gupta SK (2020) Multi-hazard disaster studies: Monitoring, detection, recovery, and management, based on emerging technologies and optimal techniques. Int J Disast Risk Reduct 47:101642

Khan N, Shahid S, Bin IT, Behlil F (2021) Prediction of heat waves over Pakistan using support vector machine algorithm in the context of climate change. Stoch Environ Res Risk Assess 35:1335–1353. https://doi.org/10.1007/s00477-020-01963-1

Kim KH, Kabir E, Ara Jahan S (2014) A review of the consequences of global climate change on human health. J Environ Sci Health C 32(3):299–318

Koutalakis PD, Tzoraki OA, Prazioutis GI, Gkiatas GT, Zaimes GN (2021) Can drones map earth cracks? Landslide measurements in north Greece using UAV photogrammetry for nature-based solutions. Sustainability 13(9):4697

Lenzen M, Li M, Malik A, Pomponi F, Sun YY, Wiedmann T, Faturay F, Fry J, Gallego B, Geschke A, Gómez-Paredes J (2020) Global socio-economic losses and environmental gains from the Coronavirus pandemic. PLoS ONE 15(7):e0235654

Leong WC, Kelani RO, Ahmad Z (2020) Prediction of air pollution index (API) using support vector machine (SVM). J Environ Chem Eng 8:103208. https://doi.org/10.1016/j.jece.2019.103208

Li H, Peng J, Jiao Y, Ai S (2022) Experiencing urban green and blue spaces in urban wetlands as a nature-based solution to promote positive emotions. Forests 13(3):473

Liu M, Lu J (2014) Support vector machine: an alternative to artificial neuron network for water quality forecasting in an agricultural nonpoint source polluted river? Environ Sci Pollut Res 21:11036–11053. https://doi.org/10.1007/s11356-014-3046-x

Liu Y, Pang Z, Karlsson M, Gong S (2020) Anomaly detection based on machine learning in IoT-based vertical plant wall for indoor climate control. Build Environ 183:107212. https://doi.org/10.1016/j.buildenv.2020.107212

Liu Y, Tian J, Liu R, Ding L (2021) Influences of climate change and human activities on NDVI changes in China. Remote Sens 13(21):4326

Liu J, Wang J, Xiong J et al (2022) Assessment of flood susceptibility mapping using support vector machine, logistic regression and their ensemble techniques in the Belt and Road region. Geocarto Int 37:9817–9846. https://doi.org/10.1080/10106049.2022.2025918

Luz AEO, Negri RG, Massi KG et al (2022) Mapping fire susceptibility in the Brazilian Amazon Forests using multitemporal remote sensing and time-varying unsupervised anomaly detection. Remote Sens 14:102429. https://doi.org/10.3390/rs14102429

Madani A, Hagage M, Elbeih SF (2022) Random forest and logistic regression algorithms for prediction of groundwater contamination using ammonia concentration. Arab J Geosci 15:1619. https://doi.org/10.1007/s12517-022-10872-2

Mallik S, Das S, Chakraborty A et al (2023) Prediction of non-carcinogenic health risk using Hybrid Monte Carlo-machine learning approach. Hum Ecol Risk Assess 29:777–800. https://doi.org/10.1080/10807039.2023.2188417

Mandal A, Majumder A, Dhaliwal SS, Toor AS, Mani PK, Naresh RK, Gupta RK, Mitran T (2022) Impact of agricultural management practices on soil carbon sequestration and its monitoring through simulation models and remote sensing techniques: a review. Crit Rev Environ Sci Technol 52(1):1–49

McDonald RI, Mansur AV, Ascensão F, Colbert ML, Crossman K, Elmqvist T, Gonzalez A, Güneralp B, Haase D, Hamann M, Hillel O (2020) Research gaps in knowledge of the impact of urban growth on biodiversity. Nat Sustain 3(1):16–24

Menon S, Bawa KS (1997) Applications of geographic information systems, remote-sensing, and a landscape ecology approach to biodiversity conservation in the Western Ghats. Curr Sci 32:134–145

Mercer J, Kelman I, Alfthan B, Kurvits T (2012) Ecosystem-based adaptation to climate change in Caribbean small island developing states: integrating local and external knowledge. Sustainability 4 (8):1908–1932

Munawar HS, Mojtahedi M, Hammad AW, Kouzani A, Mahmud MP (2022) Disruptive technologies as a solution for disaster risk management: a review. Sci Total Environ 806:151351

Naderpour M, Rizeei HM, Ramezani F (2021) Forest fire risk prediction: a spatial deep neural network-based framework. Remote Sens 13(13):2513

Nourani V, Razzaghzadeh Z, Baghanam AH, Molajou A (2019) ANN-based statistical downscaling of climatic parameters using decision tree predictor screening method. Theor Appl Climatol 137:1729–1746. https://doi.org/10.1007/s00704-018-2686-z

Park S, Park SK (2021) A micro-genetic algorithm (GA v1.7.1a) for combinatorial optimization of physics parameterizations in the Weather Research and Forecasting model (v4.0.3) for quantitative precipitation forecast in Korea. Geosci Model Dev 14:6241–6255. https://doi.org/10.5194/gmd-14-6241-2021

Passow C, Donner RV (2020) Regression-based distribution mapping for bias correction of climate model outputs using linear quantile regression. Stoch Environ Res Risk Assess 34:87–102. https://doi.org/10.1007/s00477-019-01750-7

Pecl GT, Araújo MB, Bell JD, Blanchard J, Bonebrake TC, Chen IC, Clark TD, Colwell RK, Danielsen F, Evengård B, Falconi L (2017) Biodiversity redistribution under climate change: impacts on ecosystems and human well-being. Science 355 (6332):eaai9214

Pei T, Xu J, Liu Y, Huang X, Zhang L, Dong W, Qin C, Song C, Gong J, Zhou C (2021) GIScience and remote sensing in natural resource and environmental research: status quo and future perspectives. Geogr Sustain 2(3):207–215

Pereira P, Bašić F, Bogunovic I, Barcelo D (2022) Russian-Ukrainian war impacts the total environment. Sci Total Environ 837:155865

Pettorelli N, Laurance WF, O'Brien TG, Wegmann M, Nagendra H, Turner W (2014) Satellite remote sensing for applied ecologists: opportunities and challenges. J Appl Ecol 51(4):839–848

Piao S, Liu Q, Chen A, Janssens IA, Fu Y, Dai J, Liu L, Lian XU, Shen M, Zhu X (2019) Plant phenology and global climate change: current progresses and challenges. Global Change Biol 25(6):1922–1940

Powell TWR, Lenton TM (2013) Scenarios for future biodiversity loss due to multiple drivers reveal conflict between mitigating climate change and preserving biodiversity. Environ Res Lett 8:025024. https://doi.org/10.1088/1748-9326/8/2/025024

Prüss-Ustün A et al (2016) Diseases due to unhealthy environments: an updated estimate of the global burden of disease attributable to environmental determinants of health. J Public Health 39(3):464–475

Putra FM, Sitanggang IS (2020) Classification model of air quality in Jakarta using decision tree algorithm based on air pollutant standard index. IOP Confer Ser Earth Environ Sci 528:012053

Raymond C, Breil M, Nita M, Kabisch N, de Bel M, Enzi V, Frantzeskaki N, Geneletti G, Lovinger L, Cardinaletti M, Basnou C (2017) An impact evaluation framework to support planning and evaluation of nature-based solutions projects. Report prepared by the EKLIPSE expert working group on nature-based solutions to promote climate resilience in Urban Areas. Centre for ecology and hydrology

Reid H, Huq S, Murray LA (2010) Community champions: adapting to climate challenges. IIED

Renn O (2008) Risk governance: coping with uncertainty in a complex world

Rocchini D, Marcantonio M, Da Re D, Chirici G, Galluzzi M, Lenoir J, Ricotta C, Torresani M, Ziv G (2019) Time-lapsing biodiversity: an open source method for measuring diversity changes by remote sensing. Remote Sens Environ 231:111192

Rollo F, Bachechi C, Po L (2023) Anomaly detection and repairing for improving air quality monitoring. Sensors 23:640. https://doi.org/10.3390/s23020640

Ruckelshaus M, Reguero BG, Arkema K, Compean RG, Weekes K, Bailey A, Silver J (2020) Harnessing new data technologies for nature-based solutions in assessing and managing risk in coastal zones. Int J Disast Risk Reduct 51:101795

Russo S, Lürig M, Hao W et al (2020) Active learning for anomaly detection in environmental data. Environ Model Softw 134:104869. https://doi.org/10.1016/j.envsoft.2020.104869

Schmeller DS, Courchamp F, Killeen G (2020) Biodiversity loss, emerging pathogens and human health risks. Biodiver Conserv 29:3095–3102

Shorabeh SN, Samany NN, Minaei F et al (2022) A decision model based on decision tree and particle swarm optimization algorithms to identify optimal locations for solar power plants construction in Iran. Renew Energy 187:56–67. https://doi.org/10.1016/j.renene.2022.01.011

Singh A, Sarma AK, Hack J (2020) Cost-effective optimization of nature-based solutions for reducing urban floods considering limited space availability. Environ Processes 7:297–319

Slovic P (1999) Trust, emotion, sex, politics, and science: surveying the risk-assessment battlefield. Risk Anal 19 (4):689–701

Somvanshi SS, Vashisht A, Chandra U, Kaushik G (2019) Delhi air pollution modeling using remote sensing technique. Handbook Environ Mater Manag 132:1–27

Sorek-Hamer M, Cohen A, Levy RC, Ziv B, Broday DM (2013) Classification of dust days by satellite remotely sensed aerosol products. Int J Remote Sens 34 (8):2672–2688

Sultana Z, Mallick B (2015) Adaptation strategies after cyclone in southwest coastal Bangladesh–pro poor policy choices. Am J Rural Develop 3(2):24–33

Talukdar S, Shahfahad AS et al (2023) Predicting lake water quality index with sensitivity-uncertainty analysis using deep learning algorithms. J Clean Prod 406:136885. https://doi.org/10.1016/j.jclepro.2023.136885

Tampubolon FRS, Yuwono AS, Tambunan AH, Achsani NA (2023) Renewable energy consumption and its environmental impacts: a meta-regression analysis. Environ Qual Manag 32:43–55

Tang RX, Kulatilake PHSW, Yan EC, Sen CJ (2020) Evaluating landslide susceptibility based on cluster analysis, probabilistic methods, and artificial neural networks. Bull Eng Geol Environ 79:2235–2254. https://doi.org/10.1007/s10064-019-01684-y

Taubenböck H, Wurm M, Netzband M, Zwenzner H, Roth A, Rahman A, Dech S (2011) Flood risks in urbanized areas–multi-sensoral approaches using remotely sensed data for risk assessment. Nat Hazards Earth Syst Sci 11(2):431–444

Teodoro AC, Duarte L (2022a) The role of satellite remote sensing in natural disaster management. Nanotechnology-based smart remote sensing networks for disaster prevention. Elsevier, Amsterdam, pp 189–216

Teodoro AC, Duarte L (2022b) The synergy of remote sensing and geographical information systems in the management of natural disasters. Nanotechnology-based smart remote sensing networks for disaster prevention. Elsevier, Amsterdam, pp 217–230

Torabi HA, Darabi H, Karimidastenaei Z et al (2021) Land degradation risk mapping using topographic, human-induced, and geo-environmental variables and machine learning algorithms, for the Pole-Doab watershed. Iran. Environ Earth Sci 80:1–21. https://doi.org/10.1007/s12665-020-09327-2

UNEP (2020) Global biodiversity outlook 5

Vieira RM, Tomasella J, Barbosa AA et al (2021) Land degradation mapping in the MATOPIBA region (Brazil) using remote sensing data and decision-tree analysis. Sci Total Environ 782:146900. https://doi.org/10.1016/j.scitotenv.2021.146900

Wang Y, Zheng T, Zhao Y et al (2013) Monthly water quality forecasting and uncertainty assessment via bootstrapped wavelet neural networks under missing data for Harbin, China. Environ Sci Pollut Res 20:8909–8923. https://doi.org/10.1007/s11356-013-1874-8

White GF (2018) Environmental risk. Routledge, New York

White JD, Ryan KC, Key CC, Running SW (1996) Remote sensing of forest fire severity and vegetation recovery. Int J Wildland Fire 6(3):125–136

Wu S, Hu X, Zheng W et al (2021) Effects of reservoir water level fluctuations and rainfall on a landslide by two-way ANOVA and K-means clustering. Bull Eng Geol Environ 80:5405–5421. https://doi.org/10.1007/s10064-021-02273-8

Xu K, Zhang X, Chen Z, Wu W, Li T (2016) Risk assessment for wildfire occurrence in high-voltage power line corridors by using remote-sensing techniques: a case study in Hubei Province. China. Int J Remote Sens 37(20):4818–4837

Yadav B, Gupta PK, Patidar N, Himanshu SK (2020) Ensemble modelling framework for groundwater level prediction in urban areas of India. Sci Total Environ 712:135539. https://doi.org/10.1016/j.scitotenv.2019.135539

Yan J, Tao T (2022) Unsupervised anomaly detection in hourly water demand data using an asymmetric encoder–decoder model. J Hydrol 613:128389. https://doi.org/10.1016/j.jhydrol.2022.128389

Yang X, Geng L (2022) An integrated analysis of social, economic, and environmental indicators' effects on public health and health inequality globally: from the perspective of vulnerability. Soc Indic Res 162:1261–1279. https://doi.org/10.1007/s11205-022-02877-x

Yousefi S, Pourghasemi HR, Avand M et al (2021) Assessment of land degradation using machine-learning techniques: a case of declining rangelands.

L Degrad Dev 32:1452–1466. https://doi.org/10.1002/ldr.3794
Youssef YM, Gemail KS, Sugita M, Al-Barqawy M, Teama MA, Koch M, Saada SA (2021) Natural and anthropogenic coastal environmental hazards: an integrated remote sensing, GIS, and geophysical-based approach. Surv Geophys 162:1–33
Zeng J, Guijarro M, Carrilero-Castillo A (2020) A regression discontinuity evaluation of the policy effects of environmental regulations. Econ Res Istraz 33:2993–3016. https://doi.org/10.1080/1331677X.2019.1699437

Changes in Land Cover, Forest and Environmental Degradation and Sustainability in Manipur, India

2

Lohra Salew, Balamurugan Guru, and Pradheeps Muthulingam

Abstract

Ecosystems play a prominent role in reducing the risk of hazards and are of fundamental importance for our survival. Considering these aspects, the study focuses on the issues of regional environmental degradation, land use changes, as well as their impacts carried out with satellite images, field and secondary data then integrated into the geo-informatics to analyses the various influencing parameters of environmental impacts. The study reveals that human actions account for most of the reduction of forests cover from 52.81% in 1995 to 41.96% in 2006 and the expansion of agriculture from 34.98% in 1995 to 45.57% in 2006. The "open forest cover" has increased in 2018 with 47.33%, yet the NDVI analysis shows a negative trend due to the massive reduction in dense vegetation from 28.57% in 1995 to 2.86% in 2018. The correlation analysis carried out between NDVI value (0.4–0.6), and area of coverage result shows a negative correlation with a root mean square value of 0.54, which indicates a decrease in "dense vegetation". Further, the root means square values of "built-up" and "agricultural" areas are accounted to 0.8582 and 0.0061 respectively which indicates the increasing of built-up category. Variations in the agricultural land trends indicate the shifting cultivation practices in Manipur. The environmental impacts are subsequently increasing risks among the local community, and therefore, the management of our resources is crucial to building the resilience of the ecology for long-term sustainability.

Keywords

Environmental degradation · Hazards · Natural resource · Risks · Sustainability · Mao-Maram

2.1 Introduction

The functional system and unit of the living and the non-living are interrelated, and they interact with each other and where the biotic community interacts with the abiotic environment in an ecological system (Odum and Barrett 2009).

Present Address:
L. Salew
Centre for Geo-Informatics, Tata Institute of Social Sciences, Mumbai 400088, India

B. Guru (✉)
Department of Geology, School of Earth Sciences, Central University of Tamil Nadu, Thiruvarur 610005, India
e-mail: balamurugan.guru@gmail.com

Present Address:
P. Muthulingam
Chennai Metropolitan Development Authority (CMDA), Egmore, Chennai 600008, India

S. Talukdar et al. (eds.), *Environmental Risk and Resilience in the Changing World*, GIScience and Geo-environmental Modelling, https://doi.org/10.1007/978-3-031-62442-1_2

The management of natural resources is a means to provide environmental balance, and it is unreasonable not to manage our environment to prevent it from being affected by various hazards and risk we are facing today (Bakker and Ritts 2018; Gobin et al. 2013; Krausmann et al. 2018). There needs to be sustainable development to decrease the depletion of resources and identify the future consequences (Folke et al. 2002; Sudmeier-Rieux et al. 2006; Taylor et al. 2016). If environmental sustainability is compromised, our lives are endangered, and global equity, as well as the opportunity of becoming better, are cut short (United Nations Environment 2018). Nearly 1.6 billion people in the world depend on forests as a means of livelihood and home to 80% of terrestrial species (United Nations Development Programme (UNDP) report 2018). Forests contain about 70% of the world's biodiversity (Angelsen and Kaimowitz 2001) which provides a vital ecosystem, for instance protection against soil erosion, maintaining the climate and weather conditions, purification of air, supporting the livelihoods of the people and gives shelter and its habitat to thousands of wildlife (Mahlstein and Knutti 2010; Giam 2017; Grooten 2018; Ghasemi et al. 2021; Moghanlo et al. (2021). According to UNDP annual report 2017, about 650 million are living in extreme poverty of which 16% are in the eco-sensitive fragile areas. About 24% of global health risk and 23% of deaths are due to environmental factors, of which the developing countries have 15 times higher due to exposure to environmental hazards (Anand 2013; Swati et al. 2014; Turner et al. 2018).

We not only adapt to the environment but concerted action to the ecosystems based on our biological needs (Davies et al. 2015). While some exploit the needs, we humans, more than any other organisms modify the physical environment and in doing so, increases the risks (Annisa and Eric 2016). The changes and degradation caused by human actions accelerate the risks of floods, landslides, forest fires and soil erosion (Wolff et al. 2018; Singha et al. 2020; Bera et al. 2021). The risks and hazards are subsequently increasing, and therefore, proper management of resources is crucial for the analysis of environmental processes (Eric et al. 2003; Smit and Wandel 2006). Geographical Information System (GIS) facilitate in managing, manipulating, analysing, modelling, representing and displaying geospatial data for the management of resources (Andreopoulou 2011; Balamurugan and Aravind 2015a, b). The remote sensing and GIS used as tools in analysing the complex system of the changes in the biophysical, physical and socio-economic status of the environment (Singh et al. 2004; Patra et al. 2006). The significant advantage of GIS is to be able to analyse the spatial information in the form of quantified maps and the ability to modelling in the process of development (Eta et al. 2014). The spectral bands of visible and near-infrared are commonly used to detect the health of the vegetation, and this would be reflected in the Normalized Difference Vegetation Index (NDVI) value where the degradation of the ecosystem can be quantified and monitored (Borowik et al. 2013).

Ecosystems management is of fundamental importance to our human survival and is crucial for meeting societal needs as well as to address the impending issue of climate change and environmental degradation. Some of the measures may not directly benefit us through commercial gains; however, they are of vital significance and support the endurance of those that we need as they are dependent on each other (Foley et al. 2011). For example, regional planning is essential for the reduction of biodiversity loss, proper management of land resources and sustainable use of natural resources. The ecological knowledge of the resource users and their local institutions acts as scientific knowledge and expand the information for ecosystem management and environmental feedback (Koontz et al. 2015). Strategies that can provide employment opportunities, proper utilization of the natural resources, marketing facilities, awareness among the communities and land use resource management need to be deployed through systematic improvisation and strengthening alternative land use systems with appropriate research and extension (Glamann et al. 2017). The "International Union for Conservation of Nature and Natural Resources" (IUCN) moving towards protecting people and nature shows how conservation efforts may bring benefits to economies and people, and how they reduce vulnerability to disasters. The acquisition of ecological knowledge and understanding the dynamics of ecosystems contribute to responding to the uncertainty and surprises of environmental changes (Olsson and Folke 2001). The changes have to be made at present (Schandl et al. 2018) and the absence of environmental regulations, ongoing political commitment, and the lack of institutional coordination, inadequate knowledge and public awareness needs to be addressed while moving towards sustainability (Griggs et al. 2013).

2.2 Rationale of the Study

The ecosystems offer different types of services to the community in terms of supporting, provisioning and regulating several resources. In recent years, there have been incidences of environmental degradation (Tyagi et al. 2014) as well as an increasing threat to many species; however, the consequences of human actions and various hazards have often left unexplored (Patel et al. 2013; Steffen et al. 2015; Talluto et al. 2018; Talukdar et al. 2021). The state of Manipur mainly comprises of hilly terrain where 90% falls under mountainous region and 10% of the valley. About 80% of the state population in Manipur is engaged with agriculture, and the reasons for the decrease in the forest are mainly due to deforestation, human interventions, Jhum or Shifting cultivation and timber supply for commercial purposes (Reimeingam 2017). Forest cover accounts to 66.43% of the land in the study area and the remaining utilized for agriculture and settlement (India State Forest Report (ISFR) 2017). There has been large-scale deforestation, which leads to soil erosion, landslide and flash floods. It is also compound by the increase in population, lack of development and infrastructures in the hilly areas, poor socio-economic conditions, unemployment and inconvenience of law and order. The present study focuses on identifying the environmental changes in the region, anthropogenic factors that are responsible for increasing risks and hazards prevalent in the area. Identification of land use and land cover (LULC) changes and detecting the changes of forest cover is one significant aspect, which will help us understand the several variations for a period of years in a specific extended area and contribute in management of environmental degradation. It also looks into how environmental impacts are increasing risks and hazards of the local community and how mitigation should be done in order to move towards sustainable development. While assessing the effects of environment changes, the findings will be helpful for the local administration in forming suitable and remedial measures at regional level. The effectiveness of assessing the environment will inform the public as well as improve opportunities and accountability, and it is a critical to approach whether the environmental conditions prioritized for long-term sustainability.

2.3 Study Area

The study region lies in the northern part of Senapati District, Manipur, and falls under the subdivision of Mao-Maram with its headquarters at Tadubi. It lies approximately between 25°24′N to 25°32′N latitude and 94°04′E to 94°11′E longitude and elevation ranging from 1061 to 2500 m. The Mao-Maram region is endowed with a rich landscape of blue hills ranging from Himalayan that are young and parallel to folded, valleys, streams and rivers flowing through the mountains and deep gorges. The region also comprised with endemic and endangered flora and fauna as it lies in the North East of India which falls under the global diversity hotspot (Roy et al. 2015). The climate is characterized by sub-tropical monsoon with annual rainfall ranging from 670 to 1450 mm (Manipur State Action Plan on Climate Change 2013). It comes under the classification of humid sub-tropical climate, having an average temperature of 19 °C and ranges from 4 to 34 °C.

The National Highway 39 (NH-39), now known as the Asian Highway 1 (AH-1) is the main economic lifeline of the state. This highway passes through the study region of Senapati District connecting the border state of Nagaland in the north and connects the Imphal in the south. The dominant inhabitant of the study area is the Mao-Naga community having its own rich culture and traditions and predominantly practice agriculture as their livelihood with types of farming like Jhum cultivation; terrace farming and low-lying paddy fields. The Barak River, which is one of the major river basins of the Manipur state, originates from the northern hills of Senapati District in the study area and flow westward connecting to other tributaries (Jugindro 2018). The highest peak in Manipur known as Mount Tenipu, which is 2995 m amsl (above mean sea level), situated near Mao-gate, a town is bordering the state of Nagaland. The geology consists of two major rock types, namely "Disangs and Barails" with a folded structure in continuation of the Himalayas. The Disang rock consists of "monotonous sequence with dark grey, thick splintery with minor mudstones, sandstones and siltstones" which are reported to have deposited during the Middle and lower Eocene period. The Barail rock formed during the "Oligocene and Upper Eocene" periods consists of mainly well-bedded sandstones, clay and sandy shells due to the presence of carbonaceous materials (Geological Survey of India 2011). Figure 2.1 shows the base map of the study area with essential locations and transportation networks.

2.4 Materials and Methodology

While developing the research, the materials decide what sort of information can be collected from various sources and how to be collected. Figure 2.2 shows the materials and methodology flow chart of the study. For meeting the needs of the research and gather the information, the Landsat 5, Landsat 7 and Landsat 8 satellite images of the year 1995, 2001, 2006, 2011 and 2018 with path 135 and row 42 were downloaded from USGS Earth Explorer. The satellite image during the period of January to March chosen as it gives precise information with less cloud cover (Table 2.1).

The satellite images help in assessing the land use and land cover changes (LULC) in the study area and indicate the purpose of which the land is used. The "Normalized Differential Vegetation Index" (NDVI) analysis carried out based on the satellite images to analyse the health of the vegetation and detect the changes of the vegetative area. The Survey of India (SOI) toposheet number G46Q2 and G46Q3 with the scale of 1:50,000 used as it provides the information of the important topographical features specified with legend and symbols. In this study, the toposheet was geo-referenced and the base map of the study created from the given toposheet produced by the Survey of India. The base map, watershed boundary and drainage map along with transportation networks and other important locations delineated. The Global Positioning System (GPS) coordinates taken from the sampled villages for producing accurate and real-time data from the field. The Geo-informatics software such as ERDAS Imagine, ArcMap and ENVI used for the data generation and analysis. The published documents including various journals, articles, administrative records, statistical records also referred for the study.

2.4.1 Field Data

The study uses mixed methods of both quantitative and qualitative data. The stratified random sampling method used where 11 villages sampled from the study area. The structured with open-ended questionnaires were used to conduct interviews and focus group discussions (FGDs) for the field survey done during May and June 2018. Based on the objectives of the study, the questionnaires framed and twenty interviews along with three FGDs conducted by means of personal communication during the field visits. The respondents from the local community including farmers, women societies and village authorities. The information from the community then used to validate the study as well as in analysing the information during the study process. The field observations were integrated for additional information and assess the different parameters of environmental impacts and management of the environment.

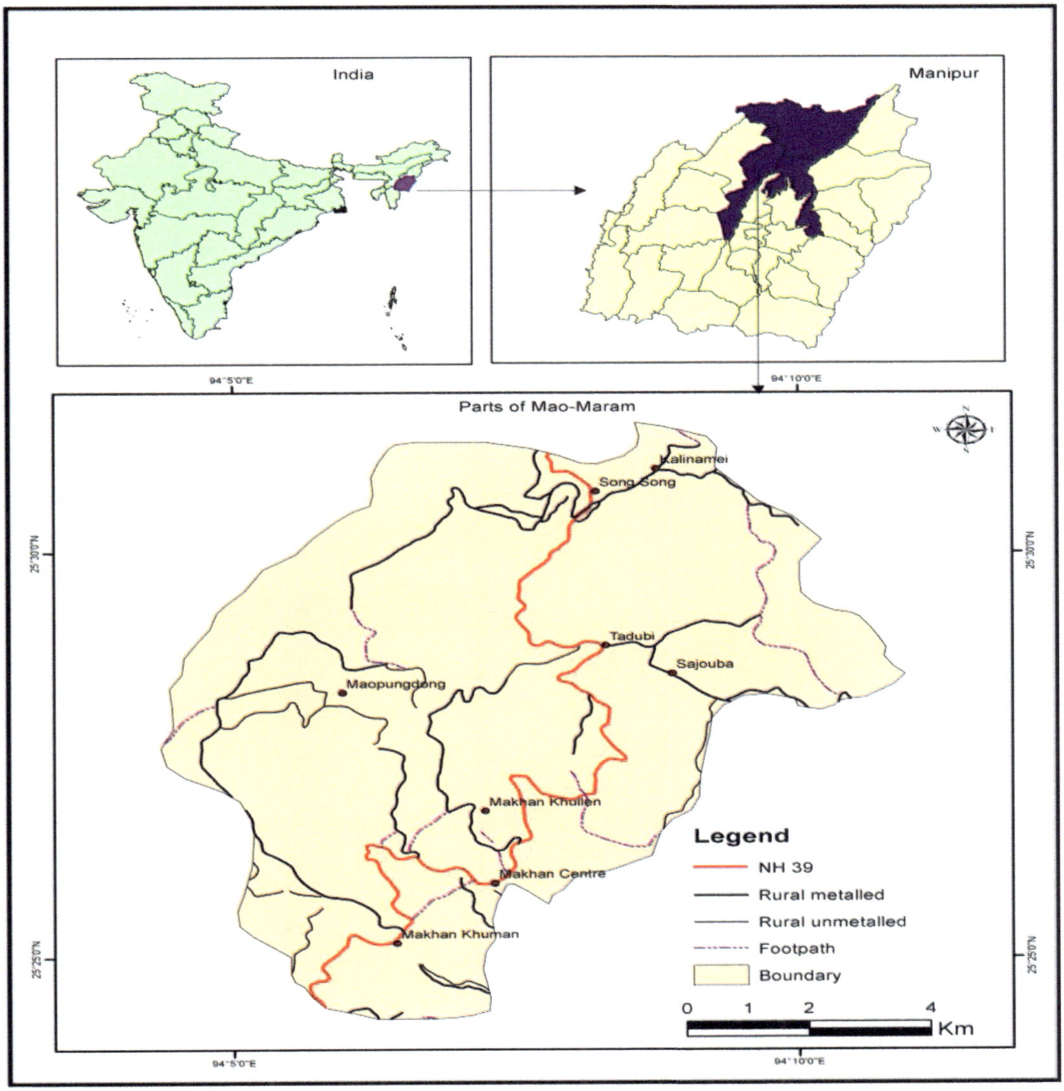

Fig. 2.1 Map of the study area, Mao-Maram, Senapati District, Manipur, with important locations and transportation networks

2.4.2 Thematic Maps Generation

The graphical representation of the quantitative data, remote sensing data are represented in the form of a thematic map and provides information on the specific theme of the geographical location, the spatial patterns and can be used to compare the different maps of different periods. The characteristics of the study region in terms of physical or social can be represented through different shades of colour, objects or numerical values. As the study area Mao-Maram region falls under the hilly terrain and situated in a less developed area, remote sensing gives many benefits by identifying the objects and detect the changes of the resources beyond many ranges of the hills. It can also monitor various types of risks and hazards that can lead to a disaster and in generating the data for systematic evaluation of environmental mapping.

Fig. 2.2 Methodology flowchart of the study

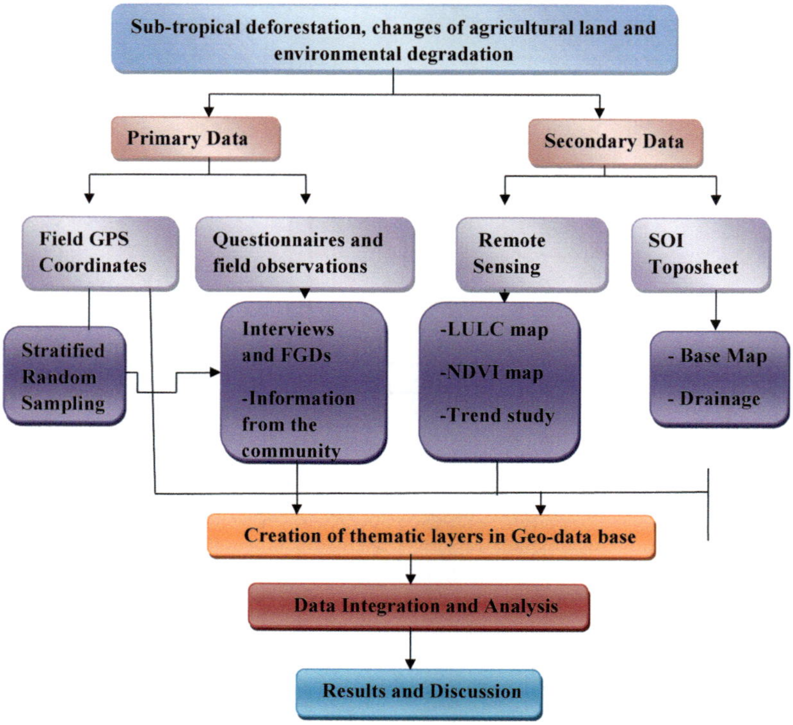

Table 2.1 Data source and its details

Sl. No.	Data source	Date and year of acquisition	Resolution/scale	Satellite type
1	USGS earth explorer	07/01/1995	30 m	LT05
2	USGS earth explorer	15/01/2001	30 m	LE07
3	USGS earth explorer	13/01/2006	30 m	LE07
4	USGS earth explorer	11/01/2011	30 m	LE07
5	USGS earth explorer	22/01/2018	30 m	LC08
6	SOI Toposheet	NA	1:50,000	No. G46Q2 and No. G46Q3

2.4.2.1 Mapping of LULC

The LULC mapping is useful for local as well as regional planning, vulnerability and risk assessment, environmental impact assessment and policy development (Balamurugan and Anubhooti 2015a, b). Generally, the LULC is done in three processes, (i) assessments through mapping and classification of data, (ii) evaluation of the productivity of vegetation through vegetation indices and (iii) studying the parameters obtained through satellite data. The fine resolution and spatially explicit data are required to predict the impact and effects of land degradation. After distinguishing the environment monitoring and assessment, Geo-informatics System tool is used to generate, analyse, manipulate and management of data. In this study, the five satellite images (1995, 2001, 2006, 2011 and 2018) were used to generate changes and analyse the LULC. The three different features, namely forest area, an agricultural area and built-up area, were classified based on their classes through ArcGIS software. The geographic surface was also compared through Google Maps in order to

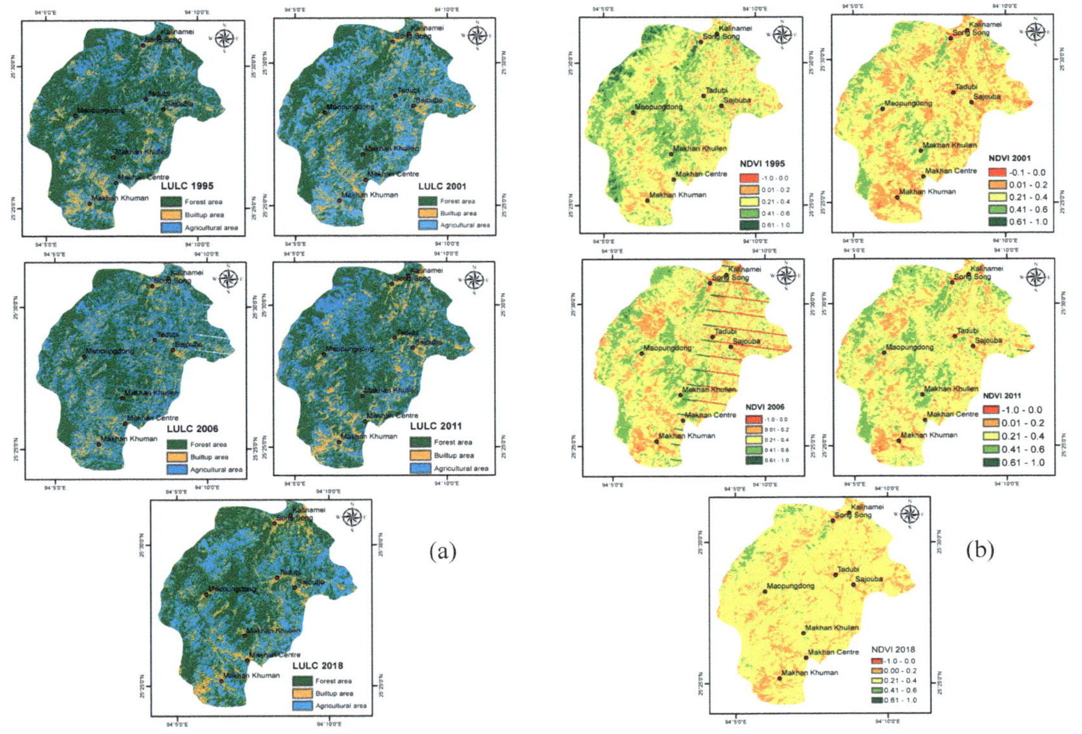

Fig. 2.3 **a** Land use land cover map with respective features. **b** Normalized Differential Vegetation Index map of the study during the period of 1995, 2001, 2006, 2011 and 2018

identify the correct pixels and features of the area. The land use land cover map generated from the study area is shown in Fig. 2.3a.

2.4.2.2 False Colour Composite

The Landsat satellite images with different spectral bands during the period of 1995, 2001, 2006, 2011 and 2018 were assigned with the bands and displayed with the colour through the multispectral images and assigned through standard false colour composite (FCC), it allows in identifying the features, and the vegetation can be detected as it appears in different shades of red due to higher reflectance of near-infrared. Depending on the type of forests and the health of the vegetation, the colours are displayed, turbid water in cyan, agricultural areas or soil in shades of blue and built-up area such as roads and buildings in yellow or grey depending on the composition.

2.4.2.3 Normalized Differential Vegetation Index (NDVI)

NDVI techniques provide quantitative information for the productivity of the vegetation extracted from the satellite imagery. It is the most widely used index of vegetation and based on the reflectance of red and near-infrared bands. Thus, the consequences of human actions and other natural phenomena on the environment can be highlighted through real-time data (Borowik et al. 2013). The standard NDVI is calculated in the following Eq. 2.1,

$$\mathrm{NDVI} = \frac{(\mathrm{NIR} - \mathrm{Red})}{(\mathrm{NIR} + \mathrm{Red})}, \qquad (2.1)$$

where

NDVI = Normalized Differential Vegetation Index.

NIR = Near-Infrared band

In this study, NDVI was mapped in order to explore the health of the photosynthetic vegetation cover. This was done through image processing techniques in a GIS environment where we can identify based on the reflectance of the green vegetation at different wavelengths. The NDVI maps during the period of 1995 to 2018 were generated using ERDAS Imagine and ArcGIS software where the changes were detected and analysed. The forest cover with the high vegetative area has higher NIR reflectance and low visible reflectance. Therefore, different years have shown different range where the higher values with upper limit +1 indicate healthier vegetation cover and lower values with a limit of −1 indicate low vegetation or absence of vegetation. The vegetation cover lying in the range of 0.2–0.4 was considered under the sparse vegetation or open forest and the values under the 0.4–0.6 come under the dense vegetation, and the NDVI map is shown in Fig. 2.3b.

2.5 Results and Discussion

The LULC changes quantified for further spatial analysis, and the relationship of the land use changes varies during the period of 1995–2001, 2001–2006, 2006–2011 and 2011–2018. The forest cover declined from 52.81% in 1995 to 42.43% in 2001 (Table 2.2), and it continued to decrease till 2006 with 36,121,538 m^2 covering 41.96% of the total land cover. The major causes in the loss of forest cover during this period were due to large exploitation of timber for commercial purposes, illicit felling of trees for firewood and agricultural expansion, especially the practice of shifting cultivation which was higher during this period. In addition, the analysis (Table 2.2) shows that there was a gradual increase from 2011 to 2018 from 41.96 to 47.33% and this was due to afforestation activity that took place around the Jhum cultivated area. However, the increase in forest cover was mainly in the open forest and the dense forest reduced. The built-up area in 1995 was 10,511,550 m^2 out of the total study area of 86,076,900 m^2 (Table 2.2). The year 2018 shows an increase in

Table 2.2 .

Sl. No	LULC features	Area in 1995 (sqm)	Area in % 1995	Area in 2001 (sqm)	Area in % 2001	Area in 2006 (sqm)	Area in % 2006	Area in 2011(sqm)	Area in % 2011	Area in 2018 (sqm)	Area in % 2018
1	Built-up area	10511550	12.21	10780087	12.52	10732512	12.47	11617813	13.50	12393000	14.40
2	Forest area	45453825	52.81	36519412	42.43	36121538	41.96	39417382	45.79	40744300	47.33
3	Agricultural land	30111525	34.98	38777401	45.05	3922285	45.57	35041705	40.71	32939600	38.27
	Total area	86076900	100.00	86076900	100.00	86076900	100.00	86076900	100.00	86076900	100.00

a built-up area with 12,393,000 m^2 indicating the expansion within the span of 23 years. It shows the increase from 12.21% in 1995 to 14.40% in 2018. In 1995, the agricultural land covered about 30,111,525 m^2, which is about 34.98% out of the total land cover; however, we observed a huge expansion of agriculture during the year 2001 and 2006 with 45.05 and 45.57%, respectively; further, it has decreased to 40.71% in 2011 and continued to decrease until the recent year 2018 with coverage of 38.27%.

2.5.1 NDVI Analysis

The NDVI analysis shown in Table 3 reveals that health of the vegetation varies for each class, values and different period of years. The class 1 lying in the range of −1.0 to 0.00 in 1995 covers around 112,500 m^2 with 0.13%. The same class in 2001 and 2006 showed an increased with 0.69 and 2.32% respectively due to the presence of water bodies and the expansion of agriculture. The class 2 lying in the range of 0.00 to 0.20 contains soil, barren and fallow land. Due to the expansion of agricultural activities and the subsequent decrease in land use patterns, the cover of the forest exploited within a short period of years and the health of vegetation values reduced. It also indicates that there was an increase in fallow land and class 3 values lying between 0.201 and 0.400 showed an increase from 13.95% in the year 1995 to 34.16% in 2001. The classes 4 and 5 lying in the range of 0.401–1.00 resulted in declining values due to the reduction of thick forests cover and dense vegetation. The result showed that the dense forests cover of class 4 reduced from 28.57% in 1995 to only 2.86% in 2018 and in class 5 with 2.02% in 1995 to 0% value in the year 2018 analysis indicating inferior dense vegetation cover.

2.5.2 Environment Impact Analysis

The analysis of the LULC features, namely built-up area, agricultural land and forest area, showed different trends during the study period from

Table 2.3 NDVI analysis of the study area from 1995 to 2018

Sl. No	NDVI values	Area in 1995 (sq.m)	Area in 1995 (%)	Area in 2001 (sq.m)	Area in 2001 (%)	Area in 2006 (sq.m)	Area in 2006 (%)	Area in 2011 (sq.m)	Area in 2011 (%)	Area in 2018 (sq.m)	Area in 2018 (%)
Class 1 (Water bodies)	−1.0 to 0.00	112,500	0.13	594,000	0.69	2,000,700	2.32	385,200	0.44	0	0
Class 2 (Soil, Rocky, Fallow land, Barren land)	0.00 to 0.200	12,014,100	13.95	29,403,900	34.16	20,949,300	24.33	18,641,700	21.65	19,344,600	22.47
Class 3 (Open forest or Sparse Vegetation)	0.201 to 0.400	47,607,300	55.3	46,690,200	54.24	46,918,800	54.5	51,480,900	59.8	64,263,600	74.65
Class 4 (Dense Vegetation or Dense forest)	0.401 to 0.600	24,597,000	28.57	9,376,200	10.89	15,424,200	17.91	15,451,200	17.95	2,468,700	2.86
Class 5 (Healthy Vegetation)	0.601 to 1.00	1,746,000	2.02	12,600	0.01	783,900	0.91	117,900	0.13	0	0

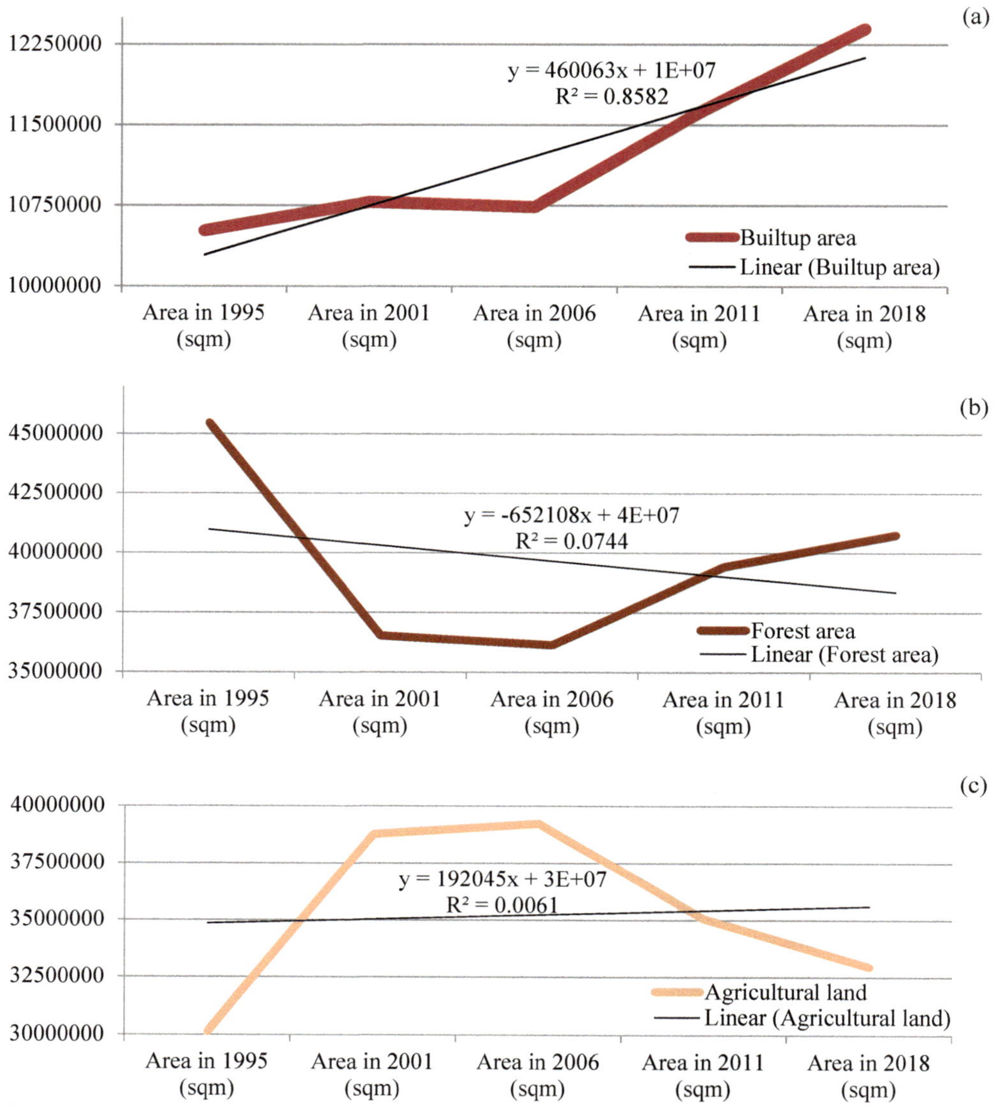

Fig. 2.4 **a** Graph showing the trend line of built-up area, **b** forest cover, **c** agricultural land changes from 1995 to 2018

1995 to 2018. The built-up area showed an increasing positive linear trend from 1995 to 2018 due to the expansion of settlement and development activities taking place (Fig. 2.4a). This indicates the positive correlation of built-up area with an R^2 value of 0.8582. On the other hand, there was a negative trend seen in forest area from the period 1995 to 2006 (Fig. 2.4b) and positive trend after 2006, however, the overall forest cover showed a negative correlation with an R^2 value of 0.0744 and this was because of the reduction in dense forest. The large agricultural expansion during the period of 1995–2006 indicated an increasing linear trend (Fig. 2.4c). This was attributed to the declining in Jhum cultivation cycle or its land use patterns from 7 to 8 years of the cycle to 2–3 years' cycle. The change in land use patterns within a

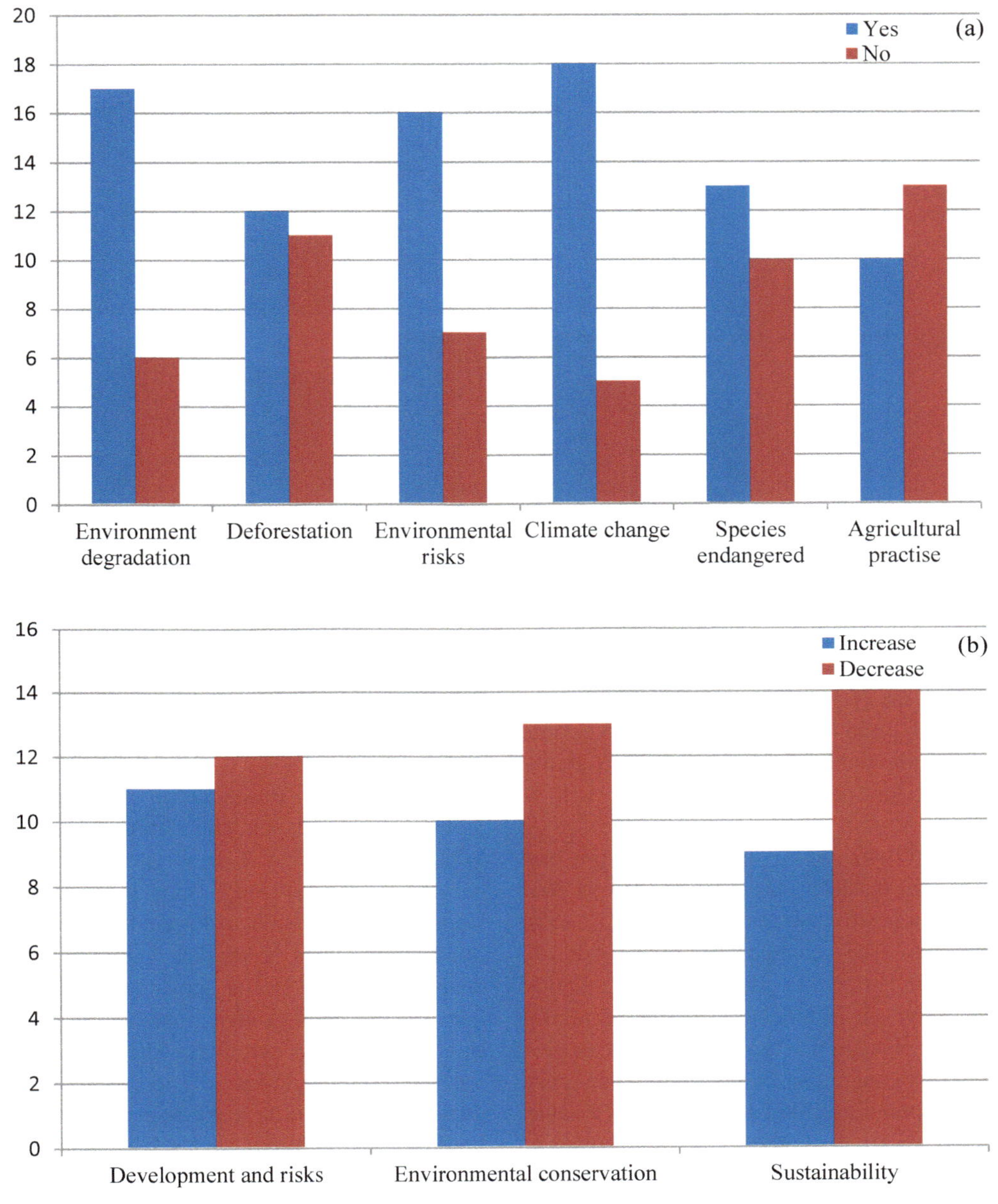

Fig. 2.5 Graph showing the respondent information from the field **a** causes of environmental degradation and **b** conservation and sustainability

short span of time was because of the degradation in soil quality, which was not able to sustain the crops for more productivity. It is noted that the expansion in agriculture on the hill slopes increased the impact of environmental degradation (Fig. 2.5a) as it occupies a wide range of land and destroys the natural habitat of many species. The community interaction from the field reveals that the species such as deer, porcupines and wild boars found in abundance during the early period; however, it is nowhere to be found now. The bird species Hornbill found

Fig. 2.6 **a** Field image showing hill slope been cleared off for cultivation, **b** showing slash and burn activity for enhancing soil fertility, **c** showing landslide blocking the village road, **d** showing paddy fields at Kashi destroyed during 2017 flash flood

mainly on the hilly areas is now considered as endangered and the high socio-economic medicinal plants like *Paris polyphylla* is getting extinct as the people are exploiting without bounds.

In addition, the agricultural practices decreased in the later stages from 2011 to 2018 indicating a positive correlation with an R^2 value of 0.0061. The change in livelihood and decreased in agriculture attribute towards the low productivity, an insufficient amount of rainfall during the plantation season, change in seasonal patterns, weather conditions and unscientific form of agriculture. Similarly, the NDVI values classified under five classes showed different trends, either negative or positive correlation. Class 1 with an R^2 value of 0.0072 shows a negative correlation, and class 2 values with an R^2 value of 0.0098 showed a positive correlation. The class 3 with an R^2 value of 0.6536 shows positive relationship because the open forest increased in several areas. The classes 4 and 5 with an R^2 value of 0.5413 and 0.5072 respectively show negative correlation indicating the exploitation and unsustainable utilization of resources.

2.5.3 Potential Hazards of the Study Area

From the field observations and responses given by the respondents from the sampled villages, the study area is prone to multi-hazards. The frequently occurring hazards are landslide, soil erosion, river bank erosion, flash flood, forest fires, earthquakes and road accidents. Landslide is the most common hazard due to its geologically unstable slopes and significant human activities. The impacts were seen in the depletion of forests cover and decrease in agricultural productivity, impact on lifestyle and traditions and subsequent impacts on the livelihoods. The widening of the village roads and National Highway (NH), now known as Asian trans-highway has also triggered several landslides (Fig. 2.5b). The high-frequency movement of heavily loaded vehicles also increases air pollution and affects those villages situated along the highway. For the women are selling food along the highway is at significant risks as the food can gets polluted with dust easily if proper measures like road repairing are not taken up. From the settlement point of view, some of the houses were built on the watershed and natural drainage systems, which make them more vulnerable to landslides and floods.

The 2004 landslide that occurred at Kalinamei Village and destroyed nearly 100 households was associated with the unsustainable practices in the forest catchment area that triggered the hazards. In a similar case, the landslide at Shajouba Village in 2006 swept away almost half of the village settlement areas, but the mitigation efforts to reduce the future impacts have been neglected. Riverbank erosion, especially during the rainy season, is another common hazard in the study area due to the surface runoff from the hill slopes triggered by deforestation practices and other anthropogenic factors. The mystical green mountains are shaved off (Fig. 2.6a, and b), and the heavy downpour during the monsoon brings down debris and mud which increases the river bank erosion, landslide (Fig. 2.6c) as well as destroys the low-lying paddy fields. One such case was seen from 2017 flash flood where the heavy rain swept away many agricultural fields. The incessant rain during the time of harvest season bought an enormous amount of debris and mud with the force of the water in the river and destroyed the fields around "Kashi" (Fig. 2.6d). The fields which were once rice bowls for many of the surrounding villages like Makhan Centre, Sajouba and Makhan Khuman are now left barren. The excessive use of chemical fertilizers in cultivation on the hill slopes also degrades the quality of the water in the streams which are a major source of water for the community. Significantly, this will likely to cause a health problem if not treated properly and if proper measures are not taken up immediately.

2.5.4 Development Schemes and Programmes

Development programmes improve the quality of life, but the narrowing gap in the level of development was seen. For example, the schemes that are meant for the local communities hardly benefitted them due to the lack of information and inaccessibility. This shows the lacuna of moving towards sustainable development, and at the same time, lack of monitoring prevails. When feasibility is taken into account, it still lacks the basic facilities such as health centres and schools. Drinking water has also to be brought from far away from water source, which are the rivers. This was not feasible for all the households. Despite the projects and schemes mentioned, there were barriers observed in the irregularity of budget distribution to the local people, long process of paperwork, inaccessibility and lack of information to the concerned households. Upon interacting with the villages, the majority of them want development since the region is still lacking behind in terms of socio-economic status, but it should come in a planned manner without affecting the cost of the ecosystem. In order to protect and preserve the environment, awareness at the local level by the students', youths, Mao-Naga community council/authorities, village authorities and women societies have been taken up however, the higher authorities from the state-district level doesn't seem to show much concern about environmental degradation of the area. For instance, the changes in the weather and climate patterns, loss of biodiversity were known; however, no significant attention was shown in almost any of the villages and these are likely to cause serious consequences to the ecosystem and human life in the study area. Therefore, efforts and promotion should function at a greater level for the welfare of society.

2.6 Conclusion

The increase in environmental impact and the dwindling of the natural resources is getting severe due to lack of development and infrastructures in the study area, poor socio-economic conditions, unemployment, inconvenience of law and order. The narrowing gap in the level of development between the valley and the hills with lack of efforts and the promotion been put up and the level of interest through the administrative support was seen. The development standpoint questions the sustainability considering all the existing risks on which the environment is shaped. The damage to the ecosystems is depressing, and therefore, the survival depends on the changes that we have to make at present with the motive of "future now". The need to question is whether the ecosystem services are taken into consideration as the foster innovation to improve the mitigation services and to enhance environmental management. The future is unpredictable, and there are many unforeseen variables like that of the anthropogenic failures and existing hazards causing sudden onset of disasters. The numerous cultural factors have an important influence on decision making of land use as the local communities have certain motivations behind, collective fond of memories and histories based on its traditional practices. The values and belief of the individuals influence a decision on land use systems and its attitude towards risk, and therefore, based on the understanding of various factors, the management of the resources could be improved. Reducing the vulnerability through community participation is extremely important, as involving the community in understanding risks and designing appropriate plan will build disaster resilient communities. The understanding of disaster risk reduction within the community's context of socio, cultural, environmental and economic context will help cope faster with self-motivated participation. The comprehensive monitoring is essential for prevention and sustainable management of the ecology while also upgrading the environment.

Sustainable use of natural resources is essential to improve the environment and its sustainability, livelihood of the local people because people depend mainly on natural resources like dependency on land and forests for several factors. Value creation is essential for the longevity of the ecological value, and the negative impacts such as landslides, erosion, floods and other hazards are to be taken into account as it deserves further attention (Colkesen et al. 2016). Database management of natural resources will help understand the type of resource availability by managing the track for environmental sustainability. Accounting of the natural resources in the study region is essential where the proper record, monitoring and information dissemination will thus improve the ability to manage future risks.

Acknowledgements The authors thank the "United States Geological Survey (USGS) Earth Explorer" for providing the satellite images used in the study, the Survey of India for its toposheet, the Indian Council of Agricultural Research, Manipur, for the rainfall and climate data, the India State of Forest Report for its annual report data, the Mao-Naga community for the field voices and everyone who have involved and contributed directly or indirectly during the study.

Funding This research did not receive any specific grant from funding agencies in the public, commercial or not-for-profit sectors.

References

Anand (2013) Global environmental issues

Andreopoulou Z (ed) (2011) Agricultural and environmental informatics, governance and management: emerging research applications. Information Science Reference, Hershey

Angelsen A, Kaimowitz D, Center for International Forestry Research (eds) (2001) Agricultural technologies and tropical deforestation. CABI Pub. in association with Center for International Forestry Research, Wallingford

Bakker K, Ritts M (2018) Smart earth: a meta-review and implications for environmental governance. Glob Environ Change 52:201–211. https://doi.org/10.1016/j.gloenvcha.2018.07.011

Balamurugan G, Aravind M (2015a) Urban growth effects for land use changes leading to natural hazards I parts of Mumbai metro city. India. Disast Adv 8(10):1–14

Balamurugan G, Aravind M (2015b) Land use and land cover changes in pre- and post-earthquake affected area using Geoinformatics–Western Coast of Gujarat, India

Bera S, Melo R, Guru B (2021) Assessment of exposed elements in a changing built environment by using an integrated model of debris flow initiation and runout (Kalimpong region, Himalaya). Bull Eng Geol Environ 80:7131–7152. https://doi.org/10.1007/s10064-021-02352-w

Borowik T, Pettorelli N, Sönnichsen L, Jędrzejewska B (2013) Normalized difference vegetation index (NDVI) as a predictor of forage availability for ungulates in forest and field habitats. Eur J Wildlife Res 59(5):675–682. https://doi.org/10.1007/s10344-013-0720-0

Colkesen I, Sahin EK, Kavzoglu T (2016) Susceptibility mapping of shallow landslides using kernel-based Gaussian process, support vector machines and logistic regression. J Afr Earth Sci 118:53–64. https://doi.org/10.1016/j.jafrearsci.2016.02.019

Davies KK, Fisher KT, Dickson ME, Thrush SF, Le-Heron R (2015) Improving ecosystem service frameworks to address wicked problems. Ecol Soc 20 (2):200237. https://doi.org/10.5751/ES-07581-200237
Eric L, Helmut G, Erika L (2003) Dynamics of land-use and land-cover changes in tropical regions
Eta JB, Adepoju MO, Halilu SA, Mohammed SO, Adeliyu SA, (2014) Geographic information systems (GIS) as an indispensable tool for environmental impact assessment (EIA). 8:32–39
Foley JA, Ramankutty N, Brauman KA, Cassidy ES, Gerber JS, Johnston M et al (2011) Solutions for a cultivated planet. Nature 478(7369):337–342. https://doi.org/10.1038/nature10452
Folke C, Carpenter S, Elmqvist T, Gunderson L, Holling CS, Walker B (2002) Resilience and sustainable development: building adaptive capacity in a world of transformations. AMBIO J Hum Environ 31(5):437–440. https://doi.org/10.1579/0044-7447-31.5.437
Geological Survey of India (2011) Geology and mineral resources of Manipur, Mizoram, Nagaland, and Tripura. Geol Surv India 1:30
Ghasemi P, Karbasi M, Zamani Nouri A et al (2021) Application of Gaussian process regression to forecast multi-step ahead SPEI drought index. Alexandria Eng J 60:5375–5392. https://doi.org/10.1016/j.aej.2021.04.022
Giam X (2017) Global biodiversity loss from tropical deforestation. Proceed Natl Acad Sci 114(23):5775–5777. https://doi.org/10.1073/pnas.1706264114
Glamann J, Hanspach J, Abson DJ, Collier N, Fischer J (2017) The intersection of food security and biodiversity conservation: a review. Reg Environ Change 17 (5):1303–1313. https://doi.org/10.1007/s10113-015-0873-3
Gobin A, Tarquis AM, Dalezios NR (2013) Preface "Weather-related hazards and risks in agriculture." Nat Hazards Earth Syst Sci 13(10):2599–2603. https://doi.org/10.5194/nhess-13-2599-2013
Griggs D, Stafford-Smith M, Gaffney O, Rockström J, Öhman MC, Shyamsundar P et al (2013) Sustainable development goals for people and planet: policy. Nature 495(7441):305–307. https://doi.org/10.1038/495305a
Grooten A (2018) World wildlife fund (WWF) living planet report 2018: Aiming higher
India State Forest Report (ISFR) (2017) Forest cover: forest survey of India, Ministry of Environment, Forestry and Climate Change (MoEFCC)
Jugindro (2018) Water resource management in Manipur: an emerging problem in the river basin environment and related issues. 7(08):78–82
Koontz TM, Gupta D, Mudliar P, Ranjan P (2015) Adaptive institutions in social-ecological systems governance: a synthesis framework. Environ Sci Policy 53:139–151. https://doi.org/10.1016/j.envsci.2015.01.003
Krausmann F, Lauk C, Haas W, Wiedenhofer D (2018) From resource extraction to outflows of wastes and emissions: the socioeconomic metabolism of the global economy, 1900–2015. Glob Environ Change 52:131–140. https://doi.org/10.1016/j.gloenvcha.2018.07.003
Mahlstein I, Knutti R (2010) Regional climate change patterns identified by cluster analysis. Clim Dyn 35:587–600. https://doi.org/10.1007/s00382-009-0654-0
Manipur State Action Plan on Climate Change (2013) Manipur State action plan on climate change, Directorate of Environment, Government of Manipur
Moghanlo S, Alavinejad M, Oskoei V et al (2021) Using artificial neural networks to model the impacts of climate change on dust phenomenon in the Zanjan region, north-west Iran. Urban Clim 35:100750. https://doi.org/10.1016/j.uclim.2020.100750
Odum EP, Barrett GW (2009) Fundamentals of ecology. Cengage Learning, Australia
Olsson P, Folke C (2001) Local ecological knowledge and institutional dynamics for ecosystem management: a study of Lake Racken Watershed, Sweden. Ecosystems 4(2):85–104. https://doi.org/10.1007/s100210000061
Patel et al (2013) Socio economic and environmental changes with transition from shifting cultivation to settled cultivation in Nort-Eastern India: an ergonomics perspective
Patra SK, Shekher M, Solanki SS, Ramachandran R, Krishnan R (2006) A technique for generating natural colour images from false colour composite images. Int J Remote Sens 27(14):2977–2989. https://doi.org/10.1080/01431160600554322
Reimeingam (2017) Usage of land and labour under shifting cultivation in Manipur
Roy A, Das SK, Tripathi AK, Singh NU, Barman HK (2015) Biodiversity in North East India and their conservation. Progress Agricult 15(2):182. https://doi.org/10.5958/0976-4615.2015.00005.8
Schandl H, Fischer-Kowalski M, West J, Giljum S, Dittrich M, Eisenmenger N, Fishman T (2018) Global material flows and resource productivity: forty years of evidence—global material flows and resource productivity. J Ind Ecol 22(4):827–838. https://doi.org/10.1111/jiec.12626
Singh RK, Prasad VH, Bhatt CM (2004) Remote sensing and GIS approach for assessment of the water balance of a watershed/Evaluation par télédétection et SIG du bilan hydrologique d'un bassin versant. Hydrol Sci J 49(1):131–141. https://doi.org/10.1623/hysj.49.1.131.53997
Singha P, Das P, Talukdar S, Pal S (2020) Modeling livelihood vulnerability in erosion and flooding induced river island in Ganges riparian corridor. India. Ecol Indic 119:106825. https://doi.org/10.1016/j.ecolind.2020.106825
Smit B, Wandel J (2006) Adaptation, adaptive capacity and vulnerability. Glob Environ Change 16(3):282–292. https://doi.org/10.1016/j.gloenvcha.2006.03.008
Steffen W, Richardson K, Rockstrom J, Cornell SE, Fetzer I, Bennett EM, Sorlin S (2015) Planetary boundaries: guiding human development on a changing planet. Science 347(6223):1259855–1259855. https://doi.org/10.1126/science.1259855

Sudmeier-Rieux K, IUCN-The World Conservation Union, International Water Management Institute, CARE/International, IUCN Commission on Ecosystem Management (eds) (2006) Ecosystems, livelihoods and disasters: an integrated approach to disaster risk management. IUCN, Gland

Talluto MV, Mokany K, Pollock LJ, Thuiller W (2018) Multifaceted biodiversity modelling at macroecological scales using Gaussian processes. Divers Distrib 24:1492–1502. https://doi.org/10.1111/ddi.12781

Talukdar S, Pal S, Singha P (2021) Proposing artificial intelligence based livelihood vulnerability index in river islands. J Clean Prod 284:124707. https://doi.org/10.1016/j.jclepro.2020.124707

Taylor L, Watkins SL, Marshall H, Dascombe BJ, Foster J (2016) The impact of different environmental conditions on cognitive function: a focused review. Front Physiol 6:372. https://doi.org/10.3389/fphys.2015.00372

Turner C, Aggarwal A, Walls H, Herforth A, Drewnowski A, Coates J, Kadiyala S (2018) Concepts and critical perspectives for food environment research: a global framework with implications for action in low- and middle-income countries. Glob Food Sec 18:93–101. https://doi.org/10.1016/j.gfs.2018.08.003

Tyagi S, Garg N, Paude R (2014) Environmental degradation: causes and consequences. Eur Res 81 (8–2), 1491. https://doi.org/10.13187/er.2014.81.1491

UN Environment (2018) United Nations environment annual report 2017

United Nations Development Programme (UNDP) report (2018) Sustainable development goals

Wolff NH, Masuda YJ, Meijaard E, Wells JA, Game ET (2018) Impacts of tropical deforestation on local temperature and human well-being perceptions. Glob Environ Change 52:181–189. https://doi.org/10.1016/j.gloenvcha.2018.07.004

Assessing the Impacts of Urbanization on Land Use Change and Carbon Dynamics in English Bazar Municipality Using Remote Sensing and CCDM Modeling

3

Swapan Talukdar, Tanmoy Das,
Somnath Bera, and Atiqur Rahman

Abstract

Land use and land cover (LULC) dynamics play a crucial role in carbon storage and emission patterns, impacting regional carbon balance and climate change mitigation efforts. This study aimed to investigate the LULC dynamics and its influence on carbon storage and emissions in English Bazar Municipality from 2001 to 2021. To achieve these objectives, a combination of remote sensing and modeling techniques was employed. LULC mapping was performed using the random forest (RF) algorithm to categorize the study area into different land cover types for the years 2001, 2011, and 2021. Carbon storage was estimated using the Integrated Valuation of Ecosystem Services and Tradeoffs (InVEST) model, providing Mega-grams of Carbon per hectare (Mg C/ha) for each LULC class. Carbon emission estimation was based on the High Absorption, Low Absorption, Balanced, Low Emission, and High Emission categories, reflecting the carbon absorption and emission capacities of each land cover type. The coupled coordination dynamic model (CCDM) further indicated the complexity of carbon dynamics, emphasizing the need for sustainable land use planning and climate change mitigation strategies. For LULC mapping, the RF model was applied to satellite imagery of three different years (2001, 2011, and 2021). The analysis revealed a decline in vegetation cover and an expansion of built-up areas and agricultural lands over time. The InVEST model was utilized to quantify carbon storage, showing a decrease in total carbon storage from 37,916.44 Mg C/ha in 2001 to 30,358.30 Mg C/ha in 2021. Carbon emissions were estimated based on High Absorption, Low Absorption, Balanced, Low Emission, and High Emission categories. The municipality shifted from being a net carbon sink in 2001 to a net carbon emitter in 2011 and 2021, attributed to urbanization and land use changes. The results of CCDM highlight the significance of conserving vegetation, promoting green spaces, and adopting climate-friendly urban development practices to enhance carbon storage and reduce emissions. Therefore, this study provides a comprehensive assessment of LULC dynamics and carbon storage and emissions in English Bazar Municipality. The findings offer valuable

S. Talukdar
Department of Geography, Asutosh College, University of Calcutta, Kolkata, India

T. Das · A. Rahman
Department of Geography, Faculty of Natural Sciences, Jamia Millia Islamia, New Delhi 110025, India

S. Bera (✉)
Department of Geography, Central University of South Bihar, Gaya 824236, India
e-mail: somnath@cusb.ac.in

S. Talukdar et al. (eds.), *Environmental Risk and Resilience in the Changing World*, GIScience and Geo-environmental Modelling, https://doi.org/10.1007/978-3-031-62442-1_3

insights for policymakers and conservationists to develop targeted strategies for sustainable land use planning and carbon management. Further research is needed to explore long-term impacts, integrate socio-economic factors, and implement nature-based solutions for effective climate change mitigation and environmental conservation.

Keywords

Land use change · Carbon storage · Carbon emissions · Remote sensing · Random forest · Climate change mitigation

3.1 Introduction

The global environmental problems induced by climate change are the most daunting issue to human lives in the present century (IPCC 2013a, b). Increasing temperatures, melting ice, rising sea levels, droughts, forest fires, and floods are recorded high in many places of the world (Wang et al. 2019). The emission of greenhouse gases (GHG), particularly CO_2 induced by human activities, is the leading cause behind the rising temperature and climate change (Han et al. 2017). In a terrestrial ecosystem, land use land cover plays a vital role in regulating climate by balancing CO_2 in the atmosphere. Forests, wetlands, and grasslands collectively store carbon significantly and, in parallel, keep CO_2 out of the atmosphere (IPCC 2000). It also contributes to sequestering additional carbon accumulated in vegetation and soil over the years (Pechanec et al. 2018).

Urbanization leads to substantial changes in land use land cover (LULC), such as the expansion of the built-up area, loss of water bodies, cropland, and green space (Lin et al. 2021). These changes in LULC ultimately disrupt the carbon balance of a region by decreasing carbon stock and increasing carbon emissions (Chuai et al. 2014; Zhang et al. 2015b). It is reported that one-third of the carbon emissions of the world are associated with changes in LULC (IPCC 2013a, b). Simultaneously, it causes changes in soil properties and the degradation of terrestrial ecosystems and biodiversity (Yogani and Thomlinson 2010; Li and Shao 2014; Yigini and Panagos 2016). Rapid urbanization also increases energy consumption and consequently gives rise to a range of environmental problems such as pollution and CO_2 emissions (Al-mulali et al. 2012). According to the International Energy Agency (IEA 2014), urbanized regions are liable for 70% of total CO_2 emissions worldwide (The International Energy Agency IEA 2014). Notably, the urbanized area is expected to include 67% of the total population in 2050 and will cover 1.86 million km^2 in 2030 (Seto et al. 2012). Hence, it is estimated that the impacts of urbanization on carbon emissions will be more significant over the next few decades. In this background, understanding urbanization and its impact on carbon storage and carbon emission is essential for building a sustainable city.

In the literature, the estimation of carbon storage has been accomplished by using field measurements (Golubiewski 2006; Ren et al. 2011) and remote sensing techniques (Pechanec et al. 2018). The field measurement methods provide detailed data by direct contact with a site (Zhang et al. 2015a). However, these methods are costly and time-consuming, especially when applied in large areas (Coomes et al. 2002; Machar et al. 2016). Therefore, numerous studies have adopted contactless remote sensing-based techniques for determining carbon storage at global (Nelson et al. 2010), national (Ni 2013; Zhang et al. 2015b), and regional scales (Zhang et al. 2017; Lyu et al. 2019; Hernández-Guzmán et al. 2019). More recently, several spatial explicitly models have been used for carbon storage estimation, such as the Carbon Exchange in the Vegetation-Soil-Atmosphere System (CESVA) model (Gao et al. 2005), Carnegie Ames Stanford Application (CASA) productivity model (Yu et al. 2009), process-based Dynamic Land Ecosystem Model (Zhang et al. 2012), and Integrated Valuation of Environmental Services and Tradeoffs (InVEST) model (Tao et al. 2015; He et al. 2016; Pechanec et al. 2018; Zhao et al. 2018; Maanan et al. 2019).

Past studies have developed several models for assessing LULC changes as well as urbanization, such as artificial neural networks, logistic regression, Cellular Automata (CA), and Markov Chain (Koranteng and Zawila-Niedzwiecki 2015; Huang et al. 2015). Further, several studies have attempted to estimate the dynamics of carbon storage by linking future LULC scenarios. For example, Jiang et al. (2017) integrated the CLUE-S model and the InVEST model to assess the impact of carbon storage under different scenarios of urban expansion (natural increase scenario, cultivated protection scenario, ecological protection scenario). Chuai et al. (2014) adopted the CLUE-S model and the InVEST model to estimate carbon storage under optimized land use structures in the future. Liang et al. (2017) modified the CLUE-S model by integrating the system dynamics (SD) model and approached coupled SD-SCLUE-S and InVEST model to assess the impact of LULC on carbon storage. He et al. (2016) assessed carbon storage under urban expansion in Beijing from 2011 to 2030 by combining the Land Use Scenario Dynamics-urban (LUSD-urban) model and the InVEST model. Zhao et al. (2013) integrated the FOREcasting SCEnarios of future land cover (FORE-SCE) model and the General Ensemble Biogeochemical Modeling System (GEMS) for estimating carbon storage from 1992 to 2050 in the southeastern USA. Lyu et al. (2019) analyzed the impact of urban expansion on carbon storage by integrating SLEUTH-3r and InVEST model.

To reduce CO_2 emissions, many countries and organizations are focusing on carbon–neutral initiatives. In November 2021, India has promised to have net-zero carbon emissions by 2070 (Li and Haneklaus 2022). At present, India is the third-largest energy consumer in the world. It emitted ~7% of the global anthropogenic greenhouse gases in 2018 (Crippa et al. 2019). It is estimated that around 416 million people will increase in the urban area of India from 2018 to 2050 (UN 2018). This massive urban growth will drive up energy consumption and CO_2 emissions (Tong et al. 2021). In this background, low carbon urbanization is one of the sustainable ways to meet India's target. Monitoring carbon storage and footprints of a city is the initial step for tracking progress toward low carbon urbanization.

3.2 Materials and Methods

3.2.1 Study Area

English Bazar block is the cultural and political hub of Malda District of West Bengal, India. It covers almost 251 km^2, including 108 villages and one municipality. The region is located on a floodplain known as "Diara," composed of younger fluvial deposits. Mahananda is one of the main rivers in the region, flowing along the eastern side of the block. The climate of the region is humid-tropical, with four main seasons dominated: hot summer (March to May), monsoon season (June to September), retreating monsoon (October to November), and winter season (December to February). The temperature ranges from 45 °C (maximum) to 8 °C (minimum) in a year. The region receives an average annual rainfall of around 1326 mm, with the maximum percentage occurring during the monsoon season. Relative humidity ranges from 59 to 91% during the monsoons and drops to 31–55% in the summer season. The region is inhabited by 274,627 people and has experienced rapid population growth due to urbanization (124.81%) between 2001 and 2011, which is the highest growth rate recorded in West Bengal according to the Census of India 2011. Furthermore, within 20 years (between 1991 and 2011), 27 new census towns emerged in this region alone. The region has experienced uncontrolled urban expansion beyond its municipal limit, significantly noticeable along NH-34, SH-10, and the meeting point of N. F. Railway and Eastern Railway (Dutta and Das 2019). Consequently, this has led to land cover transformation, vegetation loss, and various problems such as traffic density, increasing energy consumption, pollution, water crisis, and climate change (Pal and Ziaul 2017; Ziaul and Pal 2018) (Fig. 3.1).

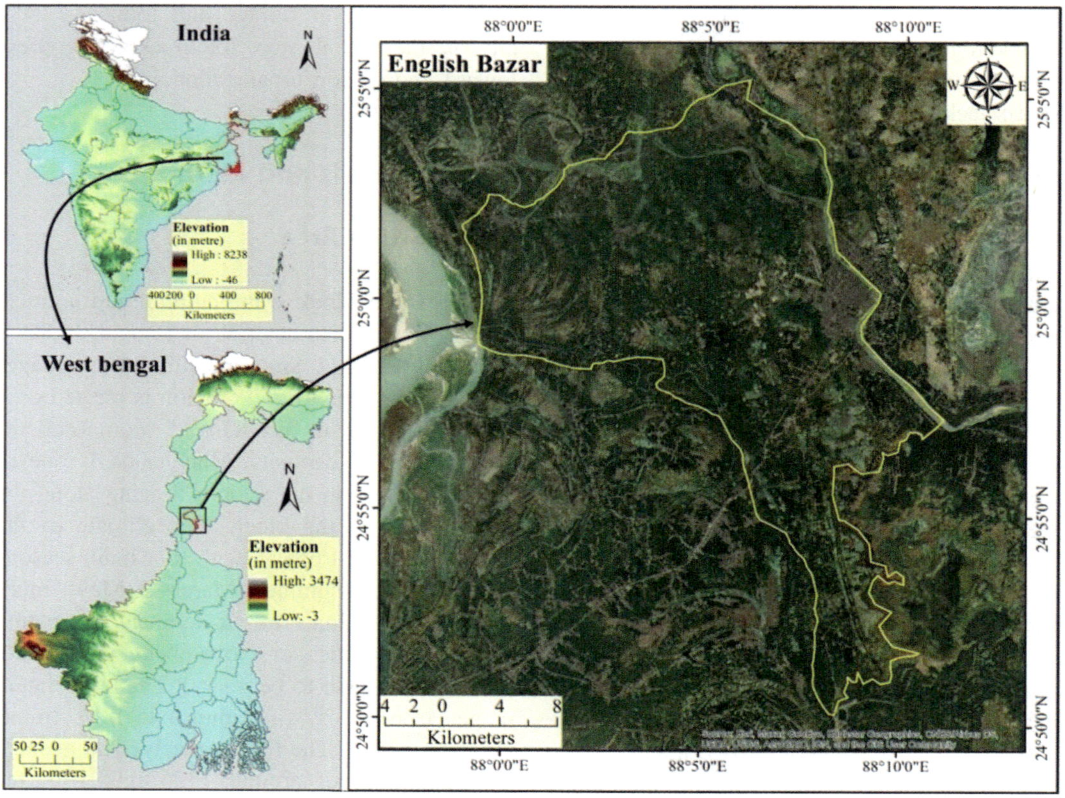

Fig. 3.1 Location of the study area map

3.2.2 Database

In this study, the LULC maps were prepared using Landsat 5 thematic mapper (TM) and the Landsat 8 operational land imager (OLI) with 30 m spatial resolution. TM was used as an input for LULC mapping of 2001 and 2011 whereas OLI was used for the same of 2021. All the satellite images were obtained from the United States Geological Survey (USGS) for December to avoid cloud coverage. In the pre-processing of LULC mapping, radiometric correction of images was carried out using radiometric rescaling coefficients. Then we extracted common three bands such as red, blue, and green bands from all satellite images. Furthermore, the field obtained points and the Google Earth image have been used for the validation of LULC maps.

3.2.3 Method for LULC Mapping (RF) and Validation

Random forest is a nonparametric and ensemble machine learning algorithm (Breiman 2001). It consists of numerous decision trees, with each tree growing based on training data, which is a subset of the original data extracted randomly following the bagging approach. The number of trees and the number of features at each node are two significant parameters of the Random Forest (RF) algorithm. Each set of training data is passed down to each decision tree, which provides individual votes to classify the new dataset. Finally, the output of the classifier is considered based on the majority of votes from all trees. RF offers multiple advantages, such as its ability to handle regression and classification functions, its applicability to both discrete and continuous data, its capacity to avoid overfitting issues, and its ability to resolve complex nonlinear relationships among datasets with higher accuracy. Recently, a host of studies have drawn attention to its potential in the field of land use and land cover (LULC) modeling (Pelletier et al. 2016; Rana and Venkata Suryanarayana 2020; Talukdar et al. 2020).

In the study, the LULC was categorized into five types: water bodies, built-up area, vegetation, agricultural land, and barren land. Kappa statistics (KC) and overall classification accuracy (OCA) were used to validate the LULC map. The accuracy of individual LULC classes was assessed using producer's accuracy (PA), user's accuracy (UA), and conditional kappa value (CKC) (Rwanga and Ndambuki 2017).

3.2.4 Method for Change Analysis

Markov model shows the change of item from one time to another using transition probability matrix. I can depict the nature of changes and as well as future forecasting of the changes. In this study, we use Markovian transition probability for analysis of changes of carbon storage and carbon emission. The Markovian transition probability has been explained through the following questions:

$$S(x+1) = P_{ij} \times S(x), \tag{3.1}$$

where $S(x)$, and $S(x + 1)$ are the status of carbon storage at time x and $x + 1$; and P_{ij} is the transition probability matrix

$$P_{ij} = \begin{bmatrix} p_{11} & p_{12} & \cdots & p_{1n} \\ p_{21} & p_{22} & \cdots & p_{2n} \\ \cdots & \cdots & \cdots & \cdots \\ p_{n1} & p_{n2} & \cdots & p_{nn} \end{bmatrix}.$$

Assuming that

$$\left(0 \le P_{ij} \le 1, \sum_{i=1}^{N} P_{ij} = 1, (i,j = 1,2,\ldots n)\right).$$

3.2.5 Method for Carbon Storage Modeling Using InVest

The InVEST model is popular open-source software which is extensively used for valuing the goods and services obtained from nature (Nelson and Daily 2010). It includes numerous algorithms and sub-modules which can simulate the changes in ecosystem service in different LULC scenarios. The study uses InVEST model for estimating carbon storage and its changes during 2001–2021. We measure the carbon storage by multiplication of the average global carbon density of each LULC feature by their corresponding areas (Table 3.1). The carbon density covers four carbon pools such as aboveground biomass (including living plant materials on the soil), belowground biomass (including living roots below ground), dead organic matter (including dead wood and litter), and soil organic matter (including organic components mixed with soil). The model measures the carbon storage on base a 30 m * 30 m spatial resolution of a cell. The overall process of carbon storage follows the equations:

Table 3.1 Global carbon pool data (IPCC Global Averages) for different land cover types

Land cover type	Aboveground biomass (Mg C/ha)	Belowground biomass (Mg C/ha)	Soil organic carbon (Mg C/ha)
Water bodies	0	0	0
Built-up area	0	0	0
Vegetation	130	62	81
Agricultural land	8	29	26
Barren land	0	0	7

Source IPCC (2014). "Climate Change 2014: Synthesis Report. Contribution of Working Groups I, II and III to the Fifth Assessment Report of the Intergovernmental Panel on Climate Change." Geneva, Switzerland

$$C = C_a + C_b + C_s + C_d \tag{3.2}$$

$$\mathrm{CT} = \sum_{k=1}^{n} A_k * C_k, \quad (k = 1, 2, \ldots n), \tag{3.3}$$

where C indicates the total carbon storage of each LULC type per unit area, C_a, C_b, C_s, C_d indicates carbon density in aboveground, belowground, soil, and dead mass for a cell respectively. A_k represents the area of LULC type and CT represent the total carbon density of a cell.

3.2.6 Method for Carbon Emission Modeling

To estimate carbon emissions from the LULC map in English Bazar Municipality, Malda, West Bengal, India, we used the following methodology, incorporating information from previous literature for carbon emission coefficients.

First, we classify the LULC into the following five categories based on their carbon characteristics: *High Absorption*: This category includes land cover types that have a high capacity for absorbing carbon from the atmosphere. Vegetation, such as tropical deciduous forests and dense tree cover, falls under this category. *Low Absorption*: This category includes land cover types with a relatively lower capacity for carbon absorption compared to High Absorption types. Agricultural lands and scattered trees are examples of Low Absorption classes. *Balanced*: This category includes land cover types that have a balanced relationship between carbon absorption and emission. Barren land and certain open spaces may fall into this category. *Low Emission*: This category includes land cover types with a low tendency to emit carbon into the atmosphere. Water bodies, like ponds and wetlands, typically have low carbon emissions. *High Emission*: This category includes land cover types that have a high potential for emitting carbon into the atmosphere. Built-up areas, including industrial, residential, and commercial spaces, usually have high carbon emissions.

Next, we calculated the area *(a_i)* of each LULC class having carbon characteristics in the study area. This was done by using GIS tools to analyze the LULC map and obtain the respective areas in square meters.

Then, we used carbon emission coefficients *(ε_i)* specific to each LULC class to estimate carbon emissions for each land cover type. The carbon emission coefficients (in kg (C) m^−2 a^−1) for each land cover type were obtained from Table 3.2 of the previous literature (Cui et al. 2018; Fang et al. 2007; Fattah et al. 2021; Hopkinson et al. 2012).

To estimate the carbon emission (LU_CE) for the entire study area, we now consider the Absorption/Emission Category for each LULC class. The calculation will be as follows:

$$\mathrm{LU\_CE} = \Sigma(a\_i * \varepsilon\_i), \tag{3.4}$$

where Σ represents the summation over all land cover classes. However, instead of just summing up the emissions, we will consider the

Table 3.2 Land use land cover features, absorption/emission category, and carbon emission coefficients

Land cover	Absorption/emission category	Description	Carbon emission coefficient (kg (C) m^−2 a^−1)
Water bodies	Low emission	Ponds, canals, lake, river, and wetlands	−0.0459
Built-up area (BU)	High emission	Industrial, residential, commercial areas, road networks, and any kinds of infrastructures	0.0742
Vegetation (VG)	High absorption	Trees and forest areas	−0.0645
Agricultural land	Low absorption	Grassland, cropland	0.0497
Barren land (VL)	Balanced	Playground, bare soils, open space, landfill sites	−0.0527

absorption/emission category of each class and adjust the net carbon emissions accordingly. For example, we will subtract the emissions from High Emission classes and add the emissions from Low Emission classes to the overall carbon balance.

By incorporating the concept of High Absorption, Low Absorption, Balanced, Low Emission, and High Emission along with the carbon emission coefficients, this improved method provides a more comprehensive understanding of the carbon dynamics in the study area. It allows us to consider both carbon emissions and absorption potentials of different land cover types, providing valuable insights for climate change mitigation and land use planning efforts.

3.2.7 Method for Carbon Excess and Surplus Modeling Using Coupling Coordination Degree Model at Pixel Scale

In the improved method, we aim to use the Coupling Coordination Degree Model (CCDM) to understand carbon dynamics through the interaction of carbon storage and carbon emission. The CCDM is a valuable tool for assessing the mutual influence and coordination between two subsystems or system elements, in our case, the carbon storage subsystem and the carbon emission subsystem. To achieve this, we first calculate the coupling degree (C) between carbon storage and carbon emission using the following equation:

$$C = \left(\frac{s(x) \times e(y)}{((s(x) + e(y)/2)^2}\right)^{1/2}, \qquad (3.5)$$

where $s(x)$ represents the index of carbon storage and $e(y)$ represents the index of carbon emission.

Next, we calculate the overall level of carbon storage and carbon emission (T) using the equation:

$$T = \alpha \times s(x) + \beta \times e(y). \qquad (3.6)$$

Here, α and β are the contributions of carbon storage and carbon emission, respectively. Since we want to give equal importance to both carbon storage and carbon emission in our analysis, we will set $\alpha = \beta = 0.5$.

Finally, we determine the degree of coupling coordination (D) using the equation:

$$D = \sqrt{C \times T}. \qquad (3.7)$$

The value of D ranges from 0 to 1, representing the degree of coupling coordination between carbon storage and carbon emission. A D value close to 0 indicates a high level of excess emission, signifying a situation where carbon emissions outweigh carbon storage, potentially leading to a net increase in atmospheric carbon dioxide concentration. On the other hand, a D value close to 1 reflects a high level of excess storage, indicating a scenario where carbon storage predominates, leading to a net decrease in atmospheric carbon dioxide.

The output of the CCDM between carbon storage and emission results in three distinct classes, namely "Excess Emission," "Balanced," and "Excess Storage." The CCDM allows us to categorize the carbon dynamics based on the degree of coupling and coordination between the carbon storage and carbon emission subsystems.

Excess Emission: This class indicates a scenario where the carbon emission subsystem dominates, and the carbon storage subsystem is relatively weaker in its ability to sequester carbon. This situation suggests a net carbon emission, where the carbon released into the atmosphere exceeds the amount being absorbed and stored.

Balanced: The "Balanced" class represents a state where there is a harmonious interaction between the carbon storage and carbon emission subsystems. The carbon sequestration capacity is relatively in equilibrium with carbon emission, resulting in a stable carbon balance.

Excess Storage: This class signifies a condition in which the carbon storage subsystem is dominant, and the carbon emission subsystem is relatively weaker. This situation indicates a net carbon absorption and storage, where the amount of carbon being sequestered exceeds the emissions, leading to a reduction in atmospheric carbon levels.

By classifying the carbon dynamics into these three distinct classes, the CCDM helps in identifying the carbon management status of the study area. It allows us to discern whether the region is acting as a net carbon source (Excess Emission), maintaining a carbon balance (Balanced), or serving as a net carbon sink (Excess Storage). Such insights are crucial for devising effective climate change mitigation and carbon sequestration strategies to combat the impacts of excess carbon emissions on the environment.

3.3 Results

3.3.1 LULC Dynamics Analysis

LULC mapping is a process of categorizing and representing the various types of land use and land cover present in a specific geographical area. This mapping technique provides valuable information about the spatial distribution and dynamics of different land cover types over time. In this study, the researchers analyzed the LULC patterns in a particular region for three different years: 2001, 2011, and 2021. Figure 3.2 displays the spatial distribution of LULC types for each of the mentioned years. The map indicates that the built-up areas are concentrated predominantly in the central-eastern parts of the study region. Surrounding the built-up areas are the vegetation and agricultural lands, which dominate the landscape. In addition to this, barren lands and water bodies are also present, but their distribution is uneven across the region. Table 3.3 presents the area coverage of each LULC class in square kilometers for the years 2001, 2011, and 2021. By analyzing this data, the researchers were able to identify the dynamics of land use and cover changes over the two-decade period.

According to Table 3.3, vegetation covered the largest portion of the study area throughout the entire period. In 2001, vegetation accounted for approximately 45% of the total area. However, over time, there was a constant decrease in vegetation cover from 2011 to 2021. Conversely, there was a decreasing trend in barren land as well. In 2001, barren land covered approximately 10.97% of the area, but by 2021, it had reduced to 4.63%. The built-up area and agricultural land, on the other hand, exhibited an increasing trend between 2001 and 2021. In 2001, built-up areas accounted for only 5.98% of the region, but by 2021, this had expanded to 14%. Similarly, agricultural land saw a substantial increase from 29.03% in 2001 to 85.83% in 2021. Water bodies experienced fluctuations over the study period. They decreased from 2001 to 2011, covering 8.75 and 6.97% of the area, respectively.

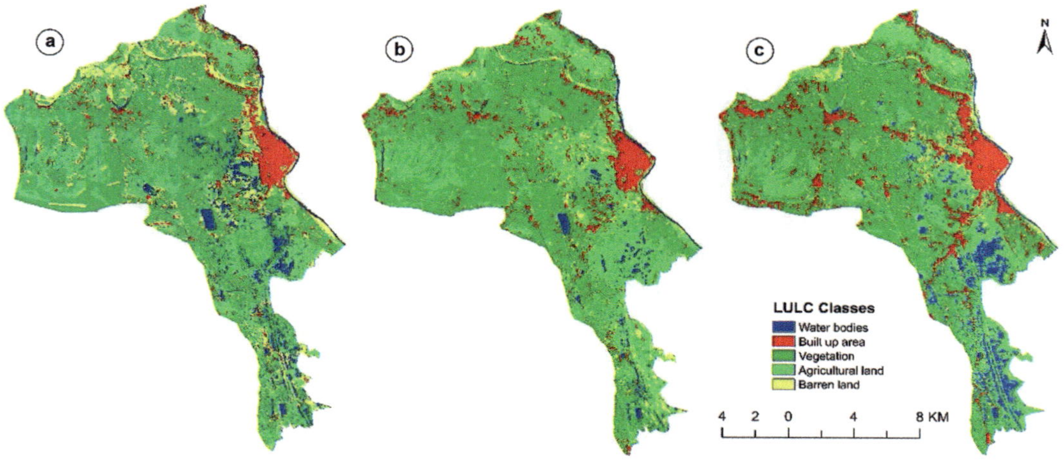

Fig. 3.2 LULC mapping using RF model for the period of **a** 2001, **b** 2011, and **c** 2021

Table 3.3 Computation of area coverages under different LULC classes for the period of 2001–2021

LULC	Area 2001 (km^2)	Area 2011 (km^2)	Area 2021 (km^2)
Water bodies	21.117301	16.82	19.498352
Built up area	14.437573	26.25	33.8
Vegetation	109.306001	99.16	91.16
Agricultural land	70.11	84.2	85.83
Barren land	26.497384	15.04	11.190303
Total area	241.47	241.47	241.47

However, water bodies increased from 2011 to 2021, covering 8.07% of the region by the end of the study period.

Overall, the net change in water bodies from 2001 to 2021 was relatively low compared to other LULC types. This indicates that while there were changes in the extent of water bodies, they were not as significant as the changes observed in vegetation, built-up area, agricultural land, and barren land.

Overall, the LULC mapping and its dynamics in this region reveal significant transformations in land use and land cover over the two-decade period. Vegetation and barren land decreased, while built-up areas and agricultural land expanded. Water bodies exhibited fluctuations, but their overall change was relatively minor.

3.3.2 Validation of LULC

The accuracy assessment of the LULC model is essential to determine the reliability of the mapped land cover types for the years 2001, 2011, and 2021. The results of the accuracy assessment are presented in Table 3.4. The overall classification accuracy of the LULC maps for 2001, 2011, and 2021 is reported to be 92, 90, and 91%, respectively. These values indicate that the LULC model's overall accuracy is quite high, confirming that the mapped land cover types are generally reliable. The Kappa statistics, which assess the agreement between the predicted LULC classes and the actual classes, show values ranging from 0.88 to 0.91 for the three maps. Kappa values above 0.8 are considered to represent excellent accuracy, and in this case, all three maps fall within this range. This demonstrates that the model's predictions closely align with the actual land cover classes.

Specifically, the accuracy of predicting different LULC types is also evaluated. The built-up area is predicted most accurately, achieving a high producer accuracy of 93.25, 96.82, and 98% for 2001, 2011, and 2021, respectively. Producer accuracy refers to the probability of correctly classifying a given LULC type. Moreover, the users' accuracy, which indicates the likelihood of

Table 3.4 Validation of LULC models for the periods of 2001, 2011, and 2021

LULC	2001			2011			2021		
	PA (%)	UA (%)	CKC	PA (%)	UA (%)	CKC	PA (%)	UA (%)	CKC
Water bodies	90	89	0.87	94.56	93	0.92	96.87	95	0.94
Built-up area	93.25	90	0.88	96.82	94	0.94	98	98	0.97
Vegetation	91.55	87	0.87	95.6	95	0.93	97.38	96	0.96
Barren land	88	91	0.91	90.33	91	0.91	95.8	95	0.93
Agricultural land	91.5	87	0.86	89.45	90	0.91	95	96	0.94
Overall classification accuracy	0.92			0.9			0.91		
Overall kappa statistics	0.88			0.9			0.91		

a pixel being correctly classified as a specific LULC class, is more than 85% for all land cover types, further highlighting the model's effectiveness. Looking at individual LULC classes, water bodies, vegetation, barren land, and agricultural land consistently demonstrate high accuracy across all 3 years. Their producer accuracy values range from 88 to 96.87%, and users' accuracy values range from 87 to 96%.

Therefore, the validation results in Table 3.4 indicate that the LULC model used in this study has shown excellent accuracy in classifying land cover types for the years 2001, 2011, and 2021. The high overall classification accuracy and Kappa statistics demonstrate the reliability of the model's predictions. The accuracy of individual land cover types, including built-up areas, vegetation, barren land, and agricultural land, also remains at a satisfactory level, making the LULC mapping and its dynamics analysis a robust and credible representation of the study region's land use changes over time.

3.3.3 Carbon Storage Modeling and Its Dynamics

The analysis of carbon storage dynamics based on the computed results for the years 2001, 2011, and 2021 reveals some interesting trends in the study area. The carbon storage is quantified in terms of Mega-grams of Carbon per hectare (Mg C/ha) for each land cover class.

Between 2001 and 2011:

During this period, there was a decrease in the total carbon storage in the study area. The total carbon storage reduced from 37,916.44 Mg C/ha in 2001 to 32,505.16 Mg C/ha in 2011. This decline can be attributed to changes in land use and land cover, including the conversion of vegetation and agricultural land to other land cover types, as well as possible carbon losses due to deforestation or land degradation. The decrease in carbon storage is particularly noticeable in the vegetation class, which experienced a decrease in area coverage during this decade. This highlights the potential impact of land use changes on carbon sequestration and storage in the region.

Between 2011 and 2021:

In the following decade (2011–2021), there was a further reduction in total carbon storage. The total carbon storage decreased from 32,505.16 Mg C/ha in 2011 to 30,358.30 Mg C/ha in 2021. This indicates a continued trend of carbon loss in the study area. The decline in carbon storage may be influenced by ongoing land use changes, human activities, and environmental factors affecting carbon accumulation and retention. It's essential to investigate the specific drivers of these changes to develop targeted strategies for carbon management and conservation.

Vegetation and Agricultural Land Dynamics:

The two primary contributors to carbon storage in the study area are the vegetation and agricultural land classes. Between 2001 and 2011, there was a noticeable reduction in the carbon storage of vegetation, from 32,505.16 to 27,094.08 Mg C/ha, indicating a significant loss of carbon stored in forests and natural ecosystems. In contrast, agricultural land experienced an increase in carbon storage from 5305.8 Mg C/ha in 2011 to 5404.29 Mg C/ha in 2021, possibly due to changes in land management practices or land cover conversions from other classes.

Barren Land Dynamics:

The carbon storage in barren land also exhibited a declining trend, decreasing from 105.28 Mg C/ha in 2011 to 78.33 Mg C/ha in 2021. This decline could be associated with changes in land management practices, vegetation encroachment, or soil degradation in barren areas.

Dynamics and Urbanization's Influence on Carbon Storage:

The dynamics of carbon storage over the studied period were significantly influenced by urbanization and land use changes. As urban areas expanded, there was an increase in built-up areas, leading to the conversion of natural vegetation and agricultural lands into impervious surfaces.

Built-up areas had negligible carbon storage (0 Mg C/ha) due to the absence of significant vegetation or soil organic carbon.

Urbanization-driven changes in land cover resulted in a decline in vegetation cover, which is evident from the reduction in carbon storage within the vegetation class from 2011 to 2021. Urban development often involves clearing natural vegetation for infrastructure and residential spaces, leading to a loss of carbon sinks. Moreover, agricultural land also experienced some reduction in carbon storage over the years, which may be attributed to changes in agricultural practices, such as increased use of mechanization and the conversion of natural habitats into farmlands.

Mapping of Carbon Dynamics:

The study conducted a comprehensive analysis of carbon storage dynamics in the study area over the period of 2001 to 2021. The carbon storage was mapped for different land use and land cover (LULC) classes, and the generated carbon storage map was further classified into three categories: high, moderate, and low storages. The analysis presented in Fig. 3.3 and Tables 3.5 and 3.6 provides valuable insights into the changes in carbon storage distribution and transitions over the study period.

Table 3.5 shows the computation of area coverage under each carbon storage category for 2001 and 2021, along with the changes (Delta) and the percentage change (%). Notably, the high storage area decreased from 19.24 km^2 in 2001 to 16.67 km^2 in 2021, indicating a reduction of 2.57 km^2 or 1.15% in high carbon storage. In contrast, the medium storage area saw a sharp increase from 19.12 km^2 in 2001 to 129.22 km^2 in 2021, representing an addition of 110.1 km^2 or a remarkable 49.113% increase. Furthermore, the low storage area decreased significantly from 185.81 km^2 in 2001 to 78.28 km^2 in 2021, experiencing a decline of 107.53 km^2 or a considerable 47.97% reduction.

Table 3.6 provides a transition matrix of carbon storage dynamics between 2001 and 2021. The matrix highlights the probabilities of transitions between different carbon storage categories over the study period. The results show that a substantial portion (44.51%) of the high storage area in 2001 transformed into the medium storage area by 2021, indicating a conversion of high carbon storage lands to moderate carbon storage conditions. Similarly, approximately 55.47% of the low storage area in 2001 transitioned into the medium storage area, indicating a shift toward increased carbon storage in these lands. Interestingly, the conversion between high storage area and low storage area was found to be negligible (less than 0.1%), suggesting that these two categories remained relatively stable and underwent minimal changes.

The overall analysis demonstrates significant changes in carbon storage distribution across the

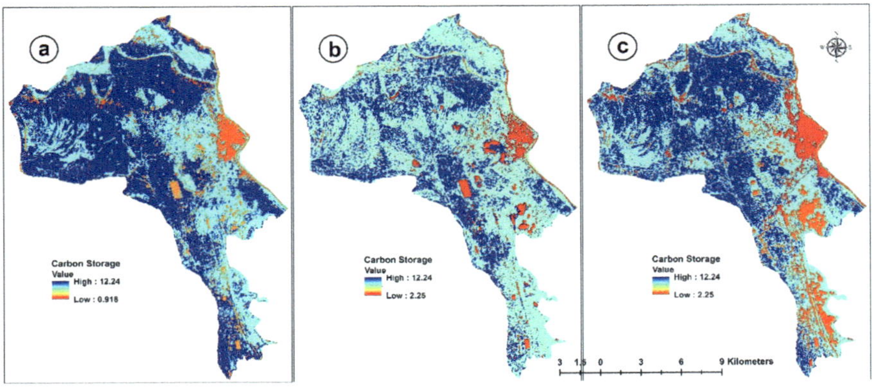

Fig. 3.3 Carbon storage mapping for the period of 2001, 2011, and 2021

Table 3.5 Computation of area coverage under different carbon storage categories and change

	2001(km^2)	2021(km^2)	Delta (km^2)	2001(%)	2021%	Delta%
High storage	19.24	16.67	− 2.57	8.58	7.4361	− 1.15
Medium storage	19.12	129.22	110.1	8.53	57.643	49.113
Low storage	185.81	78.28	− 107.53	82.89	34.92	− 47.97

Table 3.6 Carbon storage dynamics using Markovian transition probability for the period of 2001–2021

2001–2021	High storage	Medium storage	Low storage
High storage	0.445105	0.423586	0.131309
Medium storage	0.034265	0.94173	0.024005
Low storage	0.0401	0.554663	0.405237

study area from 2001 to 2021. High storage areas experienced a decline, likely influenced by factors such as urbanization, deforestation, and land use changes. On the other hand, the moderate storage category witnessed substantial growth, indicating potential carbon sequestration efforts or natural forest regeneration in these regions. The decline in low storage areas may be attributed to carbon loss due to land use practices, land degradation, or agricultural expansion. This study's findings provide valuable insights for policymakers and conservationists, highlighting the need for targeted measures to conserve and enhance carbon storage in high-priority areas, while also recognizing the potential for carbon sequestration in moderate storage regions. Sustainable land use planning, afforestation, reforestation, and climate-friendly policies are crucial to maintaining and enhancing carbon storage in the study area and contributing to climate change mitigation efforts.

3.3.4 Carbon Emission Modeling and Its Dynamics

The analysis of carbon emissions based on the concept of High Absorption, Low Absorption, Balanced, Low Emission, and High Emission provides valuable insights into the carbon dynamics of English Bazar Municipality over the period of 2001 to 2021.

In the *High Absorption* category, which includes land cover types with a high capacity for carbon absorption, the municipality had significant carbon sinks in 2001, 2011, and 2021. However, there was a gradual decline in carbon absorption over time. The total carbon absorption in this category decreased from approximately −7048 tC in 2001 to −5878 tC in 2021. This decline can be attributed to factors such as deforestation, urbanization, and changes in land management practices that may have reduced the area of dense forest and vegetation, which act as potent carbon sinks.

In contrast, the *Low Absorption* category, representing land cover types with lower carbon absorption capacities, experienced an increasing trend in carbon emissions. Carbon emissions from this category rose from around 3484 tC in 2001 to 4259 tC in 2021. This increase is likely due to the expansion of agricultural lands and scattered tree cover, which may have contributed to higher carbon emissions from these areas.

The *Balanced category*, which has a balanced relationship between carbon absorption and emission, exhibited a decline in carbon emissions over time. The total carbon emissions in this category decreased from approximately −1395 tC in 2001 to −590 tC in 2021. This reduction may be influenced by changes in land use and management practices in barren lands and open spaces, leading to a decrease in emissions.

In the *Low Emission* category, comprising land cover types with low tendencies to emit carbon, the carbon emissions remained negative, indicating a net carbon absorption. However, there was a slight increase in carbon emissions over time, with the total carbon absorption declining from approximately −970 tC in 2001 to −896 tC in 2021. This trend might be due to changes in water bodies and associated land use patterns, impacting their carbon absorption capabilities.

The *High Emission* category, representing land cover types with high potential for carbon emissions, showed a consistent increase in carbon emissions. The total carbon emissions from this category rose from around 1072 tC in 2001 to 2509 tC in 2021. The expansion of built-up areas and infrastructures, accompanied by increased human activities and urbanization, likely contributed to this escalation in carbon emissions.

Overall, the municipality experienced a shift from being a net carbon sink in 2001 to a net carbon emitter in 2011 and 2021. The total carbon emissions for the entire study area increased from approximately −4857 tC in 2001 to 1403 tC in 2021. This shift can be primarily attributed to the rapid urbanization and expansion of built-up areas, which contributed to higher carbon emissions. Additionally, changes in land use patterns, deforestation, and alterations in agricultural practices might have further influenced the carbon dynamics.

The increasing trend of urbanization in English Bazar Municipality has led to a decline in carbon absorption over time, primarily due to the reduction in forested areas and the expansion of built-up areas. Urbanization often leads to the conversion of green spaces into concrete structures, resulting in the loss of vegetation and carbon sinks. As a result, the overall carbon balance has shifted toward a net carbon emission, indicating an increased contribution to greenhouse gas concentrations and climate change. Understanding the impact of urbanization on carbon emission and absorption dynamics is crucial for sustainable land use planning and climate change mitigation strategies. Implementing measures to conserve and restore vegetation, promoting green spaces, and adopting sustainable urban planning practices can help offset carbon emissions and enhance the capacity for carbon absorption in urban areas. Moreover, it highlights the significance of balancing development goals with environmental conservation to address the challenges posed by increasing urbanization and climate change.

Dynamics Using Transition Probability Matrix:

The analysis of carbon emission mapping (Fig. 3.4) and its dynamics using the Markovian transition probability matrix for the period of 2001–2021 provides valuable insights into the transformation of carbon absorption and emission zones in English Bazar Municipality over time.

The transition matrix between 2001 and 2011 shows that a significant portion (22%) of the High Absorption area, which had a high capacity for carbon absorption, turned into a High Emission zone (Table 3.7). This change can be attributed to factors such as deforestation, land use changes, and urbanization, which might have led to the loss of forested areas and vegetation, thus reducing the carbon absorption capacity. However, a substantial portion (69%) of the High Absorption area remained stable, indicating its resilience as a carbon sink over the decade. The Low Absorption zone, which had a lower capacity for carbon absorption, also experienced transformations. About 45% of the Low Absorption area converted into the High Emission zone, indicating a shift toward increased carbon emissions. This change might be associated with changes in land use practices, such as agricultural expansion and loss of vegetative cover. The Balanced zone, which had a balanced relationship between carbon absorption and emission, was the most unstable of all categories. Nearly half (48%) of the area in the Balanced zone changed into the High Emission zone, suggesting that it became a significant source of carbon emissions over time. Among the Low Emission zone, 27% of the area converted into the High Emission zone, indicating a shift toward increased carbon emissions. On the other hand,

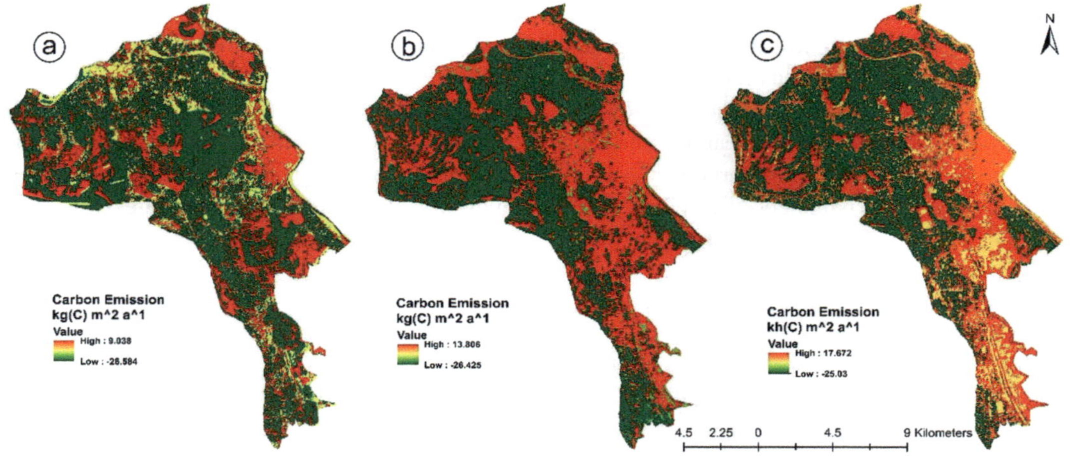

Fig. 3.4 Carbon emission modeling for the period of 2001, 2011, and 2021

Table 3.7 Transition probability matrix computation for dynamics of carbon absorption and emission for the period of 2001–2011

2001–2011	High absorption	Low absorption	Balanced	Low emission	High emission
High absorption	0.694594	0.009802	0.027229	0.040385	0.227991
Low absorption	0.148259	0.320673	0.035316	0.044853	0.4509
Balanced	0.293863	0.013052	0.112967	0.098326	0.481792
Low emission	0.190418	0.009362	0.012299	0.512697	0.275225
High emission	0.400007	0.022684	0.043086	0.067979	0.466244

Table 3.8 Transition probability matrix computation for dynamics of carbon absorption and emission for the period of 2011–2021

2011–2021	High absorption	Low absorption	Balanced	Low emission	High emission
High absorption	0.732033	0.023296	0.036487	0.061269	0.146915
Low absorption	0.009329	0.71947	0.022483	0.001493	0.247225
Balanced	0.138686	0.076149	0.255869	0.060761	0.468534
Low emission	0.215598	0.035751	0.042334	0.518231	0.188086
High emission	0.241841	0.11586	0.088561	0.105991	0.447746

40% of the High Emission zone turned into the High Absorption zone, suggesting some efforts in reforestation or afforestation during this period.

The similar trend of transformation was observed between 2011 and 2021 and 2001 and 2021 (Tables 3.8 and 3.9). The Balanced zone remained the most unstable during both time periods, with 46% of the Balanced area turning into the High Emission zone between 2011 and 2021. In 2021, 36% of the Balanced area turned into the High Absorption zone, indicating some recovery of carbon absorption capabilities.

Overall, the transition probability matrix highlights the changes in the distribution of carbon absorption and emission zones over the two-

Table 3.9 Transition probability matrix computation for dynamics of carbon absorption and emission for the period of 2001–2021

2001–2021	High absorption	Low absorption	Balanced	Low emission	High emission
High absorption	0.629302	0.036756	0.052344	0.089547	0.192051
Low absorption	0.045657	0.47336	0.04145	0.012247	0.427286
Balanced	0.365731	0.055738	0.133981	0.109175	0.335375
Low emission	0.216191	0.023314	0.05183	0.575756	0.132909
High emission	0.422529	0.069464	0.060379	0.07239	0.375238

decade period. The High Emission zone expanded, while the High Absorption zone declined, leading to an overall shift from being a net carbon sink in 2001 to a net carbon emitter in 2011 and 2021.

The increasing trend of urbanization and changes in land use patterns appear to have had a significant impact on the carbon emission and absorption dynamics in English Bazar Municipality. The expansion of built-up areas and infrastructures, along with changes in land management practices, has likely contributed to higher carbon emissions. Concurrently, deforestation, loss of vegetative cover, and reduced green spaces have led to a decline in carbon absorption capacities, resulting in a shift toward net carbon emissions. To address these challenges, sustainable land use planning and climate change mitigation strategies should be implemented. Initiatives to conserve and restore vegetation, promote green spaces, and adopt sustainable urban planning practices can help offset carbon emissions and enhance the capacity for carbon absorption in urban areas. Furthermore, reforestation and afforestation efforts in High Emission zones could aid in increasing carbon absorption capabilities and mitigate the impact of urbanization on carbon dynamics in the municipality.

3.3.5 Carbon Storage and Emission Excess and Surplus Modeling and Its Dynamics

The CCDM was applied to analyze the carbon dynamics through the interaction of carbon storage and carbon emission systems for the periods of 2001, 2011, and 2021. The results are presented in Fig. 3.5, illustrating the degree of coupling coordination between these two subsystems over time.

Figure 3.5 displays the CCDM results for 2001, 2011, and 2021, showing the transition of the system from excess emission, balanced, to excess storage. The color-coded maps provide a visual representation of the coordination between carbon storage and emission, with values ranging from 0 to 1. In these figures, areas with green shades represent a higher degree of coupling coordination, while red areas indicate a lower degree of coordination.

To further analyze the dynamics of carbon storage and emission, Markovian transition probability matrices were computed for the periods of 2001–2011, 2011–2021, and 2001–2021. Tables 3.10, 3.11, and 3.12 present the transition probability matrices for each respective time period.

Table 3.10 indicates the probabilities of transitioning between excess emission, balanced, and excess storage categories from 2001 to 2011. The results show that there was a considerable likelihood (0.599266) of remaining in the excess emission category, suggesting that areas with high carbon emissions tended to maintain this status over the decade. The balanced category had a probability of 0.481634 of remaining balanced, indicating relative stability in the carbon storage and emission relationship. Notably, there was a higher probability (0.875896) of transitioning from excess storage to excess emission, indicating that some areas shifted from being carbon sinks to carbon sources.

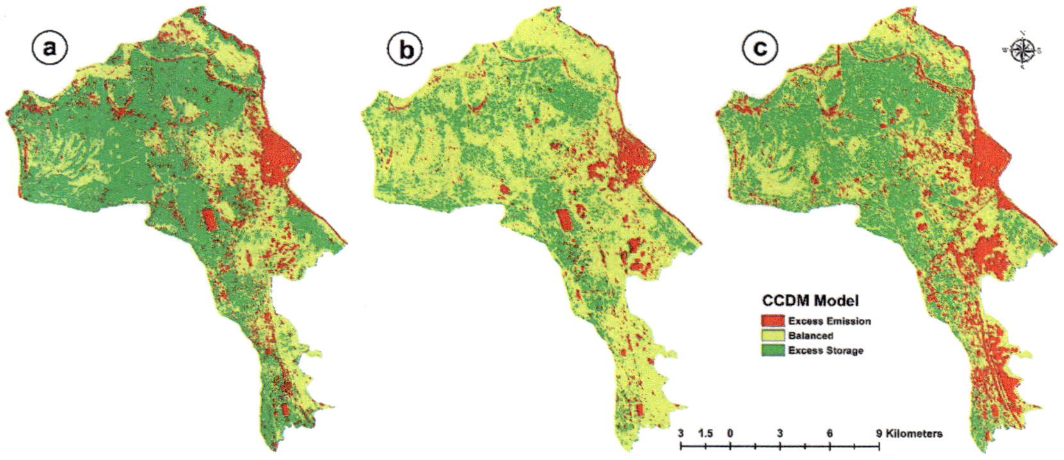

Fig. 3.5 CCDM for carbon dynamics through the interaction of carbon storage and carbon emission systems for the period of **a** 2001, **b** 2011, and **c** 2021

Table 3.10 Transition probability matrix computation for dynamics of carbon storage and emission for the period of 2001–2011

2001–2011	Excess emission	Balanced	Excess storage
Excess emission	0.599266	0.307813	0.092921
Balanced	0.111018	0.481634	0.407348
Excess storage	0.023933	0.100171	0.875896

Table 3.11 Transition probability matrix computation for dynamics of carbon storage and emission for the period of 2011–2021

2001–2011	Excess emission	Balanced	Excess storage
Excess emission	0.668153	0.313438	0.01841
Balanced	0.157664	0.557454	0.284881
Excess storage	0.036269	0.33931	0.624422

Table 3.12 Transition probability matrix computation for dynamics of carbon storage and emission for the period of 2001–2021

2011–2021	Excess emission	Balanced	Excess storage
Excess emission	0.583072	0.333323	0.083605
Balanced	0.200053	0.683581	0.116366
Excess storage	0.036385	0.377151	0.586464

Table 3.11 provides the transition probabilities for the period of 2011–2021. The results show an increase in the probability of transitioning from excess emission to balanced (0.313438) and from balanced to excess storage (0.284881). This suggests that some areas shifted

toward more balanced carbon dynamics, with a higher tendency toward carbon storage. However, there was still a significant probability (0.668153) of remaining in the excess emission category, indicating persistent carbon emission hotspots.

Table 3.12 displays the transition probabilities for the entire period of 2001–2021. The results show that the probability of remaining balanced increased (0.683581) compared to the previous time intervals, indicating a trend toward enhanced carbon coordination. However, the probabilities of remaining in the excess emission and excess storage categories remained relatively high, suggesting that both carbon emission and carbon storage dynamics were prevalent throughout the study period.

Overall, the CCDM analysis, along with the transition probability matrices, provides valuable insights into the dynamics of carbon storage and emission over time. The results indicate that the study area experienced changes in carbon coordination, with some areas transitioning from excess emission to more balanced or excess storage states. However, persistent carbon emission hotspots and carbon storage areas were observed, highlighting the complexity of carbon dynamics in the study region. These findings can be crucial for designing targeted climate change mitigation strategies and carbon management initiatives to achieve a more sustainable and balanced carbon footprint in the future.

3.4 Policy Implication

Based on the results of the study, several specific policy implications can be drawn for English Bazar Municipality to address the observed changes in land use, carbon storage, and carbon emissions:

1. Conservation and Restoration of Vegetation: Given the significant decline in vegetation cover and carbon storage over the two-decade period, conservation and restoration efforts should be prioritized. Implementing afforestation and reforestation projects can help increase the carbon sequestration potential of the region, promoting biodiversity and ecosystem services.
2. Sustainable Land Use Planning: To mitigate the impact of urbanization and expansion of built-up areas on carbon storage and emissions, municipal authorities should focus on sustainable land use planning. This may involve incentivizing green spaces, green roofs, and urban forests to maintain carbon sinks within the city.
3. Climate-Friendly Urban Development: Encouraging climate-friendly urban development practices, such as green buildings, energy-efficient infrastructure, and public transportation, can help reduce carbon emissions from the built environment. Smart growth strategies can also limit urban sprawl and preserve natural areas.
4. Carbon Offset Initiatives: Considering the increased carbon emissions in certain areas, implementing carbon offset initiatives can help balance the carbon budget. Supporting projects like carbon sequestration through improved land management or renewable energy development can help offset emissions from high-emission zones.
5. Agricultural Practices and Carbon Farming: Given the substantial increase in agricultural land and its potential for carbon storage, promoting sustainable agricultural practices and carbon farming techniques can enhance carbon sequestration in these areas.
6. Integrated Watershed Management: To address fluctuations in water bodies and their impact on carbon storage, adopting integrated watershed management practices can help maintain healthy aquatic ecosystems and preserve carbon stocks in wetlands.
7. Climate Change Mitigation and Adaptation Policies: The municipality should develop and implement climate change mitigation and adaptation policies that address both carbon storage and emissions. These policies can support initiatives such as green infrastructure, carbon trading programs, and sustainable urban development guidelines.

8. Public Awareness and Education: Raising awareness among citizens and stakeholders about the importance of carbon storage and emissions can foster support for climate-friendly initiatives. Educational campaigns can promote responsible land use, carbon-conscious behavior, and community involvement in conservation efforts.
9. Monitoring and Assessment: Continuous monitoring of land use changes, carbon storage, and emissions is essential for informed decision-making. Establishing a regular assessment mechanism can help track progress, identify areas of concern, and adapt policies accordingly.
10. Stakeholder Collaboration: Engaging various stakeholders, including government agencies, NGOs, private sector, and local communities, in carbon management efforts can lead to effective and inclusive strategies for carbon reduction and sustainable development.

In conclusion, the study's findings indicate significant transformations in land use, carbon storage, and carbon emissions dynamics in English Bazar Municipality over the studied period. Policymakers and local authorities can use these results to develop targeted strategies and action plans that promote sustainable land use, carbon sequestration, and climate change mitigation in the region. By adopting a holistic approach and collaborating with various stakeholders, the municipality can work toward achieving a more balanced carbon footprint and contribute to global efforts to combat climate change.

3.5 Conclusion

The present study conducted a comprehensive analysis of land use and land cover dynamics and its impact on carbon storage and emissions in English Bazar Municipality over the period of 2001–2021. Through LULC mapping and carbon storage modeling, the researchers observed significant changes in the spatial distribution of different land cover types and carbon storage levels. The results revealed a decline in vegetation cover and carbon storage over time, while built-up areas and agricultural lands expanded. This expansion of urbanization and changes in land use patterns were found to be major contributors to increased carbon emissions, leading to a shift from being a net carbon sink in 2001 to a net carbon emitter in 2011 and 2021. The analysis further explored the dynamics of carbon storage and emission through transition probability matrices and coupled coordination models. While some areas exhibited a shift toward balanced carbon dynamics, persistent carbon emission hotspots and storage areas were identified, emphasizing the complexity of carbon dynamics in the study region.

The significance of this study lies in providing valuable insights for policymakers and conservationists to develop targeted strategies for sustainable land use planning, carbon sequestration, and climate change mitigation. The findings emphasize the importance of conserving vegetation and promoting green spaces within the municipality to enhance carbon storage and biodiversity. Sustainable urban development practices, such as green infrastructure and energy-efficient buildings, are crucial to offsetting carbon emissions from the expanding built environment. The study also highlights the potential for carbon sequestration in agricultural lands, encouraging the adoption of sustainable agricultural practices and carbon farming techniques.

However, the study has several limitations that should be considered. Firstly, the research focused on a specific time frame of two decades, and longer-term data may provide a more comprehensive understanding of land use and carbon dynamics. Additionally, the study utilized remote sensing and modeling techniques, which may introduce uncertainties in the accuracy of results. Further field validation and ground-truthing of data could enhance the reliability of the findings. Moreover, the research focused on carbon storage and emissions at a regional level and did not account for potential variations within smaller sub-regions, which could affect policy implications.

For future research, an extended temporal analysis beyond 2021 could provide more insights into the long-term impacts of land use changes and carbon dynamics. Integrating socio-economic factors and human activities into the analysis could provide a more holistic understanding of the drivers of carbon emissions. Furthermore, exploring the potential of nature-based solutions, such as afforestation, reforestation, and wetland restoration, could offer practical solutions for carbon sequestration and climate change adaptation. Additionally, conducting scenario-based modeling to assess the effectiveness of different policy interventions and land management strategies could aid in formulating evidence-based decisions.

References

Al-mulali U, Binti Che Sab CN, Fereidouni HG (2012) Exploring the bi-directional long run relationship between urbanization, energy consumption, and carbon dioxide emission. Energy 46:156–167. https://doi.org/10.1016/j.energy.2012.08.043

Breiman L (2001) Random forests. Mach Learn 45:5–32. https://doi.org/10.1023/A:1010933404324

Chuai X, Huang X, Wang W et al (2014) Spatial simulation of land use based on terrestrial ecosystem carbon storage in Coastal Jiangsu, China. Sci Rep 4:5667. https://doi.org/10.1038/srep05667

Coomes DA, Allen RB, Scott NA et al (2002) Designing systems to monitor carbon stocks in forests and shrublands. For Ecol Manag 164:89–108. https://doi.org/10.1016/S0378-1127(01)00592-8

Crippa M, Oregionni G, Guizzard D, et al (2019) Fossil CO_2 and GHG emissions of all world countries—2019 Report

Cui Y, Li L, Chen L, Zhang Y, Cheng L, Zhou X, Yang X (2018) Land-use carbon emissions estimation for the Yangtze River Delta Urban Agglomeration using 1994–2016 Landsat image data. Remote Sens 10(9):1334

Dutta I, Das A (2019) Application of geo-spatial indices for detection of growth dynamics and forms of expansion in English Bazar Urban Agglomeration, West Bengal. J Urban Manag 8:288–302. https://doi.org/10.1016/j.jum.2019.03.007

Fang J, Guo Z, Piao S, Chen A (2007) Terrestrial vegetation carbon sinks in China, 1981–2000. Sci China Ser D Earth Sci 50(9):1341–1350

Fattah MA, Morshed SR, Morshed SY (2021) Multi-layer perceptron-Markov chain-based artificial neural network for modelling future land-specific carbon emission pattern and its influences on surface temperature. SN Appl Sci 3:1–22

Gao Z, Liu J, Cao M et al (2005) Impacts of land-use and climate changes on ecosystem productivity and carbon cycle in the cropping-grazing transitional zone in China. Sci China Ser D Earth Sci 48:1479–1491. https://doi.org/10.1360/03yd0372

Golubiewski NE (2006) Urbanization increases grassland carbon pools: effects of landscaping in Colorado's front range. Ecol Appl 16:555–571. https://doi.org/10.1890/1051-0761(2006)016[0555:UIGCPE]2.0.CO;2

Han J, Meng X, Zhou X et al (2017) A long-term analysis of urbanization process, landscape change, and carbon sources and sinks: a case study in China's Yangtze River Delta region. J Clean Prod 141:1040–1050. https://doi.org/10.1016/j.jclepro.2016.09.177

He C, Zhang D, Huang Q, Zhao Y (2016) Assessing the potential impacts of urban expansion on regional carbon storage by linking the LUSD-urban and InVEST models. Environ Modell Softw 75:44–58. https://doi.org/10.1016/j.envsoft.2015.09.015

Hernández-Guzmán R, Ruiz-Luna A, González C (2019) Assessing and modeling the impact of land use and changes in land cover related to carbon storage in a western basin in Mexico. Remote Sens Appl Soc Environ 13:318–327. https://doi.org/10.1016/j.rsase.2018.12.005

Hopkinson CS, Cai WJ, Hu X (2012) Carbon sequestration in wetland dominated coastal systems—a global sink of rapidly diminishing magnitude. Curr Opin Environ Sustain 4(2):186–194

Huang J, Wu Y, Gao T et al (2015) An integrated approach based on Markov chain and cellular automata to simulation of urban land use changes. Appl Math Inform Sci 9:769–775

IPCC (2000) Land use, land-use change, and forestry: a special report of the IPCC

IPCC (2013a) Working group I contribution to the IPCC fifth assessment report, climate change 2013: the physical science basis. IPCC AR5:2014

IPCC (2013b) International panel of climate change. Climate change 2013: the physical science basis contribution of working group I to the fifth assessment report of the intergovernmental panel on climate change 33

Jiang W, Deng Y, Tang Z et al (2017) Modelling the potential impacts of urban ecosystem changes on carbon storage under different scenarios by linking the CLUE-S and the InVEST models. Ecol Modell 345:30–40. https://doi.org/10.1016/j.ecolmodel.2016.12.002

Koranteng A, Zawila-Niedzwiecki T (2015) Modelling forest loss and other land use change dynamics in Ashanti Region of Ghana. Folia Forestalia Polonica A 57:96–111. https://doi.org/10.1515/ffp-2015-0010

Li B, Haneklaus N (2022) The potential of India's net-zero carbon emissions: analyzing the effect of clean energy, coal, urbanization, and trade openness. Energy Rep 8:724–733. https://doi.org/10.1016/j.egyr.2022.01.241

Li D, Shao M (2014) Soil organic carbon and influencing factors in different landscapes in an arid region of northwestern China. CATENA 116:95–104. https://doi.org/10.1016/j.catena.2013.12.014

Liang Y, Liu L, Huang J (2017) Integrating the SD-CLUE-S and InVEST models into assessment of oasis carbon storage in northwestern China. PLoS ONE 12:72494. https://doi.org/10.1371/journal.pone.0172494
Lin Q, Zhang L, Qiu B et al (2021) Spatiotemporal analysis of land use patterns on carbon emissions in China. Land 10:1–13. https://doi.org/10.3390/land10020141
Lyu R, Mi L, Zhang J et al (2019) Modeling the effects of urban expansion on regional carbon storage by coupling SLEUTH-3r model and InVEST model. Ecol Res 34:380–393. https://doi.org/10.1111/1440-1703.1278
Maanan M, Maanan M, Karim M et al (2019) Modelling the potential impacts of land use/cover change on terrestrial carbon stocks in north-west Morocco. Int J Sustain Develop World Ecol 26:560–570. https://doi.org/10.1080/13504509.2019.1633706
Machar I, Simon J, Rejsek K et al (2016) Assessment of forest management in protected areas based on multidisciplinary research. Forests 7:285. https://doi.org/10.3390/f7110285
Nelson E, Sander H, Hawthorne P et al (2010) Projecting global land-use change and its effect on ecosystem service provision and biodiversity with simple models. PLoS ONE 5:14327. https://doi.org/10.1371/journal.pone.0014327
Nelson EJ, Daily GC (2010) Modelling ecosystem services in terrestrial systems. F1000 Biol Rep 2
Ni J (2013) Carbon storage in Chinese terrestrial ecosystems: approaching a more accurate estimate. Clim Change 119:905–917. https://doi.org/10.1007/s10584-013-0767-7
Pal S, Ziaul S (2017) Detection of land use and land cover change and land surface temperature in English Bazar urban centre. Egy J Remote Sens Space Sci 20:125–145. https://doi.org/10.1016/j.ejrs.2016.11.003
Pechanec V, Purkyt J, Benc A et al (2018) Modelling of the carbon sequestration and its prediction under climate change. Ecol Inform 47:50–54. https://doi.org/10.1016/j.ecoinf.2017.08.006
Pelletier C, Valero S, Inglada J et al (2016) Assessing the robustness of random forests to map land cover with high resolution satellite image time series over large areas. Remote Sens Environ 187:156–168. https://doi.org/10.1016/j.rse.2016.10.010
Rana VK, Venkata Suryanarayana TM (2020) Performance evaluation of MLE, RF and SVM classification algorithms for watershed scale land use/land cover mapping using sentinel 2 bands. Remote Sens Appl Soc Environ 19:351. https://doi.org/10.1016/j.rsase.2020.100351
Ren Y, Wei X, Wei X et al (2011) Relationship between vegetation carbon storage and urbanization: a case study of Xiamen, China. For Ecol Manag 261:1214–1223. https://doi.org/10.1016/j.foreco.2010.12.038
Rwanga SS, Ndambuki JM (2017) Accuracy assessment of land use/land cover classification using remote sensing and GIS. Int J Geosci 8:611–622. https://doi.org/10.4236/ijg.2017.84033
Seto KC, Güneralp B, Hutyra LR (2012) Global forecasts of urban expansion to 2030 and direct impacts on biodiversity and carbon pools. Proceed Natl Acad Sci USA 109:16083–16088. https://doi.org/10.1073/pnas.1211658109
Talukdar S, Singha P, Mahato S et al (2020) Land-use land-cover classification by machine learning classifiers for satellite observations: a review. Remote Sens 12:526–593
Tao Y, Li F, Liu X et al (2015) Variation in ecosystem services across an urbanization gradient: a study of terrestrial carbon stocks from Changzhou, China. Ecol Modell 318:210–216. https://doi.org/10.1016/j.ecolmodel.2015.04.027
The International Energy Agency IEA (2014) Key world energy statistics 2014. Int Energy Agency 82:e11012
Tong K, Nagpure AS, Ramaswami A (2021) All urban areas' energy use data across 640 districts in India for the year 2011. Sci Data 8:853. https://doi.org/10.1038/s41597-021-00853-7
UN (2018) World urbanization prospects 2018
Wang S, Wang J, Fang C, Li S (2019) Estimating the impacts of urban form on CO_2 emission efficiency in the Pearl River Delta, China. Cities 85:117–129. https://doi.org/10.1016/j.cities.2018.08.009
Yigini Y, Panagos P (2016) Assessment of soil organic carbon stocks under future climate and land cover changes in Europe. Sci Total Environ 557–558:838–850. https://doi.org/10.1016/j.scitotenv.2016.03.085
Yogani G, Thomlinson JR (2010) Changes in landuse/landcover affect distribution and habitat of the land crab, *Cardisoma guanhumi* (Gecarcinidae, decapoda) in two estuaries in Puerto Rico. Caribbean J Sci 46:258–266
Yu Y, Guo Z, Wu H et al (2009) Spatial changes in soil organic carbon density and storage of cultivated soils in China from 1980 to 2000. Glob Biogeochem Cycles 23:3428. https://doi.org/10.1029/2008GB003428
Zhang C, Tian H, Chen G et al (2012) Impacts of urbanization on carbon balance in terrestrial ecosystems of the Southern United States. Environ Pollut 164:89–101. https://doi.org/10.1016/j.envpol.2012.01.020
Zhang C, Ju W, Chen JM et al (2015a) Disturbance-induced reduction of biomass carbon sinks of China's forests in recent years. Environ Res Lett 10:114021. https://doi.org/10.1088/1748-9326/10/11/114021
Zhang M, Huang X, Chuai X, et al (2015b) Impact of land use type conversion on carbon storage in terrestrial ecosystems of China: a spatial-temporal perspective. Sci Rep 5:10233
Zhang F, Zhan J, Zhang Q et al (2017) Impacts of land use/cover change on terrestrial carbon stocks in Uganda. Phys Chem Earth 101:195–203. https://doi.org/10.1016/j.pce.2017.03.005
Zhao S, Liu S, Sohl T et al (2013) Land use and carbon dynamics in the southeastern United States from 1992

to 2050. Environ Res Lett 8:044022. https://doi.org/10.1088/1748-9326/8/4/044022
Zhao Z, Liu G, Mou N et al (2018) Assessment of carbon storage and its influencing factors in Qinghai-Tibet plateau. Sustainability 10:61864. https://doi.org/10.3390/su10061864
Ziaul S, Pal S (2018) Analyzing control of respiratory particulate matter on land surface temperature in local climatic zones of English Bazar municipality and surroundings. Urban Clim 24:34–50. https://doi.org/10.1016/j.uclim.2018.01.006

4 Quantifying the Impact of Urban Expansion on Land Resources Using the Land Resource Impact Assessment Framework: A Case Study of Aligarh City, India

Md Kaikubad Ali and Shahab Fazal

Abstract

This paper aims to assess the changes in land use and land cover caused by the urban expansion in one of the rapidly growing urban centers in Western Uttar Pradesh. The prime objective of the research is to evaluate the impact of urban expansion and urban development on the land resources in the urban fringe of Aligarh city using seven Land Resource Impact Assessment (LRIA) Indicators. The Land Resource Impact Assessment (LRIA) is a framework that analytically examines the per capita consumption of land due to urban expansion and resultant new development by integrating demographic data. The study has been carried out in the GIS environment incorporating satellite and analog maps of different times with population data of similar census periods. The results show a significant increase in urban built-up (307.3% increase) during 1980–2018. The LRIA reveals urban expansion in Aligarh city has consumed 179.73 ha per annum during 1980–2000 and 293.35 ha per year during 2000–2018. When integrated with population change, the per capita consumption of land to new urban growth was 103.23 m^2 during 1980–2000 and 180.17 m^2 during 2000–2018.

Keywords

Urbanization · Urban expansion · Land resource · Land use change · Impact Assessment

4.1 Introduction

Urbanization is not only a change in society and the economy but also an important geospatial process (Jiao et al, 2015), where the increase in population numbers is compensated by the spatial growth of the built-up area. Urban expansion, therefore, is a phenomenon of urban space increase and an important measuring index for the urbanization process (Longyu et al. 2009). India shows an average 4.84% urban land expansion growth rate between 1970 and 2000 (Seto et al. 2011). Census data reveal the country experienced a 31% increase in urban land between 2001 and 2011. In the 2001 census, India had 5161 urban settlements spread over 78,163.26 km^2 (2.38% of the total land). The number of urban settlements increased to 7935 in the 2011 census covering 102,252.03 km^2 area (3.11% land). This increase in the urban built-up area can be attributed to the result of, firstly, the identification of more settlements as urban (Census Towns and Statutory Towns) in every

M. K. Ali (✉) · S. Fazal
Department of Geography, Aligarh Muslim University, Aligarh, India
e-mail: md.kaikubadali@gmail.com

S. Talukdar et al. (eds.), *Environmental Risk and Resilience in the Changing World*, GIScience and Geo-environmental Modelling, https://doi.org/10.1007/978-3-031-62442-1_4

census and, secondly, the expansion of the existing cities, especially the big cities, on the surrounding rural hinterland.

Researchers found that urban expansion in India, as in the case of many developing countries, is taking place on the prime agricultural land (Kavitha et al. 2017; Del Mar López et al. 2001; Fazal 2000; Gumma et al. 2017; Jiang et al. 2013; Ali 2021; Pandey and Seto 2015). There is more scope for urban expansion around small cities than large ones, and it is likely to result in more agricultural land conversions (Pandey and Seto 2015). Besides, the expansion of cities and the resulting invasion of the rural land by urban forces have caused a succession of land uses in a discordant and dissociated way. The growing pressure of population explosion on land resources can be seen in the fact that the per capita availability of arable land has declined from 0.34 ha in 1961 to 0.11 ha in 2021 (World Bank 2021). Therefore, the immediate impact of urbanization is on the land in the periphery. The alteration of land use and land cover due to the increasing population and urban expansion is receiving attention from urban researchers and planners.

The land is a basic natural resource. Land resource plays an essential role in supporting our day-to-day life and activities. It provides us with the living space, the primary products to support our material needs, and opportunities that satisfy our ways of life (Baarlowe 1986). Therefore, land acts as a resource base providing the stage for other resource-generating opportunities and space for all kinds of human activities. As it is a finite resource, its physical supply is limited. But as the population is increasing rapidly, especially in urban areas, the demand for housing space, industrial space, infrastructural space, and so on is bound to increase. The direct implication of urbanization is the changes in land use and land cover within and around the settlement. The urban fringe that, in the immediate past, was a rural hinterland, experiences rapid and dynamic changes in the landscape.

Studies on urban expansion and consequent land use and land cover (LULC) changes are abundant in urban research both in the global and Indian contexts. Much progress has occurred in predicting urban growth or sprawl. Cellular automata (CA) and agent-based modeling (ABM) are two commonly employed techniques of urban growth modeling (Shafizadeh Moghadam and Helbich 2013). However, there was an inadequate focus on assessing the impact of urban expansion on land resources. Few studies have examined the impact of urban expansion on individual land use classes. For example, agricultural land loss due to urban sprawl (Del Mar López et al. 2001; Fazal 2000; Gumma et al. 2017; Jiang et al. 2013; Pandey and Seto 2015); loss of arable land (Huang et al. 2015; Tan et al. 2005); farmland loss (Fazal 2001; Jim 1997; Narducci et al. 2019; Seto et al. 2000; Song et al. 2015; Theobald 2001) impact of urban sprawl on forest land (Bengston et al. 2005; MacDonald and Rudel 2005; Miller 2012; Novak and Wang 2004; Nowak and Walton 2005); and on wetlands (Holland et al. 1995; Kuchay and Bhat 2014; Mao et al. 2018; Mondal et al. 2017).

Land transformation due to urban expansion has an immense impact on different types of land based on regional spatial variability in particular (as studied by scholars) and on the land resources as a whole. This study, thus, in a general context, tries to assess the impact of urban expansion on land resources in Aligarh city using different indicators of land resources.

4.2 Study Area

Aligarh city lies at the frontier of the National Capital Region of Delhi. It is connected to the capital by railway (Howrah-Delhi Railway Line) and roadway (Kanpur to Delhi Grand Trunk Road), which helped the city to emerge as one of the rapidly growing urban centers in Western Utter Pradesh. This historical city has experienced a three-time increase in population during the last four decades, unequally spreading over a much larger area (Farooq and Ahmad 2008; Fazal 2013; Ali 2021; Fazal et al. 2015). Due to recent industrial expansion (e.g., the development of Talanagari—a planned industrial zone devoted to the lock industries), this growth in the population is expected to be accentuated. In 1971, the city had 252,314 population, which increased to 874,408 in 2011. At present, it has a population of 1.2 million (Population Census 2022). The pressure of overgrowing population, the demand for housing and infrastructure, and the expansion of industrial units and commercial centers have forced the city to expand beyond the municipal limit along the fringes, resulting in unexpected land use transformation (Banu and Fazal 2013). Therefore, urban expansion in the city during the last few decades impacted the altercation of land resources. The Government of India has identified Aligarh city to develop under the Smart City Mission. Before starting the implementation of this project, there is an urgent need to assess the urban expansion and its impact on land resources in the rapidly transforming city surroundings.

4.3 Objective of the Study

The present study is aimed at the following objectives:

- To assess the change in land use land cover due to urban expansion in Aligarh city
- To evaluate the impact of urban expansion and urban development on the land resources using land resource impact indicators.

4.4 Database and Methodology

The present study has been carried out by integrating satellite and analog maps for different periods and incorporating population data from different census periods. The land use/land cover maps of three periods viz. 1980, 2000, and 2018 were prepared using remote sensing imageries and topographical maps. Table 4.1 shows the details of the images used. All the maps were brought on the same scale and coordinate system in the GIS environment using ArcGIS 10.5. The NRSA Level II LU/LC scheme was adopted, and 13 level LU/LC classes were interpreted and used to prepare land use maps. The validation and accuracy assessment was done with 30 sample points through ground checks, cross-verified with Google Earth images.

The 1980 LULC map was integrated with the 1981 census population data; the 2000 LULC map was integrated with the 2001 census population data. However, there are no data available for the current year. Therefore, the population in 2018 was calculated by simple linear interpolation using an exponential growth rate from the 2001 to 2011 censuses. The detailed schematic expression of the methodologies followed in this study is shown in Fig. 4.1.

4.4.1 Land Resource Impact Assessment Indicators

The Land Resource Impact Assessment (LRIA) is a framework to analytically examine the per capita consumption of land due to urban expansion and resultant new development by integrating population growth with land resource loss (Hasse and Lathrop 2003). The LRIA of urban expansion has been developed using different land resource impact indicators for fulfilling the objective. The LRIA in this study has been modified and redefined as per the regional and spatial differences of the study area and exhibition of different geographical settings.

Table 4.1 Details of data sets used

Images/Maps	Type	Year	Resolution/Scale
Sentinel 2A	Multi-spectral	2018	10 m
IRS 1D	PAN	2000	5.8 m
Topographical sheet	54 I/1	1978	1:50,000
Aligarh city guide map	Guide map	1980	1:20,000

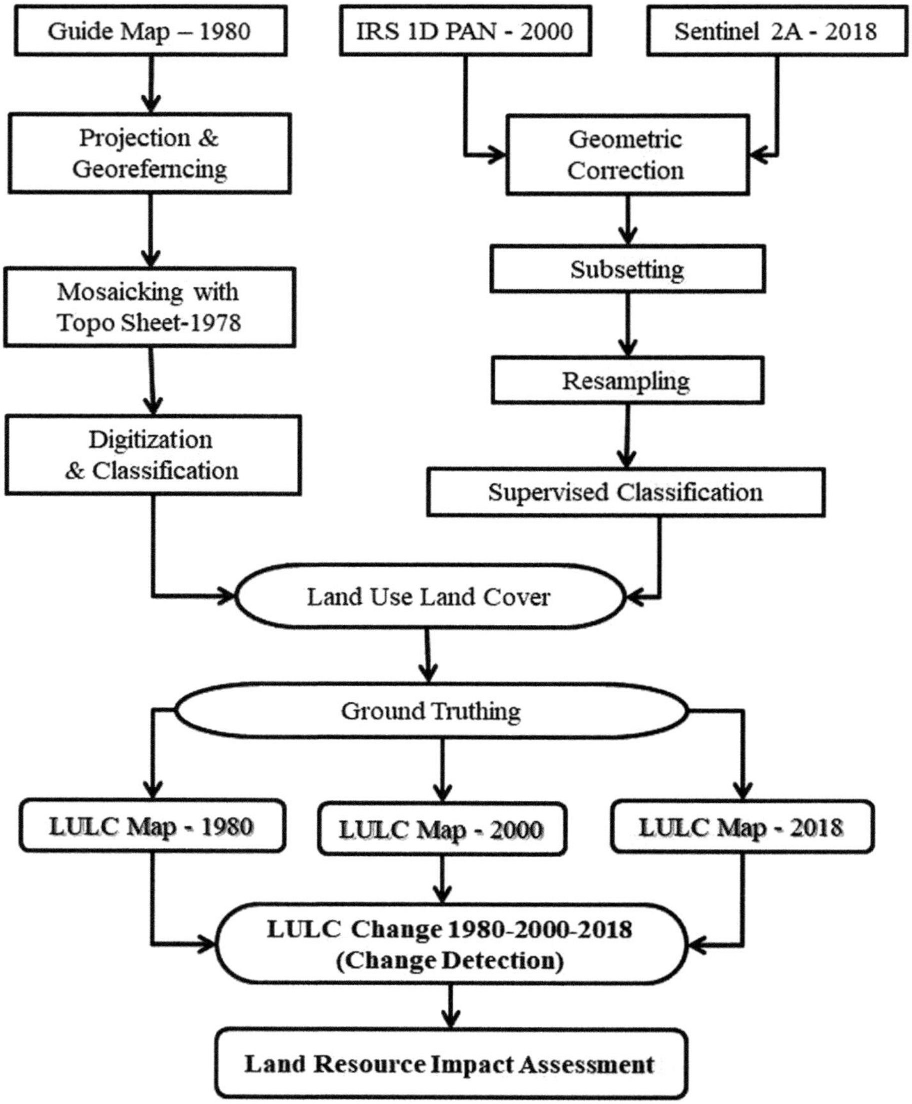

Fig. 4.1 Flow chart for methodology

The land resource impact indicators examine the impact of new urban development on five specific critical land resources, including (a) efficiency of land utilization (i.e., urban growth density); (b) prime agricultural land; (c) vacant land; (d) tree plantation; (e) waterlogged bodies. The indicators taken for the study based on these five land resources and their impact status are described in Table 4.2.

4.4.1.1 New Urban Growth (UG)

New urban growth (UG) measures the land consumption for new urban development in this region. The UG indicator calculates the area of all non-urban land use categories in the initial year that became urban built-up in the proceeding year of study. The percent growth of urban land (UG*pct*) was calculated by comparing the new growth of urban land to the previously urban

Table 4.2 Land resource impact indicators

LRI indicators	Impact	Recoverable
1. New urban built-up growth (UG)	Gain	No
2. New vacant land growth (SVG)	Gain	Yes
3. Loss of agricultural land to urban growth (AG-U)	Critical loss	No
4. Loss of agricultural land to vacant land (AG-V)	Loss	Yes
5. Loss of vacant land to urban growth (VAC-U)	Loss	No
6. Loss of tree plantation to urban growth (TP-U)	Critical loss	No
7. Loss of water bodies to urban growth (WAT-U)	Critical loss	No

land uses. The per capita new urban growth (UG*pct*) was calculated by normalizing the area of UG by the concurrent change in population.

4.4.1.2 Speculative Vacant Land Growth (SVG)

The SVG indicator measures the increase in vacant lands. It was calculated by tabulating the area of all non-vacant land uses (except urban) in the initial year converted to vacant lands in the proceeding study year. The percent growth of vacant land (SVG*pct*) was calculated by comparing the new growth of vacant land to the previously vacant land uses. The per capita vacant land growth (SPG*percap*) was calculated by normalizing the area of new vacant land by the concurrent change in population.

4.4.1.3 Loss of Agricultural Land to Urban Development (AG-U)

The loss of agricultural land to new urban growth (AG-U) indicator includes the area of all agricultural land uses in the initial year that became urbanized by the final year. The percent AG-U (AG-U*pct*) was calculated by comparing the area of agricultural land uses that was converted to urban land with the total agricultural land in the initial year. The per capita (AG-U*percap*) was generated by normalizing the amount of agricultural land loss to urban development by the concurrent increase in population.

4.4.1.4 Loss of Agricultural Land to Vacant Land (AG-V)

The loss of agricultural land to vacant land (AG-V) measures the amount of agricultural land from the previous study period converted to vacant in the proceeding study year. The percent AG-V (AG-V*pct*) was calculated by comparing the area of agricultural land converted to vacant land with the area of agricultural land in the initial year. The per capita agricultural land loss to vacant land (AG-V*percap*) was generated by normalizing the amount of agricultural land loss to vacant land by the concurrent increase in population during the study period.

4.4.1.5 Upgradation of Vacant Land to Urban Development (VAC-U)

This indicator measures the vacant land lost to new urban growth. The percent vacant land loss to urban development (VAC-U*pct*) was calculated by comparing the area of vacant land converted to urban land uses with the area under vacant land in the initial year. The per capita loss of vacant to urban development (VAC-U*percap*) was generated by normalizing the amount of vacant land converted to new urban built-up by the concurrent increase in population during the study period.

4.4.1.6 Loss of Tree Plantation to Urban Development (TP-U)

This indicator measures the area under tree plantation lost to new urban growth (TP-U). The percent loss of tree plantation to urban development (TP-U*pct*) was calculated by comparing the area under tree plantation converted to urban land uses by the area under tree plantation in the initial year. The per capita loss of tree plantation to urban development (TP-U*percap*) was calculated by normalizing the amount of tree plantation lost to urban land uses by the concurrent increase in population.

4.4.1.7 Loss of Water Bodies to Urban Development (WAT-U)

This indicator measures the loss of water bodies to new urban growth (WAT-U). The percent loss of water bodies to urban development (Wat-U*pct*) was calculated by comparing the area under waterlogged bodies that were converted to urban land uses by the area under water-logged bodies in the initial year. The per capita loss of water bodies (Wat-U*percap*) was generated by normalizing the amount of water-logged bodies lost to urban built-up by the concurrent increase in population.

4.4.1.8 Population Change

Urban spatial expansion is driven by the increase in the population. New urban development around the city occurs to compensate for the pressure of the growing population. Therefore, it is also essential to integrate population data with the LULC change data to measure the per capita amount of urban growth and change in select land resource indicators. Table 4.3 depicts the change in population in the study area over the past four decades.

4.5 Results and Discussion

4.5.1 Features of Land Use Changes

4.5.1.1 Expansion of Built-Up Area

The increase in population along with modernization-driven socio-economic transformation has resulted in increase and expansion of built-up area in the city. The built-up area in the city was increased by 4234.9 ha (307.3% increase) during the study period 1980 and 2018. The notable land use classes of built-up area included residential, industrial, commercial, etc. (Table 4.4). Since the city is located in the fertile agricultural tract between the Ganga and the Yamuna rivers and agriculture is the mainstay of the economy, all this built of extension occurred on city peripheral agricultural lands (Fig. 4.2).

4.5.1.2 Increase in Residential Land

The significant growth in population in the city has led to necessary demand of residential land uses. As a result residential land use has increased from 745.3 ha (3.7% of the total study area) in 1980 to 3947.9 ha (19.3% of total study area) in 2018, adding a total of 3202.6 ha during the same period (an increase of 429.7%).

4.5.1.3 Increase in Vacant Land

The farmers under the influence of urban forces, first, stop farming on agricultural lands and offer their land for urban uses and until they receive desired revenue the land remains vacant. The analysis of land use in Aligarh city also finds that there was significant increase in vacant land during the study period. The vacant land has increased 2918.3 ha in during the study period between 1980 and 2018, an increase of 479.3%. These vacant lands are speculative land holdings.

4.5.1.4 Increase in Industrial and Commercial Land

Aligarh city, during the study period, has emerged as an important industrial and trading center in Western Uttar Pradesh. The city growth and modernization have fuelled the economic transformation in the city. The city has witnessed an increase of 258.4 ha (470.5% increase) in commercial lands and 277.3 ha (468.7% increase) in industrial lands during the study period.

Table 4.3 Change in population in the study area

Census	Population	Change in population
1981	320,861	–
2001	669,087	348,226
2018[a]	962,160	293,073

[a] Population projected through exponential growth rate

Table 4.4 Land use changes in the Aligarh city during 1980, 2000, and 2018

Land use	1980 percent	2000 percent	2018 percent	Change (1980–2000)	Change (2000–2018)	Change (1980–2018)
Residential	745.3 (3.7)	2432.8 (11.9)	3947.9 (19.3)	1687.6 (226.4)	1515.0 (62.3)	3202.6 (429.7)
Commercial	54.9 (0.3)	119.0 (0.6)	313.3 (1.5)	64.1 (116.7)	194.3 (163.2)	258.4 (470.5)
Industrial	59.2 (0.3)	106.6 (0.5)	336.4 (1.6)	47.4 (80.2)	229.8 (215.6)	277.3 (468.7)
Community services	34.2 (0.2)	50.3 (0.2)	106.3 (0.5)	16.1 (47.1)	56.0 (111.2)	72.1 (210.8)
Government office	60.1 (0.3)	130.6 (0.6)	226.9 (1.1)	70.4 (117.1)	96.4 (73.8)	166.8 (277.3)
Local park	29.0 (0.1)	33.5 (0.2)	65.6 (0.3)	4.5 (15.6)	32.1 (95.7)	36.6 (126.2)
Other	45.2 (0.2)	94.2 (0.5)	266.4 (1.3)	49.0 (108.4)	172.2 (182.7)	221.2 (489.3)
University	350.1 (1.7)	350.1 (1.7)	350.1 (1.7)	–	–	–
Vacant	608.9 (3.0)	1887.9 (9.2)	3527.2 (17.3)	1279.0 (210.1)	1639.3 (86.8)	2918.3 (479.3)
Village	276.7 (1.4)	356.2 (1.7)	497.7 (2.4)	79.5 (28.7)	141.5 (39.7)	221.1 (79.9)
Agriculture	17,802.1 (87.2)	14,548.1 (71.3)	10,481.2 (51.3)	−3254.0 (−18.3)	−4066.9 (−28.0)	−7320.9 (−41.1)
Shrubs	178.7 (0.9)	160.7 (0.8)	137.0 (0.7)	−18.0 (−10.1)	−23.8 (−14.8)	−41.8 (−23.4)
Tree plantation	103.4 (0.5)	49.6 (0.2)	68.5 (0.3)	−53.9 (−52.1)	18.9 (38.2)	−34.9 (−33.8)
Waterlogged	66.7 (0.3)	94.7 (0.5)	89.9 (0.4)	28.1 (42.1)	−4.8 (−5.1)	23.3 (34.9)
Total	20,414.4	20,414.4	20,414.4	–	–	–
Built-up	1378.0 (6.8)	3317.2 (16.2)	5612.9 (27.5)	1939.2 (140.7)	2295.7 (69.2)	4234.9 (307.3)

4.5.2 Land Resource Impact Assessment

The results of the LRI assessment are presented in Table 4.5.

4.5.2.1 New Urban Growth

The study area added 348,226 persons during 1980–2000. During the same time, 1912.34 ha of land was converted to new urban areas. The new urban development experienced a 148.93% increase since 1980. The annual increase in population was 17,411 persons, while the annual increase in built-up area was 95.61 ha. The per capita new urban growth was 54.92 m^2 per person added to the city.

During the 2000–2018 period, the city added 293,073 persons. 2296.88 ha of land was converted to different urban land uses to compensate for the population growth. The new urban development experienced a 72.88% increase since 2000. The annual increase in population was 16,282 persons, while the annual increase in urban land uses was 127.60 ha. The per capita

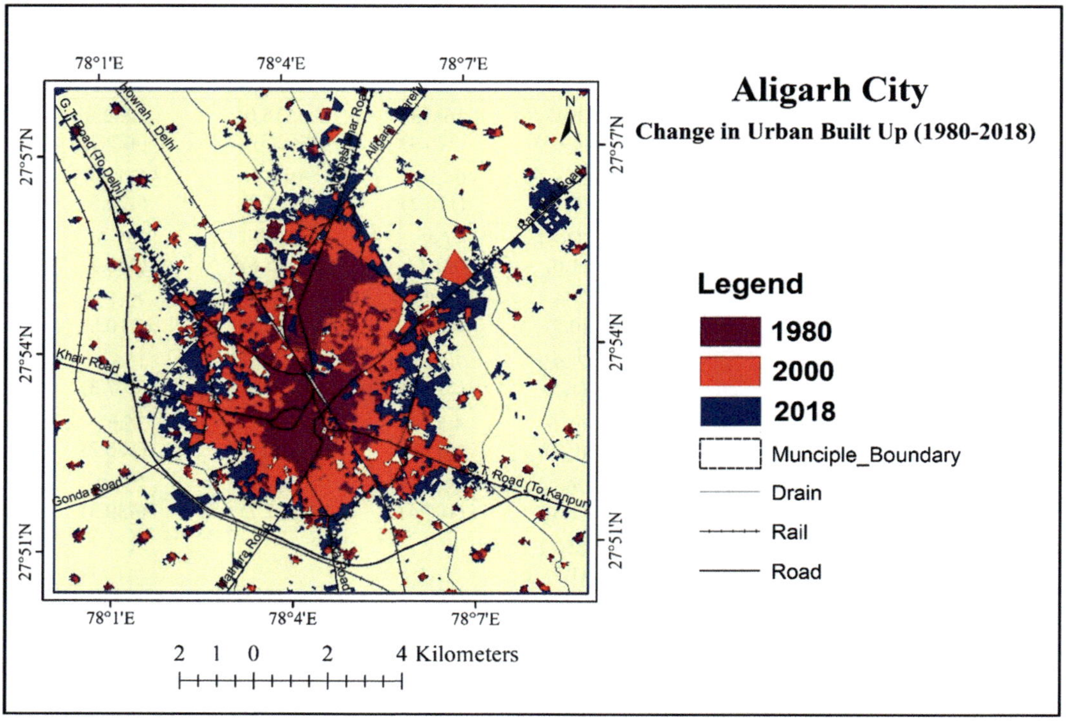

Fig. 4.2 Urban built-up change in Aligarh city (1980–2018)

Table 4.5 Land Resource Impact Assessment of urban expansion in Aligarh city

LRI indicators	1980–2000				2000–2018			
	Area (Ha)	PercapΔ (m²)	PercentΔ	AnnualΔ (ha)	Area (Ha)	PercapΔ (m²)	PercentΔ	AnnualΔ (ha)
New urban growth (UG)	1912.34	54.92	148.93	95.62	2296.88	78.37	72.88	127.60
New vacant land (SVG)	1682.26	48.31	257.07	84.11	2983.46	101.80	153.31	165.75
Loss of agricultural land to urban (AG-U)	1464.26	42.05	8.20	73.21	983.10	33.54	6.70	54.62
Loss of agricultural land to vacant (AG-V)	1667.73	47.89	9.34	83.39	2924.69	99.79	19.94	162.48
Vacant land to urban (VAC-U)	344.51	9.89	52.65	17.23	1190.22	40.61	61.16	66.12
Loss of tree plantation to urban (TP-U)	32.47	0.93	11.53	1.62	19.35	0.66	9.68	1.08
Loss of water body to Urban (WAT-U)	15.52	0.45	23.24	0.78	26.53	0.91	28.00	1.47
Total loss of land to urban	3524.49	101.21		176.22	5143.89	175.52		285.77

new urban growth was 78.37 m^2 per person added to the city, which is higher than the previous study period.

The spatial analysis of urban expansion in Aligarh city reveals new urban growth occurred in the continuity of existing built-up areas sprawling over the fringe during 1980–2000 (Fig. 4.3). There has been a radial pattern of urban expansion around the city. Most of the new urban growth was at the cost of agricultural land. During this period, clusters of new urban development can be seen linearly along the major roadways. But, as the city has a centrifugal transport network, the space between the two roads remained closed, resulting in an overlap in the linear growth and resembles the radial urban expansion. Such a new urban growth cluster expanded along the Kanpur G.T. Road, Agra Road, Khair Road, and Anupshahar Road. Dispersed new urban development occurred along the Ramghat Road and Delhi G.T. Road. The development of the Provincial Armed Constabulary (PAC) area along the Ramghat Road has ushered new urban growth during 2000–2018. During 2000–2018, new urban growth in Aligarh city can be characterized by dispersed urban growth along the major roadways. A new industrial area emerged along the Aligarh New Bypass on the city's Western periphery.

4.5.2.2 Growth in Speculative Vacant Land

The urban fringe of the study area has experienced rapid growth in vacant lands. It reflects increasing cases of land speculation. A total of 1682.26 ha of land were converted to vacant land for speculative purposes during 1980–2000 (a 257% of increase). The annual growth in vacant land was 84.11 ha per year. The per capita increase in vacant land was 48.31 m^2 per person added to the city.

Compared to this, the study area added 2983.46 ha to vacant land during 2000–2018. 98.03% (2924.69 ha) of this came from the

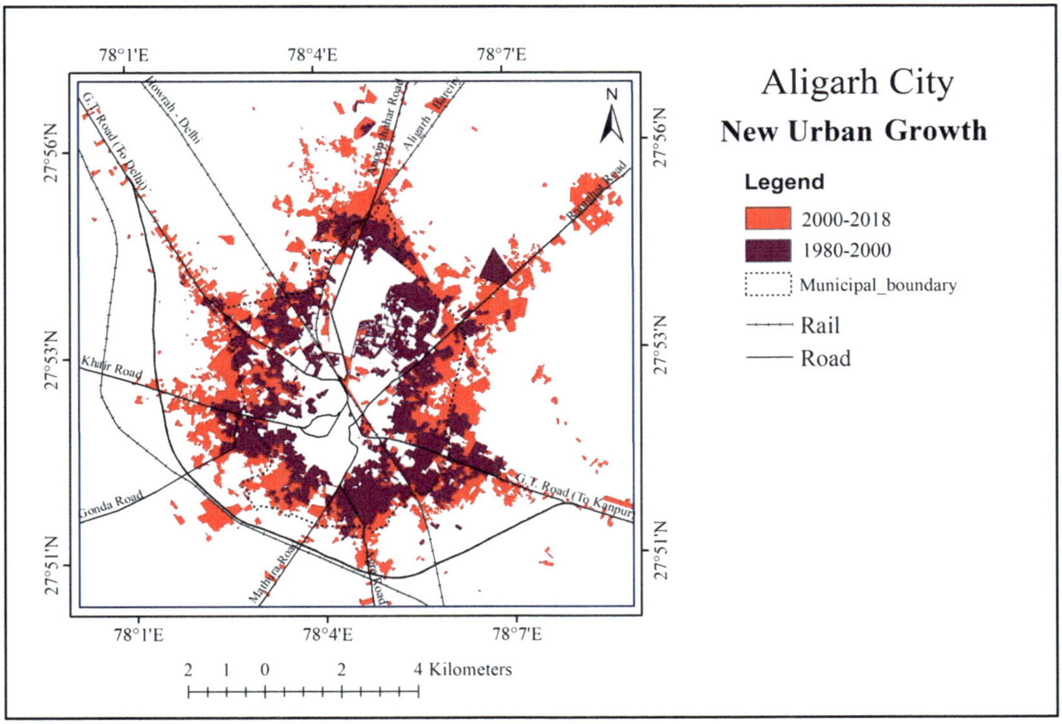

Fig. 4.3 New urban growth in Aligarh city (1980–2018)

conversion of agricultural lands. There has been a 153.31% increase in vacant land since 2000. The annual growth in vacant land was 165.75 ha during this time. The per capita increase in vacant land was 101.80 m^2 per person added to the city during the same period.

The spatial analysis of the growth in vacant land during 1980–2000 exhibits extensive development of vacant land along the Ramghat Road and Anupshahar Road. In the surrounding of Kisanpur village, along the Ramghat Road, there has been a considerable growth of vacant land due to the presence of *ushar* (infertile) land in this area and the inclusion of this region for urban development in the first master plan by Aligarh Development Authority (ADA). The ADA proposed the development of Avantika colony and Swarnjayanti Nagar in this region in master plan 2000. The proposal has led to extensive land speculation among real estate companies and builders. Secondly, a new industrial area, i.e., Talanagri, was proposed to be developed along the Ramghat Road. The vacant land grew along the Anupshahar Road due to the increasing demand for residential spaces for the growing population whose life and work are directly or indirectly dependent on the university and secondly due to the proposal of a new industrial area in the master plan. In the proceeding years, these areas experienced new urban growth. Besides, along the Kanpur Road in the Dhanipur village, vacant land emerged for the proposed sites of the airport. There had been dispersed growth of vacant land along the Delhi G.T. Road, Khair Road, and Agra Road. These small dispersed patches of vacant land are recently converted agricultural land adjacent to the existing built-up (Fig. 4.4).

During 2000–2018, the growth of speculative vacant land is observable all around the

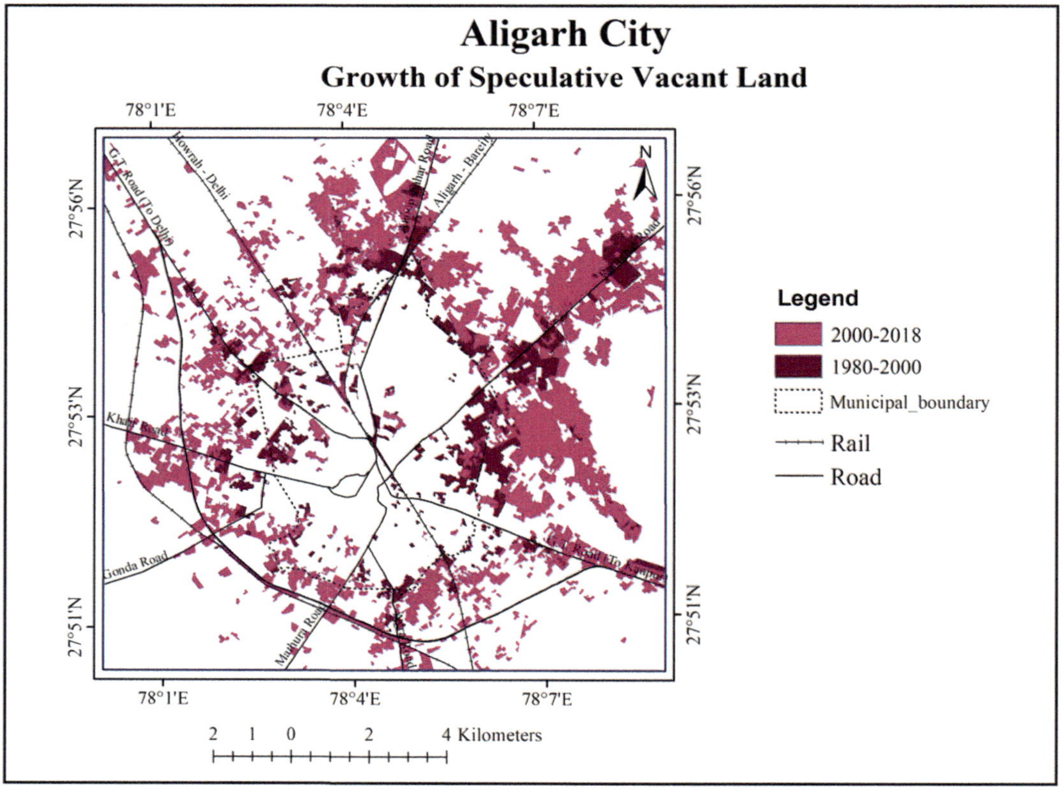

Fig. 4.4 Growth of speculative vacant land in Aligarh city (1980–2018)

city built-up. The land around Manjurgarhi and Chherat villages along the Anoopshahar Road experienced significant growth in new vacant land during this period. The proposal of an industrial area in this region by ADA in their master plan has resulted in extensive land speculation, which made way for agricultural to vacant land conversion. The vacant land also emerged along the newly constructed Aligarh Bypass connecting the Delhi-Kanpur G.T. Road at the northwest and southeast parts of the city.

4.5.2.3 Loss of Prime Agricultural Land to Urban

The urban expansion in the study area has mainly been at the cost of agricultural land. Agricultural lands were lost either to different urban land uses or speculative vacant lands. A total of 1667.73 ha of agricultural lands were lost to the new urban growth in the fringe areas during the 1980–2000 study period. This was 8.20% of the total agricultural land in 1980. The annual loss of agricultural land to urban land uses was 73.21 ha per annum. The per capita loss of agricultural land to new urban growth was 42.05 m^2 per person added to the city during this period.

Contrary to this, 983.1 ha of agricultural land was lost to different urban land uses during 2000–2018. This was 6.70% of the total agricultural land available in 2000. The annual loss of agricultural land to new urban growth was 54.61 ha per annum. The per capita loss of agricultural land to new urban uses during this period was 33.54 ha per person added to the city.

Loss of agricultural land to urban built-up during 1980–2000 occurred all along the existing built-up. But during 2000–2018, much of the loss of agricultural land to the urban growth occurred along the Western margin of the city, on the area between Mathura Road and Khair Road, Khair Road, and Delhi G.T. Road (Fig. 4.5).

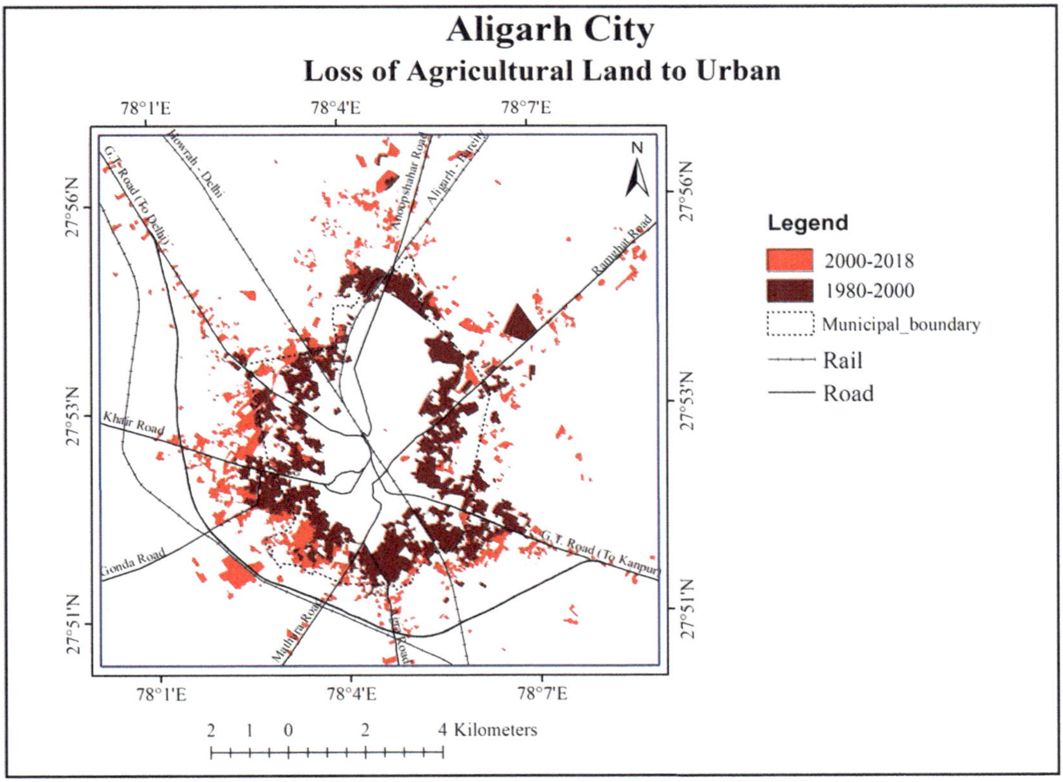

Fig. 4.5 Loss of agricultural land to urban built-up in Aligarh city (1980–2018)

4.5.2.4 Loss of Agricultural Land to Vacant Land

Similarly, while analyzing the land speculation and conversion of agricultural lands to vacant land, the data reveals a total of 1667.73 ha of agricultural land was lost to vacant land, which was 9.34% of the total agricultural land in 1980. The annual loss of agricultural land to the vacant land was 83.38 ha per annum. The per capita loss of agricultural land to vacant land was 47.89 m^2 per person during 1980–2000.

During 2000–2018, a total of 2924.69 ha of agricultural land were lost to vacant land, which was 19.94% of the total agricultural land in 2000. The annual loss of agricultural land to the vacant land was 162.48 ha per annum. The per capita loss of agricultural land to vacant land was 99.79 m^2 per person during 2000–2018.

During 1980–2000, loss of agricultural land to vacant land growth occurred mostly along the Anoopshahar Road, Ramghat Road, and the area between Ramghat Road and Kanpur G.T. Road. The loss of agricultural land to vacant growth continued to occur in these areas during 2000–2018. Also, this loss happened along Khair Road, Mathura Road, and Agra Road. This is due to the development of the Aligarh Bypass and Railway Line along the Western margin of the city (Fig. 4.6).

4.5.2.5 Upgradation of Vacant Land to Urban Development

The land use map obtained from satellite imageries exhibits vacant land around the built-up during all the study periods. Analysis of loss of vacant land to urban growth reveals 344.51 ha of vacant land was developed for urban uses during 1980–2000. This was 52.65% of total vacant land in 1980. The annual loss of vacant land to urban land uses during this period was 17.23 ha per annum. The per capita loss of vacant land to urban growth was 52.65 m^2 per person during this period.

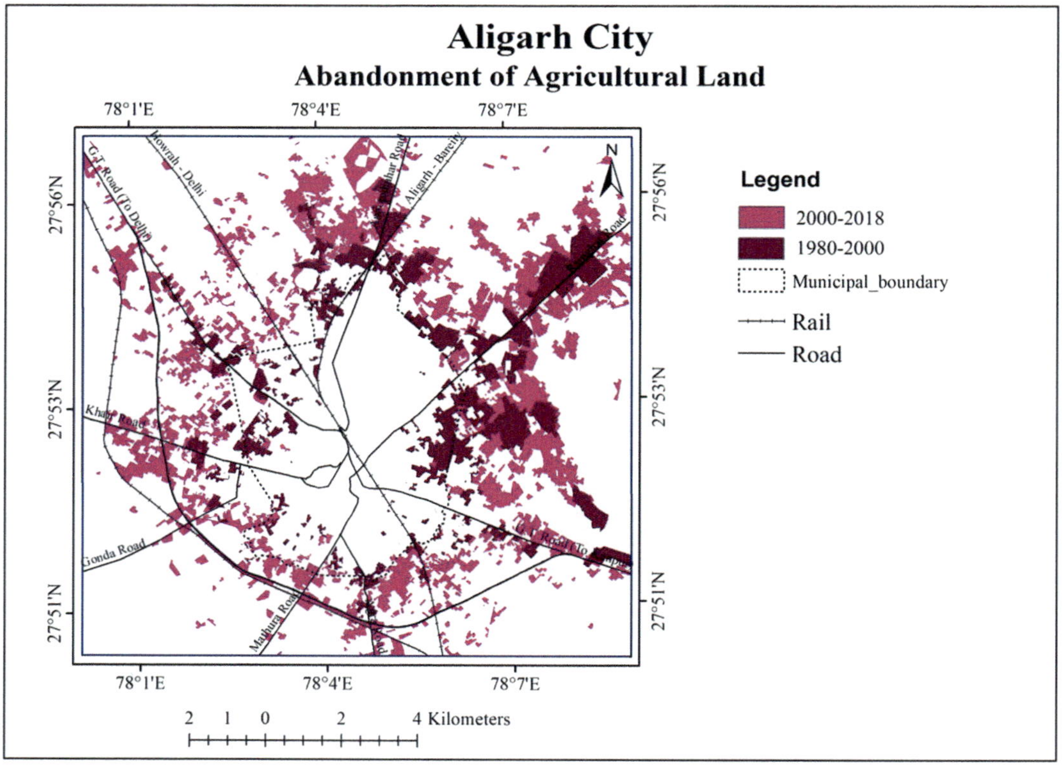

Fig. 4.6 Abandonment of agricultural land in Aligarh city (1980–2018)

During 2000–2018, a total of 1190.22 ha of vacant land was developed for urban uses. This was 61.16% of the total vacant land in the year 2000. The annual loss of vacant land to urban land uses during this period was 66.12 ha per year. The per capita loss of vacant land to urban development was 40.61 m^2 per person during this period.

Vacant land is the first kind of land that is subject to immediate consumption for urban expansion. The conversion of vacant land to urban uses occurred along Kanpur G.T. Road and Mathura Road during 1980–2000 (Fig. 4.7). Besides, much of the vacant land between Anoopshahar Road and Ramghat Road area was converted to urban land uses during this period. Contrary to this, during 2000–2018, the conversion of vacant land to urban land uses was observed around the urban fringes. However, the extent was much higher along the Anoopshahar Road and Ramghat Road. Besides, the area between Ramghat Road and Kanpur G.T. Road observed massive conversion of vacant land to urban development during the same period.

4.5.2.6 Loss of Tree Plantation

The study area is mainly devoid of any presence of natural vegetation. The old maps of the city (topographical map and guide map) exhibit some patches of tree plantations or orchards on the outskirts. Urban expansion in the city has also impacted tree plantations. The study area lost 32.47 ha of tree plantation to urban growth during 1980–2000. It was 11.53% of the total area under tree plantation in 1980. The annual loss of tree plantation to urban expansion was 1.62 ha per annum. The per capita loss of tree plantation was 0.93 m^2 during the same period.

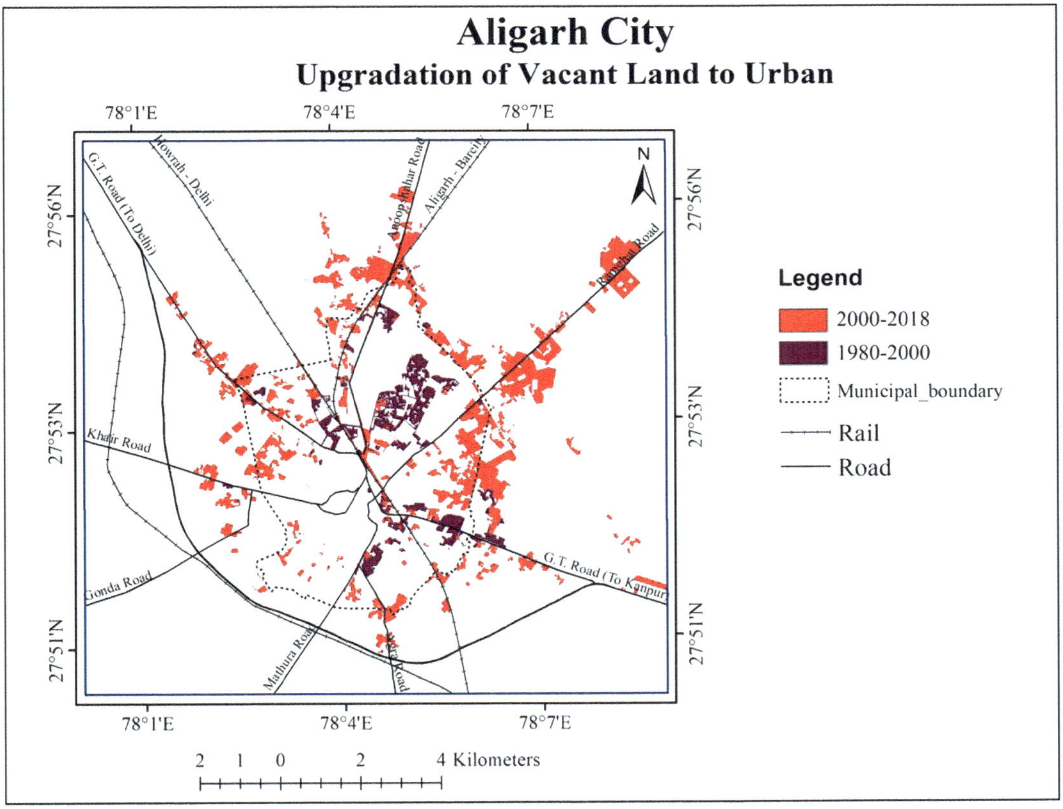

Fig. 4.7 Upgradation of vacant land to urban built-up in Aligarh city (1980–2018)

During 2000–2018, the loss of tree plantations to urban expansion was slower and 19.35 ha of the area under tree plantations was lost to urban land uses. This was 9.68% of the total area under tree plantation in 2000. This trend can be attributed to the urban expansion in the city during the 1980–2000 period had already converted existing tree plantations of the fringes to urban land uses; whatever remained were the shrubs and orchards far from the city. The annual loss of tree plantation to urban land uses was 1.07 ha per annum. The per capita loss of tree plantation to urban expansion was 0.66 m^2 per person added to the city population during 2000–2018.

As the study area did not have a significant share of land cover under tree plantation, the loss of tree plantation is not that much high. But the impact of urban expansion on the tree plantation is quite noticeable and has far-reaching implications on urban ecology. During 1980–2000, urban development consumed tree covers at the Western fringe of the then city built-up, between Delhi G.T. Road and Khair Road (Fig. 4.8). And during 2000–2018, the study area lost tree covers along the Anupshahar Road, where there has been mushrooming growth of residential land uses.

4.5.2.7 Loss of Water-Logged Bodies

Urban expansion has also impacted the water bodies as the study area lost 15.52 ha of the area under water bodies to urban growth during 1980–2000. This was 23.24% of the area under water bodies in 1980. The annual loss of water bodies area to urban land uses was 0.78 ha per annum, and the per capita loss of water bodies to urban expansion was 0.45 m^2 per person by the concurrent increase in population.

Contrary to this, a total of 26.53 ha area under water bodies was lost to urban land uses during 2000–2018. This was 28% of the total area under water bodies in the study area in 2000. The annual loss of water bodies area to urban land uses was 1.47 ha per year. The per capita loss of water bodies was 0.91 m^2 per person by the concurrent increase in population.

Although the city lacks an adequate share of land under water bodies, what was present in the

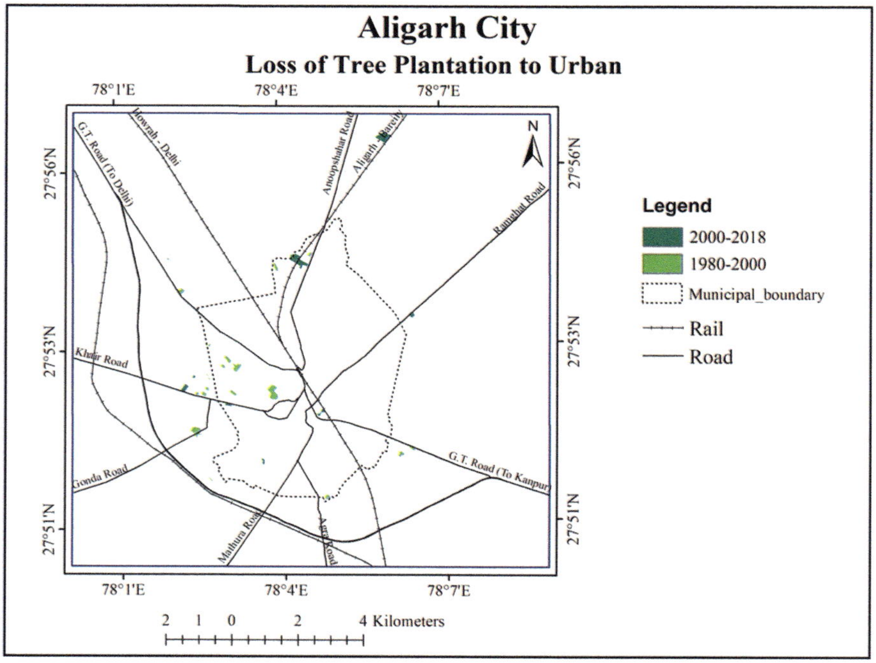

Fig. 4.8 Loss of tree plantation to urban built-up in Aligarh city (1980–2018)

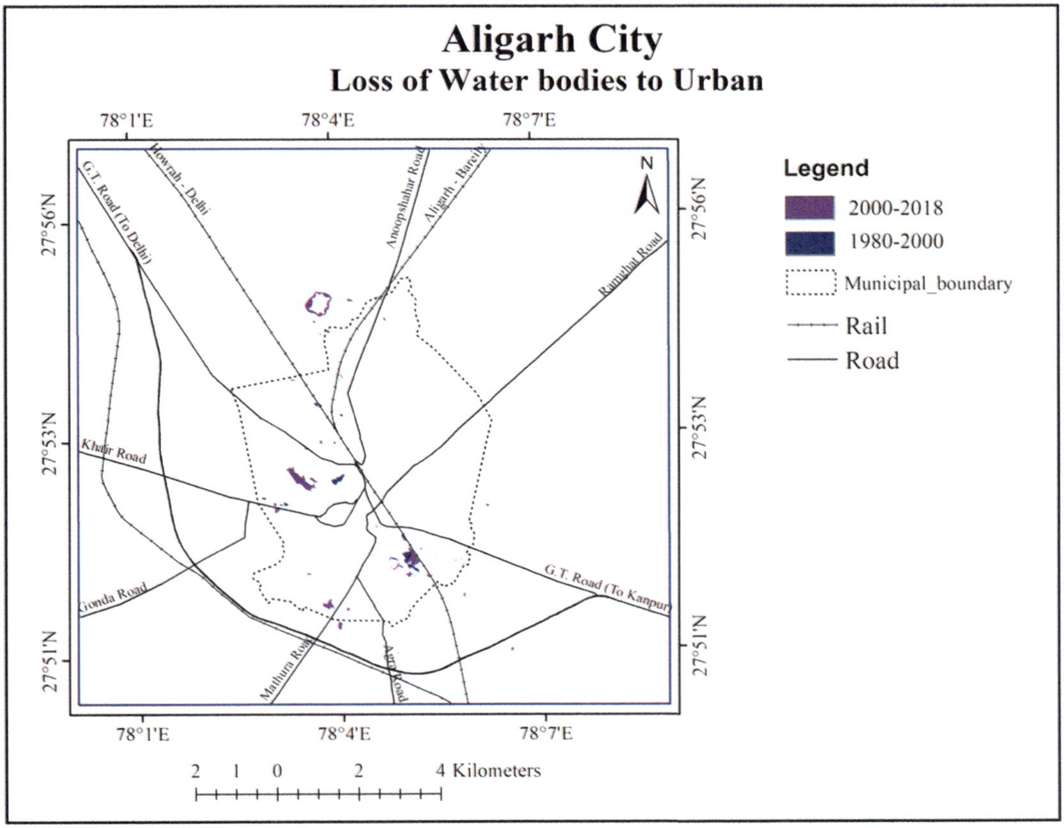

Fig. 4.9 Loss of water bodies to urban built-up in Aligarh city (1980–2018)

form of water-logged bodies has also been consumed by urban expansion. During 1980–2000, the loss of water areas was observed along the Delhi-Howrah rail line in the Bagwan Nagar area. During 2000–2018, a large water-logged area between Khair Road and Delhi G.T. Road (Fig. 4.9) was filled up to vacant land and later developed into residential uses.

4.6 Conclusion

Urban expansion in Aligarh city has consumed considerable land resources to make way for urban development. This has been done both in a planned and unplanned way. The ADA failed to manage the pace and extent of urban growth and the resultant impact on land and the environment in the region (Fazal and Ali 2022). The above discussion comes with the conclusion that urbanization in the city has consumed surrounding land in two ways-

1. Substantial growth in vacant land: Urbanization has resulted in land speculation in the urban fringe, where agricultural lands have been abandoned for future urban development aiming to get higher land values in the subsequent years.
2. A rapid increase in urban built-up: There is a rapid increase in residential, commercial, industrial, and other urban land uses along with the concurrent population growth. The growth in new urban built-up has come at the cost of vacant land, agricultural land, tree plantation, and water-logged areas.

Urban expansion in Aligarh city has come mainly at the cost of peripheral agricultural land. Agricultural land in the fringes has transformed in two ways: firstly, it transformed into vacant land, and further, the transformation converted the vacant land into urban land uses. Secondly, the agricultural lands were directly converted into urban land uses. Therefore, urbanization in Aligarh city has resulted in the extensive loss of prime agricultural land to vacant land growth and urban growth.

The impact of urban expansion on land resources can thus be assessed by the fact that the study area experienced massive transformations in land uses in both periods. The rate and extent of growth were more during 2000–2018. The per capita loss of land to urban expansion was 101.21 m^2 per person added to city population during 1980–2020. It increased to 175.52 m^2 per person added to the city population during 2000–2018. Similarly, the annual rate of land loss to urban expansion increased from 176.22 ha during 1980–2000 to 285.77 ha during 2000–2018. This reflects horizontal spilling of the urban built-up, which is one of the main characteristics of urbanization in Indian cities (Mahtta et al. 2019; Ghosh 2019; Pathak 2019). The urban expansion and consumption of land resources in Aligarh city were observed around the built-up in the urban fringes.

The overall assessment of urban expansion and its impact on land resources in the urban fringe of Aligarh city, therefore, reveals a critical scenario of rapidly growing urban centers in India where urban built-up growth is occurring at the expense of productive agricultural land. The per capita land consumption is significantly high and has increased with the increasing population. However, urban housing shortages remain a prime challenge to urban planners. Land allocation and management of urban land in the fringes are thus matters of concern for implementing master plans in the city, making way to accomplish the Smart City Mission.

Acknowledgements The authors acknowledge the fellowship grant from the Indian Council of Social Science Research (ICSSR) (F.No.3-146/2021-22/PDF/GEN) for completing this research.

References

Ali MK (2021) A GIS-based analysis of urban expansion and land use land cover changes in Aligarh city. Ind J Spat Sci 12(2)

Banu N, Fazal S (2013) Development of infrastructural facilities in public sector on the urban Fringe of Aligarh City. J Infrastruct Develop 5(2):151–168. https://doi.org/10.1177/0974930614527946

Baarlowe R (1986) Land resource economics: the economics of real property

Bengston DN, Potts RS, Fan DP, Goetz EG (2005) An analysis of the public discourse about urban sprawl in the United States: monitoring concern about a major threat to forests. For Policy Econ 7(5):745–756. https://doi.org/10.1016/j.forpol.2005.03.010

Del Mar López T, Aide TM, Thomlinson JR (2001) Urban expansion and the loss of prime agricultural lands in Puerto Rico. Ambio 30(1):49–54. https://doi.org/10.1579/0044-7447-30.1.49

Farooq S, Ahmad S (2008) Urban sprawl development around Aligarh city: a study aided by satellite remote sensing and GIS. J Indian Soc Rem Sens 36(1):77–88. https://doi.org/10.1007/s12524-008-0008-0

Fazal S (2000) Urban expansion and loss of agricultural land—a GIS based study of Saharanpur City, India. Environ Urban 12(2):133–149. https://doi.org/10.1630/095624700101285343

Fazal S, Banu N, Sultana S (2015) Expanding cities, contested land: role of actors in the context of peri-urban interface. Curr Urban Stud 03(03):187–198. https://doi.org/10.4236/cus.2015.33016

Fazal S (2001) The need for preserving farmland. A case study from a predominantly agrarian economy (India). Landsc Urban Plan 55(1):1–13. https://doi.org/10.1016/S0169-2046(00)00134-1

Fazal S (2013) Land use dynamics in a developing economy regional perspectives from India. Springer International Publishing, Aligarh. https://doi.org/10.1007/978-94-007-5255-9

Fazal S, Ali MK (2022) The realities of urban land use planning in India: case study of Aligarh City. Urban India 42(1):88–112. National Institute of Urban Affairs (NIUA)

Ghosh S (2019) A city growth and land-use/land-cover change: a case study of Bhopal, India. Model Earth Syst Environ 5(4):1569–1578. https://doi.org/10.1007/s40808-019-00605-y

Gumma MK, Mohammad I, Nedumaran S, Whitbread A, Lagerkvist CJ (2017) Urban sprawl and adverse impacts on agricultural. Rem Sens 9:1–16. https://doi.org/10.3390/rs9111136

Hasse JE, Lathrop RG (2003) Land resource impact indicators of urban sprawl. Appl Geogr 23(2–3):159–175. https://doi.org/10.1016/j.apgeog.2003.08.002

Holland CC, Honea J, Gwin SE, Kentula ME (1995) Wetland degradation and loss in the rapidly urbanizing area of Portland, Oregon. Wetlands 15(4):336–345. https://doi.org/10.1007/BF03160888

Huang D, Jin H, Zhao X, Liu S (2015) Factors influencing the conversion of arable land to urban use and policy implications in Beijing, China. Sustainability (switzerland) 7(1):180–194. https://doi.org/10.3390/su7010180

Jiang L, Deng X, Seto KC (2013) The impact of urban expansion on agricultural land use intensity in China. Land Use Policy 35:33–39. https://doi.org/10.1016/j.landusepol.2013.04.011

Jiao L, Mao L, Liu Y (2015) Multi-order landscape expansion index: characterizing urban expansion dynamics. Landsc Urban Plan 137:30–39. https://doi.org/10.1016/j.landurbplan.2014.10.023

Jim CY (1997) Land use problems associated with farmland abandonment. Geography 82(3):277–280

Kavitha A, Somashekar RK, Nagaraja BC (2017) Urban expansion and loss of agriculture land—a case of Bengaluru city. Int J Geomat Geosci 5(3):492–498

Kuchay NA, Bhat MS (2014) Urban Sprawl of Srinagar City and its Impact on Wetlands—a spatio-temporal analysis. Int J Environ Bioenergy 9(2):122–129

Longyu S, Guofan S, Shenghui C, Xuanqi L, Tao L, Kai Y, Jingzhu Z (2009) Urban three-dimensional expansion and its driving forces—A case study of Shanghai China. Chin Geogr Sci 19(4):291–298. https://doi.org/10.1007/s11769-009-0291-x

MacDonald K, Rudel TK (2005) Sprawl and forest cover: what is the relationship? Appl Geogr 25(1):67–79. https://doi.org/10.1016/j.apgeog.2004.07.001

Mahtta R, Mahendra A, Seto KC (2019) Building up or spreading out? Typologies of urban growth across 478 cities of 1 million+. Environ Res Lett 14(12):124077. https://doi.org/10.1088/1748-9326/ab59bf

Mao D, Wang Z, Wu J, Wu B, Zeng Y, Song K, Yi K, Luo L (2018) China's wetlands loss to urban expansion. Land Degrada Develo 29(8):2644–2657. https://doi.org/10.1002/ldr.2939

Miller MD (2012) The impacts of Atlanta's urban sprawl on forest cover and fragmentation. Appl Geogr 34:171–179. https://doi.org/10.1016/j.apgeog.2011.11.010

Mondal B, Dolui G, Pramanik M, Maity S, Biswas SS, Pal R (2017) Urban expansion and wetland shrinkage estimation using a GIS-based model in the East Kolkata Wetland, India. Ecol Indicat 83(July):62–73. https://doi.org/10.1016/j.ecolind.2017.07.037

Narducci J, Quintas-Soriano C, Castro A, Som-Castellano R, Brandt JS (2019) Implications of urban growth and farmland loss for ecosystem services in the western United States. Land Use Policy 86(April):1–11. https://doi.org/10.1016/j.landusepol.2019.04.029

Novak AB, Wang YQ (2004) Effects of suburban sprawl on Rhode Island's forests: a Landsat view from 1972 to 1999. Northeast Nat 11(1):67–74

Nowak DJ, Walton JT (2005) Projected urban growth (2000–2050) and its estimated impact on the US forest resource. J Forest 103(8):383–389. https://doi.org/10.1093/jof/103.8.383

Pandey B, Seto KC (2015) Urbanization and agricultural land loss in India: comparing satellite estimates with census data. J Environ Manage 148:53–66. https://doi.org/10.1016/j.jenvman.2014.05.014

Pathak CR (2019) Challenges of smart cities in India. Contemp South Asian Stud, 261–269. https://doi.org/10.1007/978-3-030-23796-7_16

Population Census (2022) Aligarh city population census 2011–2023 | Uttar Pradesh. Retrieved 18 Jan 2023, from https://www.census2011.co.in/census/city/112-aligarh.html

Seto KC, Kaufmann RK, Woodcock CE (2000) Landsat reveals China's farmland reserves, but they're vanishing fast [1]. Nature 406(6792):121. https://doi.org/10.1038/35018267

Seto KC, Fragkias M, Guneralp B, Reilly MK (2011) A Meta-analysis of global urban land expansion Karen. PLoS ONE 6(8):e23777. https://doi.org/10.1371/journal.pone.0023777

Shafizadeh Moghadam H, Helbich M (2013) Spatiotemporal urbanization processes in the megacity of Mumbai, India: a Markov chains-cellular automata urban growth model. Appl Geogr 40:140–149. https://doi.org/10.1016/j.apgeog.2013.01.009

Song W, Pijanowski BC, Tayyebi A (2015) Urban expansion and its consumption of high-quality farmland in Beijing, China. Ecol Ind 54:60–70. https://doi.org/10.1016/j.ecolind.2015.02.015

Tan M, Li X, Xie H, Lu C (2005) Urban land expansion and arable land loss in China—A case study of Beijing-Tianjin-Hebei region. Land Use Policy 22(3):187–196. https://doi.org/10.1016/j.landusepol.2004.03.003

Theobald DM (2001) Land-use dynamics beyond the American urban fringe. Geogr Rev 91(3):544. https://doi.org/10.2307/3594740

World Bank (2021) Arable land (hectares per person) - India, World Bank Open Data, https://data.worldbank.org/indicator/AG.LND.ARBL.HA.PC?locations=IN

Analysing Seasonal Variability of Land Surface Temperature with Land Use Land Cover Pattern in Jammu District, India

5

Naveen Kumar, Habibullah, Manisha Kumari, Mansoor Ahmad, Neha Thappa, Shafiq Ahmed, Shahfahad, and Md. Sarfaraz Asgher

Abstract

One of the most significant environmental parameters is land surface temperature (LST) which represents the skin temperature of land surface. It is measured remotely using thermal sensors on satellites or in situ using ground-based instruments. It is applied to investigate urban heat island (UHI) and thermal environment in a region. Hence, this research seeks to investigate the seasonal variation of LST with LULC pattern and the relation between LULC pattern and LST. For this, Landsat datasets have been used for different seasons, i.e. pre-monsoon, monsoon, as well as post-monsoon for mapping of LULC and LST retrieval. In addition, surface biophysical parameters were computed to inspect the connection amid LST and LULC pattern. The findings show that the highest LST is observed in monsoon season while it reaches to its lowest the pre-monsoon. Gully area, built-up area, and river bed area registered high LST throughout all three seasons, whereas water bodies, dense forests, scrub forests, and agricultural areas exhibited low LST throughout all three seasons. Additionally, the relation between LST and spectral indices parameters indicates that while all surface biophysical parameters displayed a substantial correlation with LST. But, the normalized difference moisture index (NDMI) and normalized difference built-up index (NDBI) had highest correlation relationship with LST throughout all seasons (correlation > 0.50). These outcomes may assist in land use planning, thermal wave heat wave mitigation in the region.

Keywords

Land use land cover pattern · Surface biophysical parameters · Land surface temperature · Seasonal variation · Landsat datasets

5.1 Introduction

The land surface temperature (LST) is a fundamental parameter in the field of Earth science, which denotes the temperature of the uppermost layer of the terrestrial surface (Kant et al. 2009; Kumari et al. 2020). Specifically, LST is defined as the temperature of the land surface as observed by remote sensing instruments, such as satellites or airborne sensors (Ma et al. 2022; Sarif et al.

N. Kumar · Habibullah · M. Kumari · M. Ahmad · N. Thappa · S. Ahmed · Md.S. Asgher (✉)
Department of Geography, University of Jammu, Jammu, Jammu & Kashmir 180006, India
e-mail: sasgher20@gmail.com

Shahfahad
Department of Geography, Faculty of Sciences, Jamia Millia Islamia, New Delhi 110025, India
e-mail: g.shahfahad@jmi.ac.in

S. Talukdar et al. (eds.), *Environmental Risk and Resilience in the Changing World*, GIScience and Geo-environmental Modelling, https://doi.org/10.1007/978-3-031-62442-1_5

2023). The measurement of LST is of great significance as it provides valuable information regarding the thermal state of the Earth's surface and its surrounding environment (Weng et al. 2004). It is just one of the numerous environmental factors that are used to estimate how much energy and material are exchanged between the Earth's surface and the atmosphere (Zhou et al. 2011; Thakur et al. 2021). The LST has the capability to depict the thermal characteristics of the land surface configuration, which exhibits a significant association with the air temperature (Schwarz et al. 2012; Ullah et al. 2019). It is a crucial tool for tracking the urban heat island (UHI), determining thermal comfort, and identifying risks associated with heat (Abrar et al. 2022; AlDousari et al. 2023).

LST varies with LULC types, surface materials, surface roughness, topography, etc. (Song et al. 2020; Hashim et al. 2022). It is due to the fact that different LULC types and building materials have different thermal properties. For example, impervious surfaces like barren rocks and built-up surfaces can retain higher amount of heat from the incoming solar energy and emit it at a steady rate (Bindajam et al. 2022; Shahfahad et al. 2022a; Sarif et al. 2023). Furthermore, the pervious surfaces like vegetation cover, scrubland, and agricultural land cannot hold the energy for long time and reflect it quickly into atmosphere (Myint et al. 2010; García and Díaz 2023). As a result, there exists a marked contrast in the land surface temperature (LST) between impervious and pervious surfaces, with impervious surfaces exhibiting notably higher LST values than pervious surfaces (Dutta et al. 2021; Waleed and Sajjad 2022). In urban areas, this elevated temperature because of high proportion of impervious surfaces is termed as UHI phenomenon (Yuan and Bauer 2007; Mohajerani et al. 2017).

Satellite-derived LST has been proven to be one of the most effective means of assessing the surface thermal properties and UHI in across the globe (Bindajam et al. 2022; Waleed and Sajjad 2022; Talukdar et al. 2022). Consequently, several researches have been done to examine the LST variation and its association with LULC pattern (Bindajam et al. 2022; Shahfahad et al. 2022a; Tariq et al. 2023), surface biophysical parameters (Guo et al. 2015; Kumari et al. 2020), urban landscape heterogeneity (Estoque et al. 2017; Dutta et al. 2021; Tali et al. 2023), green and blue space spatial configuration (Chen et al. 2022), using thermal satellite datasets. The surface biophysical parameters calculated using remote sensing datasets such as normalized difference moisture index (NDMI, normalized difference bareness index (NDBaI), normalized difference vegetation index (NDVI), normalized difference built-up index (NDBI), normalized difference water index (NDWI), etc. have been widely used to assess the influence of landscape composition on LST (Tran et al. 2017; Wu et al. 2019; Kumari et al. 2020; Guha et al. 2022).

Although several researches have investigated the discrepancy in LST with LULC pattern and surface biophysical parameters, very few researches have been done to assess the seasonal variation of LST (Wu et al. 2021; Chen et al. 2023; Mohammad and Goswami 2022). Moreover, the impression of seasonality is well understood on the environment and climate in tropics, especially in the countries of South Asia (Feng et al. 2013; Subash and Sikka 2014; Bindajam et al. 2023). In the case of India, only a few researches have been done to assess the seasonal variation in LST. The analysis of seasonal pattern of LST helps in identifying the areas of concern for mitigation of UHI and reducing the vulnerability to heat waves in a region. The current research aims to evaluate the seasonal fluctuations of land surface temperature (LST) in the Jammu district of India in response to the land use and land cover (LULC) pattern.

5.2 Materials and Methods

5.2.1 Study Area

Located on the foothills of north-western Himalayas, Jammu district is part of Jammu and Kashmir union territory. Geographically, the district is located at 33° 07′ 30″ north latitude and 75° 08′ 52″ east longitude, and the area of district

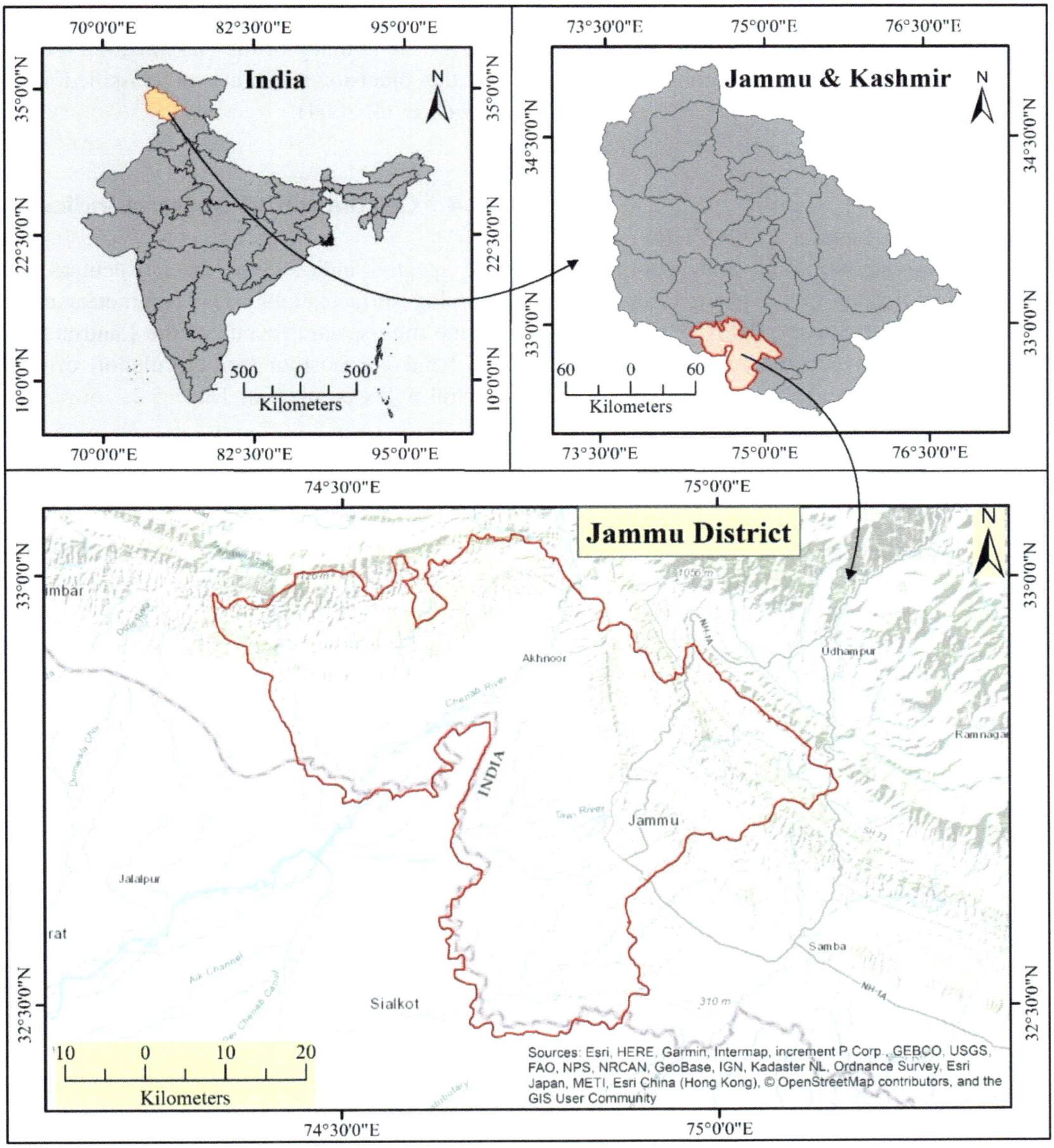

Fig. 5.1 Location of Jammu district

is about 2336 km^2 (Fig. 5.1). Jammu is surrounded by the districts of Reasi and Udhampur in the north and north-east, Rajouri in north-west, Pakistan in south and south-west, and Pakistan Occupied Kashmir (POK) in the north. With a population of 1,529,958 and population density of 653 persons/ km^2, Jammu is the district of Jammu and Kashmir. Nearly, 72% population of Jammu is urban living in cities and towns while only 28% population is rural (Census of India 2011). The district has warm temperate type (Koppen: Cwa) climate with dry and hot summers and cold winters. The minimum temperature in district goes to about 4 °C during winter months (December–January) while maximum temperature reaches up to 47 °C during summers

(May–June). Further, Jammu receives about 895–1500 mm rainfall each year, nearly 75% of which occurs from south-west monsoon.

5.2.2 Data Used

This study uses Landsat 8 (OLI/TIRS) datasets for pre-monsoon, monsoon, and post-monsoon seasons, downloaded from website of the United States Geological Survey (USGS) (Table 5.1). Landsat 8 data have eight multispectral bands, two thermal bands, and a panchromatic band. To retrieve LST, the thermal bands (band 10 & 11) are utilized. Band 10 has been utilized to retrieve the LST in this study due to its better calibration to band 11. For the LULC classification, multispectral bands, i.e. band 5, 4, and 3 were utilized. While, for the analysis of surface biophysical parameters, several spectral indices like NDVI, NDMI, NDWI, NDBaI, and NDBI have been calculated using the multispectral bands.

5.2.3 Method for LULC Mapping

For the mapping of LULC, maximum likelihood classifier-based supervised classification technique has been applied. The process involves several steps such as collection of training samples, classification of LULC pattern, recoding, and accuracy assessment. Based on the NRSC classification scheme (level-I), seven LULC classes have been identified in the Jammu district namely water bodies, agricultural land, scrubland, dense forest, built-up area, river bed, and gullies. Using the kappa coefficient, accuracy has been evaluated following the LULC classification. An important step in LULC classification is the accuracy assessment, which is used to evaluate the consistency of the classification method and the precision/reliability of classified maps (Powell et al. 2004).

5.2.4 Calculation of Spectral Indices

Five spectral indices have been calculated for assessing surface biophysical parameters utilizing the multispectral bands of the Landsat data. The band composition for calculation of each spectral index is given in Table 5.2.

5.2.5 Land Surface Temperature Retrieval

The LST retrieval from Landsat 8 data, several steps are performed. Firstly, spectral radiance ($L\lambda$) has been computed by utilizing the DN value of thermal band (Eq. 5.1).

$$L\lambda = ML \cdot Q_{\text{cal}} + AL \tag{5.1}$$

where $L\lambda$ represents spectral radiance while ML represents rescaling multiplicative factor. The Q_{cal} refers to calibrated value of quantized-pixel while AL refers to rescaling additive factor of band 10.

The brightness temperature is then derived by the calibration constant using Eq. 5.2 (Chander et al. 2009).

$$TB = \frac{K2}{\ln\left(\frac{K1}{L\lambda} + 1\right)} \tag{5.2}$$

where $K1$ as well as $K2$ refers to calibration constant while TB refers to brightness temperature.

Table 5.1 Details of the Landsat datasets used

Satellite/Sensor	Season	Path/Row	Date of acquisition	Spatial resolution (m)	Cloud cover (%)
Landsat 8 (OLI/TIRS)	Pre-monsoon	149/37	03-04-2020	30	0.20
	Monsoon	149/37	23-07-2020	30	2.60
	Post-monsoon	149/37	28-10-2020	30	0.00

Table 5.2 Details of the spectral indices used

S. No.	Spectral indices	Bands combination	References
1	NDVI	Band 5- Band 4/Band 5 + Band 4	Kriegler et al. (1969)
2	NDWI	Band 3- Band 5/Band 3 + Band 5	McFeeters (1996)
3	NDBI	Band 6- Band 5/Band 6 + Band 5	Zha et al. (2003)
4	NDMI	Band 5- Band 6/Band 5 + Band 6	Jin and Sader (2005)
5	NDBaI	Band 6- Band 10/Band 6 + Band 10	Zhao and Chen (2005)

The correction of emissivity is performed using an improved equation (Eq. 5.3) of Yu et al. (2014).

$$\varepsilon = 0.00149Pv + 0.98481 \qquad (5.3)$$

where ε refers to corrected emissivity while Pv represents vegetation proportion calculated by utilizing Eq. 5.4.

$$Pv = \left[\frac{\text{NDVI} - \text{NDVI}_{\text{min}}}{\text{NDVI}_{\text{max}} - \text{NDVI}_{\text{min}}}\right]^2 \qquad (5.4)$$

where NDVI_{max} is maximum while NDVI_{min} is minimum value of NDVI.

The LST was then retrieved utilizing the Eq. 5.5 (Artis and Carnahan 1982).

$$T = \frac{TB}{1 + (\lambda.TB/P)\ln(\varepsilon)} \qquad (5.5)$$

where T represents retrieved LST while TB refers to brightness temperature. Further, λ refers to emitted radiance wavelength while ε represents corrected emissivity.

5.2.6 Correlation Between Spectral Indices and LST

Pearson correlation technique was applied using Python to assess the association amid LST and spectral indices as given in Eq. 5.6.

$$r = \frac{\Sigma(xi - \underline{x})\left(yi - \underline{y}\right)}{\sqrt{\Sigma(xi - \underline{x})^2 \sum\left(yi - \underline{y}\right)^2}} \qquad (5.6)$$

where r is the correlation between variable x and y, while $\bar{x}$ and $\bar{y}$ refers to the mean of variable x and y, respectively.

5.3 Results

5.3.1 LULC Pattern in Jammu

The LULC pattern of Jammu district is characterized by marked heterogeneity, with variations in land cover being attributable to both natural and anthropogenic factors. The result shows that built-up area is predominantly found in east-central parts of the district, primarily in the vicinity of the Jammu city (Fig. 5.2). Moreover, the southern plains of the district feature scattered patches of built-up areas, comprising smaller villages, and hamlets. The northern hilly region of the district is primarily under scrubland and vegetation cover, owing to the dense forests and orchards that cover the Pir Panjal range. A few pockets of built-up areas are noticeable over the northern hills, while some croplands and gullies are present in the same region. In contrast, the southern plains of the district feature extensive croplands. Additionally, the northern part of the district is marked by existence of the Chenab River, while eastern part of district is traversed by the Tawi River.

Table 5.3 provides a comprehensive overview of the LULC pattern of Jammu district. It shows that maximum area in Jammu is occupied by agricultural land, covering an area of 729.45 km^2. Further, the water bodies and gullies comprise lowest proportion of the total area of district, i.e. about 18.61 and 21.92 km^2, respectively.

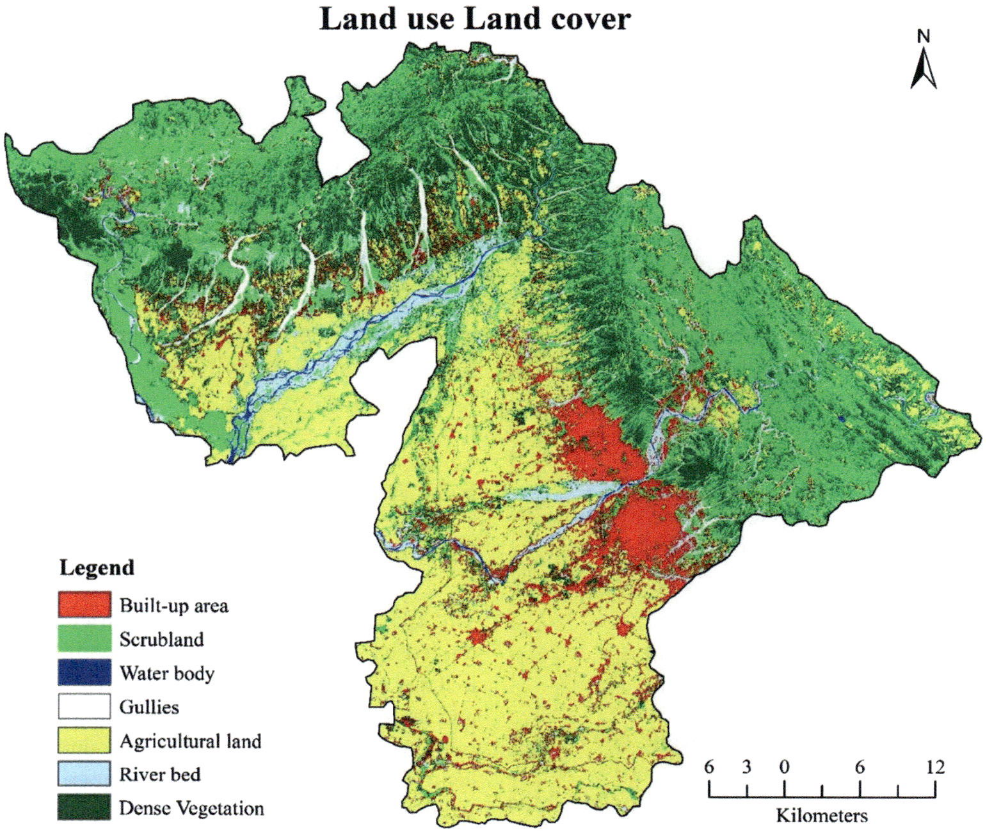

Fig. 5.2 LULC pattern of Jammu district

The table also shows that scrubland and dense vegetation have a significant proportion of the area in Jammu district, covering 615.82 km^2 and 541.49 km^2, respectively. These LULC classes are important ecological resources that provide habitat for wildlife, support soil conservation, and contribute to the local climate. Furthermore, the table indicates that the built-up area and river bed have a comparatively low proportion of area under them.

Table 5.3 Area under different LULC types in Jammu district

S. No.	LULC type	Area (in sq. km)
1	Built-up area	214.05
2	Scrubland	615.82
3	Water bodies	18.61
4	Gullies	21.92
5	Agricultural land	729.45
6	River bed	194.88
7	Dense vegetation	541.49
	Total	2336.22

5.3.2 Seasonality of Surface Biophysical Parameters

Figure 5.3 shows that biophysical parameters have a significant seasonal variation in Jammu district. Analysis of figure shows that NDVI is higher in the southern parts of Jammu district in pre-monsoon season while low in the central and northern. This may be because of the presence of croplands in the southern parts and pre-monsoon season is generally the growing season of rabi crops. On the other hand, the northern part is mostly hilly which generally has low vegetation growth in pre-monsoon. During monsoon as well as post-monsoon seasons, the NDVI shows an increase in the northern parts of district. This may be because the rainfall during monsoon season gives rise to the scrubs and increases the vegetation density. At the same time, NDBI shows higher values in northern parts during pre-monsoon and a decline in monsoon and post-monsoon. It may be because the hilly parts in north are mostly uncovered during pre-monsoon season exhibiting similar spectral signatures as built-up surfaces. However, in the monsoon and post-monsoon, the growth of vegetation cover led to decline in NDBI in northern parts.

Like NDVI, NDMI also shows higher values in the southern plains during pre-monsoon season. But, during monsoon and post-monsoon seasons, the rainfall led to increase in soil moisture in the northern parts which has resulted in higher NDMI values in the north and central parts during monsoon as well as post-monsoon seasons. The NDBaI and NDWI on the other hand show low seasonal variation in comparison to other biophysical parameters, implying that the bareness and water bodies have low seasonal variation in the district (Fig. 5.3). The NDBaI is moderate to low throughout the Jammu district except some patches in central and northern parts where bareness is high. At the same time, there is a clear pattern of high and low NDWI in the Jammu district during all seasons. The NDWI is low throughout the district except in the central parts where Tawi and Chenab Rivers flow.

The statistics of surface biophysical parameters shows a significant seasonal variation in Jammu (table 5.4). For instance, maximum NDVI is highest during pre-monsoon season (0.60) in Jammu district while the mean NDVI is maximum during monsoon season (0.36), implying that the vegetation density is maximum during monsoon season. Similarly, the mean NDMI is also very high during monsoon (0.20) while it is low during post-monsoon (0.09). On the other hand, mean NDBI is lowest during the monsoon (-0.20) and highest during post-monsoon season (-0.09). The NDBaI and NDWI show low seasonal variation with mean NDBaI varying between −0.39 to −0.43 between pre-monsoon and post-monsoon and mean NDWI varies between −0.22 to −0.31 between post-monsoon and monsoon season.

5.3.3 Seasonality of LST

The present study undertakes an analysis of the seasonal variation in land surface temperature (LST) and discerns a discernible pattern in its temporal behaviour. The results of the investigation demonstrate that the LST exhibits a low value in pre-monsoon, followed by an increase in the monsoon season. However, the LST displays a decline in the post-monsoon season, thereby revealing a consistent pattern of variation across the three seasons (Fig. 5.4). In pre-monsoon season, LST was high in the central parts, especially over the Jammu city and northern in northern parts over hills. Contrary to this, LST is low in most of the southern and western parts of Jammu district during pre-monsoon season. In the southern and central parts of the district, LST significantly increased in the monsoon season, while it decreased in the northern parts. However, during post-monsoon season, LST exhibited a substantial decline in most parts of the district, except over the hills. An interesting finding is that LST was noted high during all seasons over the gullies on hills. This may be because the data selected for the analysis was for the cloud-free dry period to avoid atmospheric errors. During dry periods, the gullies get quickly heated as they are mostly barren and are composed of rocks and bare sands. However, the

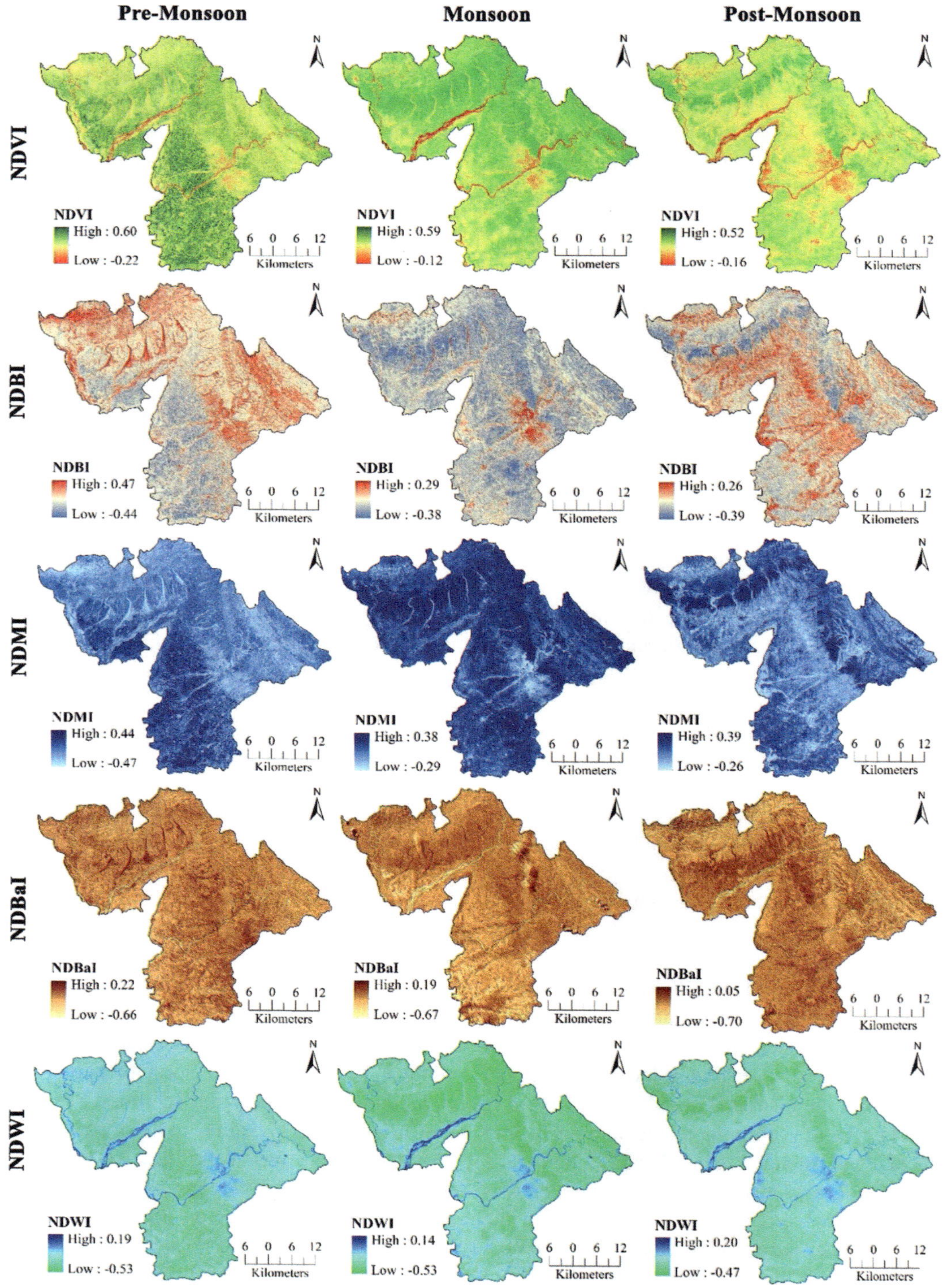

Fig. 5.3 Seasonal pattern of surface biophysical parameters

Table 5.4 Seasonal variation in statistics of surface biophysical parameters

Indices		Pre-monsoon	Monsoon	Post-monsoon
NDVI	Maximum	0.60	0.59	0.52
	Minimum	−0.22	−0.12	−0.16
	Mean	0.28	0.36	0.23
NDBI	Maximum	0.47	0.29	0.26
	Minimum	−0.44	−0.38	−0.39
	Mean	−0.12	−0.20	−0.09
NDMI	Maximum	0.44	0.38	0.39
	Minimum	−0.47	−0.29	−0.26
	Mean	0.12	0.20	0.09
NDBaI	Maximum	0.22	0.19	0.05
	Minimum	−0.66	−0.67	−0.70
	Mean	−0.39	−0.40	−0.43
NDWI	Maximum	0.19	0.14	0.20
	Minimum	−0.53	−0.53	−0.47
	Mean	−0.25	−0.31	−0.22

increase in vegetation cover and scrubs during post-monsoon season led to decline in LST in most parts of the district.

Table 5.5 shows that maximum LST in the Jammu district was observed in the pre-monsoon but the minimum LST was found in post-monsoon season. Further, mean LST was maximum during monsoon season (23.44 °C) while minimum and maximum LST was recorded highest during post-monsoon, i.e. 14.65 and 34.35 °C, respectively. Similarly, the range of LST (gap between minimum and maximum LST was noted highest during post-monsoon season (19.7 °C) and lowest during pre-monsoon season (16.09 °C). Also, the data indicate that the LST in the post-monsoon deviates more from mean LST while the pre-monsoon season LST deviates less from the mean.

5.3.4 Seasonal Variation of LST with LULC Pattern

For assessing the seasonal pattern of LST over different LULC types, seven LULC classes have been identified in the Jammu district. Among all LULC types, river bed and gullies show very high LST during all three seasons (Table 5.6). During pre-monsoon season, maximum and minimum LST was noted over river bed, i.e. 30.07 and 9.90 °C while mean LST was highest over gullies (23.77 °C). Again, in monsoon season, the minimum and maximum LST has been found over river bed, i.e. 31.05 and 7.72 °C while mean LST has been noted again maximum over gullies (18.38 °C). However, in post-monsoon season, the maximum and minimum LST has been noted over agricultural land (35.64 °C) and water bodies (13.76 °C), respectively. However, the mean LST has been again noted maximum over gullies (25.70 °C). The higher mean LST over gullies may be because the gullies are made of bare sands and rocks (Amin and Romshoo 2019) which get heated quickly.

5.3.5 Relationship Between LST and Surface Biophysical Parameters

To examine the association between surface biophysical parameters and LST, the Pearson correlation techniques have been applied in this

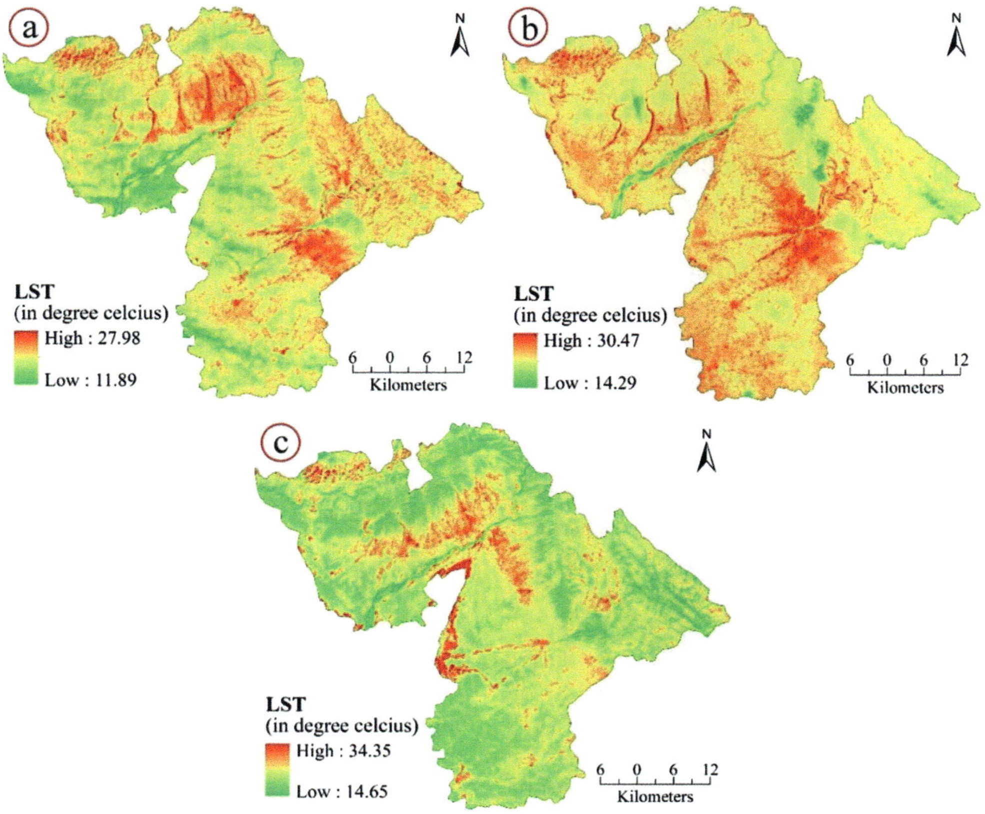

Fig. 5.4 Pattern of LST during **a** pre-monsoon, **b** monsoon, and **c** post-monsoon season

Table 5.5 Seasonal variation in descriptive statistics of LST

Season	LST (Mean)	LST (Maximum)	LST (Minimum)	LST (Standard deviation)
Pre-monsoon	20.76	27.98	11.89	1.79
Monsoon	23.44	30.47	14.29	1.22
Post-monsoon	22.98	34.35	14.65	2.23

study using Python scatter plots. For each biophysical parameter and LST for all seasons, scatter plots were made in Python. The benefit of using Python for correlation analysis is that it also provides error when computing the correlation by means of in terms of root mean square error. According to the findings, there exists a negative association of LST with NDMI and NDVI. Conversely, the investigation reveals a positive correlation of LST with NDBI and NDBaI (Fig. 5.5). This means, built-up surface as well as bare/open lands increases LST while soil moisture and vegetated areas reduce the LST. At the same time, NDWI did not have a significant relationship with LST. This is possibly because the Jammu district has only two major water bodies in the form of rivers while most of its areas do not have water bodies. Moreover, the error in correlation between NDWI and LST was also noted higher than other biophysical parameters. The NDMI ($R < -0.50$) and NDBI ($R > 0.50$) have the most consistent

Table 5.6 Seasonal variation of LST over different LULC types

Season	LULC	LST (in °C)			
		LST Max.	LST Min.	Mean	Overall mean
Pre-monsoon	Water bodies	28.44	9.98	17.49	20.76
	River bed	30.07	9.90	21.55	
	Agricultural land	29.19	11.11	19.96	
	Gullies	29.75	16.15	23.77	
	Built-up area	29.62	13.37	21.90	
	Scrubland	29.41	11.60	21.16	
	Dense vegetation	28.81	12.19	20.6	
Monsoon	Water bodies	30.56	16.75	21.69	23.44
	River bed	31.05	7.72	23.98	
	Agricultural land	30.57	10.56	23.53	
	Gullies	30.09	18.38	25.41	
	Built-up area	30.51	7.92	24.63	
	Scrubland	30.51	6.50	23.16	
	Dense vegetation	30.00	6.51	22.89	
Post-monsoon	Water bodies	32.85	13.76	20.93	22.98
	River bed	34.55	13.93	23.97	
	Agricultural land	35.64	15.00	23.04	
	Gullies	32.19	19.81	25.70	
	Built-up area	33.22	17.50	23.93	
	Scrubland	34.52	14.27	22.74	
	Dense vegetation	35.26	14.99	22.41	

negative relationship with LST with correlation coefficient for all seasons while water bodies witnessed consistently low correlation with LST ($R < 0.30$) for all seasons. Further, NDVI and NDBaI showed fluctuation in correlation coefficient with LST during pre- and post-monsoon.

5.4 Discussion

Present research has been intended to inspect the seasonality of LST with LULC pattern in Jammu district which is in the foothills of Himalayas. Study shows that the highest LST is found in pre-monsoon season in Jammu district while lowest LST has been recorded in post-monsoon season. The rationale behind this observation is that the pre-monsoon season is characterized by a significant decrease in the soil moisture content and vegetation density (Varikoden and Revadekar 2018; Sharma et al. 2021), resulting in a heightened capacity of the land surface to absorb heat (Mohammad et al. 2019; Sharma et al. 2021). Hence, the LST tends to increase, leading to a negative correlation of LST with NDMI and NDVI. Contrary to this, in the post-monsoon, the soil moisture and vegetation cover are maximum because the substantial rainfall gave rise to abundance of vegetation cover (Mathew et al. 2022; Samuel et al. 2023). This leads to a contrasting difference between the seasonal LST in Indian sub-continent. Moreover, local topographic condition and climate pattern also have crucial influence on the LST (He et al. 2019).

The investigation of seasonal pattern of LST over LULC typologies shows that the highest LST has been recorded over gullies during all three seasons followed by built-up area and river

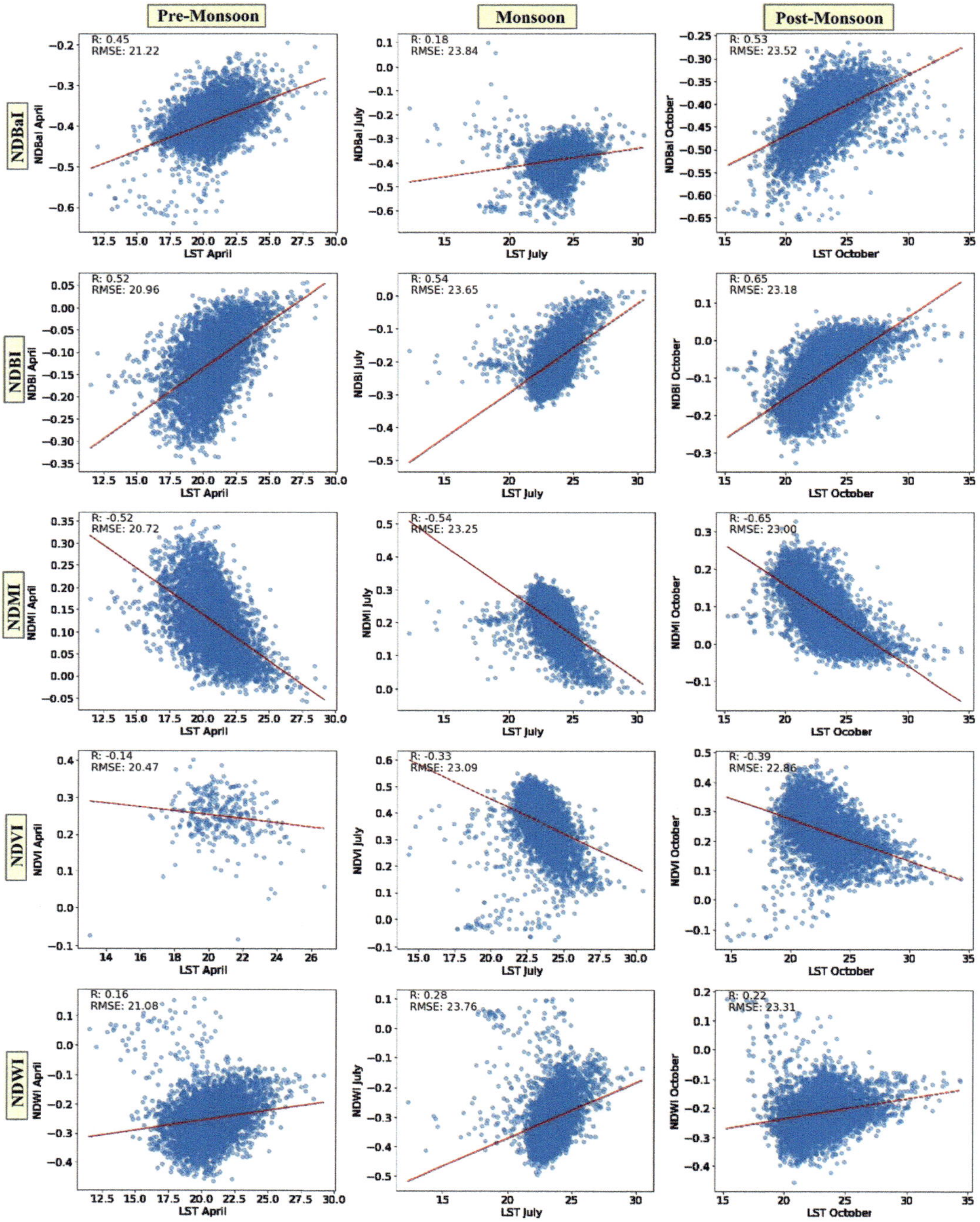

Fig. 5.5 Correlation between surface biophysical parameters and LST

bed. This may be attributed to the presence of sand, gravel, and stones in gullies which absorb higher amount of solar energy and get quickly heated than the other LULC categories (Hereher 2017; Ayanlade et al. 2021). Further, among all the LULC typologies, water bodies recorded the lowest LST during all seasons. It is because of the higher heat capacity and thermal conductivity of water bodies, which enables efficient dissipation of energy and consequent cooling of the surrounding environment (Gunawardena et al., 2017). The dense vegetation and agricultural land also show low LST in comparison to the gullies, built-up area, and river bed. This is because a significant amount of heat is transferred back to the outer atmosphere by transpiration process from vegetated surfaces like forests and croplands (Mildrexler et al. 2011; Hesslerová et al. 2019; Rihan et al. 2021). During all three seasons, the LST over scrubland is found to be more than that of dense forest. Further, in pre-monsoon period, the LST of scrubland is more than the mean LST of Jammu district because the canopy cover of scrubland is low which allows more sunlight to penetrate and reach ground surface directly, ultimately heating up the surface. In their study, Hardwick et al. (2015) noted that the lower canopy density in scrublands and forest areas allows sunlight to penetrate to ground surface and hence, the surfaces get heated quickly.

For examining the impact of LULC pattern of seasonality of LST, surface biophysical parameters are calculated, and then correlation between biophysical parameters and LST has been calculated. The study shows negative correlation of LST with NDMI as well as NDVI while positive correlation with the NDBI as well as NDBaI. The findings of Tariq and Mumtaz (2022), Shahfahad et al. (2022b), and Sannigrahi et al. (2018) are comparable to this result. Nevertheless, the NDWI did not show a significant relationship with the LST. This is because the Jammu district has only two major water bodies which are crossing through the district from north to south-west. Studies have demonstrated that the significance of spatial association between depends on the spatial pattern of variables; more dispersed the variables are, more significant will be the association and more clustered the variables are, less significant will be the association (Wu et al. 2019; Rihan et al. 2021). Moreover, positive relationship of LST with NDBI as well as NDBaI indicates that the built-up surfaces as well as bare/open land exhibit higher LST and can store and hold heat for long time. Nevertheless, negative correlation of NDMI and NDVI with LST shows that the soil moisture and vegetation cover put diminishing impacts on LST and cannot hold the temperature for long time.

Among all the biophysical parameters, NDMI and NDBI showed a consistent relationship with the LST while other showed either variation in correlation coefficient or did not show a significant correlation. This is because the spatial pattern built-up surface and soil moisture mostly remain same throughout the year while vegetation cover and bareness vary with time, location, and season (Zhou et al. 2014; Shahfahad et al. 2022a). Likewise in Jammu district, the bareness and vegetation cover varies from one season to other. For instance, the bare surfaces in pre-monsoon season get covered with grasses and scrubs in the post-monsoon season due to consistent moisture supply during monsoon season (Tse-Ring et al. 2010). Similarly, the vegetation density also varies from season to season. Moreover, the higher LST noted over gullies and river beds indicates that bareness causes higher temperature in the region. The increase bareness and deterioration of natural vegetation may increase the heat waves in the Jammu district. Thus, it is suggested to increase the vegetated areas in the region, especially in the outskirts of Jammu city and other small towns, Jammu region has a long history of heat waves and extreme temperature conditions. Furthermore, the local governing bodies and planning authorities may also look after planning some artificial water bodies which will not only mitigate the KST but also be beneficial for tourism as region is an important tourist destination.

5.5 Conclusion

This research is intended for assessing the seasonal pattern of LST in response to the LULC pattern in the Jammu district. As per the results, the mean LST has been recorded highest in monsoon season while pre-monsoon season recorded the lowest temperature. Further, it is noted that the LULC pattern varies with the LULC typologies which indicates that LULC pattern is the most important controlling factor of LST. For instance, gullies, built-up areas, and river beds recorded high LST in all the three seasons while the water body, dense forest, scrub forest, and agricultural area exhibited low LST in all the three seasons. The mean LST over gullies varies from 23.77 °C in pre-monsoon to about 25.70 °C in post-monsoon season. On the other hand, the mean LST over water bodies varies from 17.49 °C in pre-monsoon to about 20.93 °C in post-monsoon. Moreover, a contrasting difference in LST has been found during the three seasons in the Jammu district with monsoon season being warmest while pre-monsoon being the coldest. The correlation analysis showed that all the biophysical parameters have significant relationship with LST except NDWI. Nevertheless, the NDBI and NDMI have consistently high relationship with correlation coefficient > 0.50 (LST) during all three seasons while NDBaI and NDVI have variation in relationship from one season to other. The Jammu region has frequently witnessed heat waves and extreme temperatures in the past decades. Therefore, there is an urgent need for the strategies to mitigate the heat waves and maintain the thermal comfort in the region. This study may provide insights towards the thermal comfort management and heat waves mitigation in Jammu district as well as other region with similar climatic and geographical settings. For this, focus should be given to the social forestry and water bodies conservation programmes. Fallow lands and unutilized barren lands may be transformed into green spaces and artificial water bodies/wetlands to preserve the region's environmental health.

References

Abrar R, Sarkar SK, Nishtha KT, Talukdar S, Rahman A, Islam ARMT, Mosavi A (2022) Assessing the spatial mapping of heat vulnerability under urban heat island (UHI) effect in the Dhaka metropolitan area. Sustainability 14(9):4945

AlDousari AE, Kafy AA, Saha M, Fattah MA, Bakshi A, Rahaman ZA (2023) Summertime microscale assessment and prediction of urban thermal comfort zone using remote-sensing techniques for Kuwait. Earth Syst Environ, pp 1–22

Amin M, Romshoo SA (2019) Comparative assessment of soil erosion modelling approaches in a Himalayan watershed. Model Earth Syst Environ 5:175–192

Artis DA, Carnahan WH (1982) Survey of emissivity variability in thermography of urban areas. Rem Sens Environ 12(4):313–329

Ayanlade A, Aigbiremolen MI, Oladosu OR (2021) Variations in urban land surface temperature intensity over four cities in different ecological zones. Sci Rep 11(1):20537

Bindajam AA, Mallick J, Talukdar S, Shohan AAA, Rahman A (2022) Modeling the spatiotemporal heterogeneity of land surface temperature and its relationship with land use land cover using geostatistical techniques and machine learning algorithms. Environ Sci Pollut Res, pp 1–19

Bindajam AA, Naikoo MW, Horo JP, Mallick J, Rihan M, Malcoti MD, Talukdar S, Rahman MR, Rahman A (2023) Response of soil moisture and vegetation conditions in seasonal variation of land surface temperature and surface urban heat island intensity in sub-tropical semi-arid cities. Theor Appl Climatol 153(1):367–395

Census of India (2011). Jammu District—Population 2011–2023. https://www.census2011.co.in/census/district/639-jammu.html

Chander G, Markham BL, Helder DL (2009) Summary of current radiometric calibration coefficients for Landsat MSS, TM, ETM+, and EO-1 ALI sensors. Rem Sens Environ 113(5):893–903

Chen L, Wang X, Cai X, Yang C, Lu X (2022) Combined effects of artificial surface and urban blue-green space on land surface temperature in 28 major cities in China. Remote Sensing 14(3):448

Chen Y, Yang J, Yu W, Ren J, Xiao X, Xia JC (2023) Relationship between urban spatial form and seasonal land surface temperature under different grid scales. Sustain Cities Soc 89:104374

Dutta D, Rahman A, Paul SK, Kundu A (2021) Impervious surface growth and its inter-relationship with vegetation cover and land surface temperature in peri-urban areas of Delhi. Urban Climate 37:100799

Estoque RC, Murayama Y, Myint SW (2017) Effects of landscape composition and pattern on land surface temperature: An urban heat island study in the megacities of Southeast Asia. Sci Total Environ 577:349–359

Feng X, Porporato A, Rodriguez-Iturbe I (2013) Changes in rainfall seasonality in the tropics. Nat Clim Change 3(9):811–815

García, D. H., & Díaz, J. A. (2023). Space–time analysis of the earth's surface temperature, surface urban heat island and urban hotspot: relationships with variation of the thermal field in Andalusia (Spain). *Urban Ecosystems*, 1–22.

Guha S, Govil H, Taloor AK, Gill N, Dey A (2022) Land surface temperature and spectral indices: a seasonal

study of Raipur City. Geodesy and Geodynamics 13 (1):72–82
Gunawardena KR, Wells MJ, Kershaw T (2017) Utilising green and bluespace to mitigate urban heat island intensity. Sci Total Environ 584:1040–1055
Guo G, Wu Z, Xiao R, Chen Y, Liu X, Zhang X (2015) Impacts of urban biophysical composition on land surface temperature in urban heat island clusters. Landsc Urban Plan 135:1–10
Hardwick SR, Toumi R, Pfeifer M, Turner EC, Nilus R, Ewers RM (2015) The relationship between leaf area index and microclimate in tropical forest and oil palm plantation: Forest disturbance drives changes in microclimate. Agric for Meteorol 201:187–195
Hashim BM, Al Maliki A, Sultan MA, Shahid S, Yaseen ZM (2022) Effect of land use land cover changes on land surface temperature during 1984–2020: a case study of Baghdad city using landsat image. Nat Hazards 112(2):1223–1246
He J, Zhao W, Li A, Wen F, Yu D (2019) The impact of the terrain effect on land surface temperature variation based on Landsat-8 observations in mountainous areas. Int J Remote Sens 40(5–6):1808–1827
Hereher ME (2017) Retrieving spatial variations of land surface temperatures from satellite data–Cairo region, Egypt. Geocarto Int 32(5):556–568
Hesslerová P, Pokorný J, Huryna H, Harper D (2019) Wetlands and forests regulate climate via evapotranspiration. In: Wetlands: ecosystem services, restoration and wise use, pp 63–93
Jin S, Sader SA (2005) Comparison of time series tasselled cap wetness and the normalized difference moisture index in detecting forest disturbances. Remote Sens Environ 94(3):364–372
Kant Y, Bharath BD, Mallick J, Atzberger C, Kerle N (2009) Satellite-based analysis of the role of land use/land cover and vegetation density on surface temperature regime of Delhi, India. J Indian Soc Remote Sens 37:201–214
Kriegler FJ, Malila WA, Nalepka RF, Richardson W (1969), Preprocessing transformations and their effects on multispectral recognition. In: Proceedings of the sixth international symposium on remote sensing of environment. University of Michigan, Ann Arbor, MI, pp 97–131
Kumari B, Tayyab M, Ahmed IA, Baig MRI, Khan MF, Rahman A (2020) Longitudinal study of land surface temperature (LST) using mono-and split-window algorithms and its relationship with NDVI and NDBI over selected metro cities of India. Arab J Geosci 13:1–19
Ma J, Shen H, Wu P, Wu J, Gao M, Meng C (2022) Generating gapless land surface temperature with a high spatio-temporal resolution by fusing multi-source satellite-observed and model-simulated data. Remote Sens Environ 278:113083
Mathew A, Sarwesh P, Khandelwal S (2022) Investigating the contrast diurnal relationship of land surface temperatures with various surface parameters represent vegetation, soil, water, and urbanization over Ahmedabad city in India. Energy Nexus 5:100044
McFeeters SK (1996) The use of the normalized difference water index (NDWI) in the delineation of open water features. Int J Remote Sens 17(7):1425–1432
Mildrexler DJ, Zhao M, Running SW (2011) A global comparison between station air temperatures and MODIS land surface temperatures reveals the cooling role of forests. J Geophys Res Biogeosci 116(G3)
Mohajerani A, Bakaric J, Jeffrey-Bailey T (2017) The urban heat island effect, its causes, and mitigation, with reference to the thermal properties of asphalt concrete. J Environ Manage 197:522–538
Mohammad P, Goswami A (2022) Predicting the impacts of urban development on seasonal urban thermal environment in Guwahati city, northeast India. Build Environ 226:109724
Mohammad P, Goswami A, Bonafoni S (2019) The impact of the land cover dynamics on surface urban heat island variations in semi-arid cities: a case study in Ahmedabad City, India, using multi-sensor/source data. Sensors 19(17):3701
Myint SW, Brazel A, Okin G, Buyantuyev A (2010) Combined effects of impervious surface and vegetation cover on air temperature variations in a rapidly expanding desert city. Gisci Remote Sens 47(3):301–320
Powell RL, Matzke N, de Souza Jr C, Clark M, Numata I, Hess LL, Roberts DA (2004) Sources of error in accuracy assessment of thematic land-cover maps in the Brazilian Amazon. Remote Sens Environ 90 (2):221–234
Rihan M, Naikoo MW, Ali MA, Usmani TM, Rahman A (2021) Urban heat island dynamics in response to land-use/land-cover change in the coastal city of Mumbai. J Indian Soc Remote Sens 49(9):2227–2247
Samuel JB, Chakraborty A, Paleri A (2023) Deciphering the relationship between vegetation and Indian summer monsoon rainfall. Environ Res Lett
Sannigrahi S, Bhatt S, Rahmat S, Uniyal B, Banerjee S, Chakraborti S, Jha S, Lahiri S, Santra K, Bhatt A (2018) Analyzing the role of biophysical compositions in minimizing urban land surface temperature and urban heating. Urban Clim 24:803–819
Sarif MO, Gupta RD, Murayama Y (2023) Assessing local climate change by spatiotemporal seasonal lst and six land indices, and their interrelationships with suhi and hot-spot dynamics: a case study of Prayagraj City, India (1987–2018). Remote Sens 15(1):179
Schwarz N, Schlink U, Franck U, Großmann K (2012) Relationship of land surface and air temperatures and its implications for quantifying urban heat island indicators—An application for the city of Leipzig (Germany). Ecol Ind 18:693–704
Shahfahad, Naikoo MW, Islam ARMT, Mallick J, Rahman A (2022a) Land use/land cover change and its impact on surface urban heat island and urban thermal comfort in a metropolitan city. Urban Clim 41:101052
Shahfahad, Talukdar S, Rihan M, Hang HT, Bhaskaran S, Rahman A (2022b) Modelling urban heat island (UHI) and thermal field variation and their relationship with land use indices over Delhi and Mumbai metro cities. Environ Dev Sustain 24:3762–3790

Sharma M, Bangotra P, Gautam AS, Gautam S (2021) Sensitivity of normalized difference vegetation index (NDVI) to land surface temperature, soil moisture and precipitation over district Gautam Buddh Nagar, UP, India. Stoch Env Res Risk Assess 36:1779–1789

Song J, Chen W, Zhang J, Huang K, Hou B, Prishchepov AV (2020) Effects of building density on land surface temperature in China: spatial patterns and determinants. Landsc Urban Plan 198:103794

Subash N, Sikka AK (2014) Trend analysis of rainfall and temperature and its relationship over India. Theoret Appl Climatol 117:449–462

Tali JA, Ganaie TA, Naikoo MW (2023) Assessing the link between land use/land cover changes and land surface temperature in Himalayan urban center, Srinagar. Environ Sci Pollut Res, pp 1–15

Talukdar S, Ali R, Nguyen KA, Naikoo MW, Liou YA, Islam ARMT, Mallick J, Rahman A (2022) Monitoring drought pattern for pre-and post-monsoon seasons in a semi-arid region of western part of India. Environ Monit Assess 194(6):396

Tariq A, Mumtaz F (2022) Modeling spatio-temporal assessment of land use land cover of Lahore and its impact on land surface temperature using multispectral remote sensing data. Environ Sci Pollut Res 30:23908–23924

Tariq A, Mumtaz F, Majeed M, Zeng X (2023) Spatio-temporal assessment of land use land cover based on trajectories and cellular automata Markov modelling and its impact on land surface temperature of Lahore district Pakistan. Environ Monit Assess 195(1):114

Thakur S, Maity D, Mondal I, Basumatary G, Ghosh PB, Das P, De TK (2021) Assessment of changes in land use, land cover, and land surface temperature in the mangrove forest of Sundarbans, northeast coast of India. Environ Dev Sustain 23:1917–1943

Tran DX, Pla F, Latorre-Carmona P, Myint SW, Caetano M, Kieu HV (2017) Characterizing the relationship between land use land cover change and land surface temperature. ISPRS J Photogramm Remote Sens 124:119–132

Tse-Ring K, Sharma E, Chettri N, Shrestha AB (2010) Climate change vulnerability of mountain ecosystems in the Eastern Himalayas. International centre for integrated mountain development (ICIMOD). http://lib.riskreductionafrica.org/bitstream/handle/123456789/485/climate%20change%20vulnerability%20of%20mountain%20ecosystems%20in%20the%20Eastern%20Himalayas.pdf?sequence=1

Ullah S, Ahmad K, Sajjad RU, Abbasi AM, Nazeer A, Tahir AA (2019) Analysis and simulation of land cover changes and their impacts on land surface temperature in a lower Himalayan region. J Environ Manage 245:348–357

Varikoden H, Revadekar JV (2018) Relation between the rainfall and soil moisture during different phases of Indian monsoon. Pure Appl Geophys 175:1187–1196

Waleed M, Sajjad M (2022) Leveraging cloud-based computing and spatial modeling approaches for land surface temperature disparities in response to land cover change: evidence from Pakistan. Remote Sens Appl Soc Environ 25:100665

Weng Q, Lu D, Schubring J (2004) Estimation of land surface temperature–vegetation abundance relationship for urban heat island studies. Remote Sens Environ 89(4):467–483

Wu C, Li J, Wang C, Song C, Chen Y, Finka M, La Rosa D (2019) Understanding the relationship between urban blue infrastructure and land surface temperature. Sci Total Environ 694:133742

Wu W, Li L, Li C (2021) Seasonal variation in the effects of urban environmental factors on land surface temperature in a winter city. J Clean Prod 299:126897

Yu X, Guo X, Wu Z (2014) Land surface temperature retrieval from Landsat 8 TIRS—Comparison between radiative transfer equation-based method, split window algorithm and single channel method. Rem Sens 6(10):9829–9852

Yuan F, Bauer ME (2007) Comparison of impervious surface area and normalized difference vegetation index as indicators of surface urban heat island effects in Landsat imagery. Remote Sens Environ 106 (3):375–386

Zha Y, Gao J, Ni S (2003) Use of normalized difference built-up index in automatically mapping urban areas from TM imagery. Int J Remote Sens 24(3):583–594

Zhao HM, Chen XL (2005) Use of normalized difference bareness index in quickly mapping bare areas from TM/ETM+. In: Proceedings of the 2005 IEEE international geoscience and remote sensing symposium, Seoul, Korea, vol 3, pp 1666–1668

Zhou W, Huang G, Cadenasso ML (2011) Does spatial configuration matter? Understanding the effects of land cover pattern on land surface temperature in urban landscapes. Landsc Urban Plan 102(1):54–63

Zhou D, Zhao S, Liu S, Zhang L, Zhu C (2014) Surface urban heat island in China's 32 major cities: Spatial patterns and drivers. Remote Sens Environ 152:51–61

Part II

Environmental Risk, Vulnerability and Health Assessment: Methods, Innovations, and Technologies

Evaluating the Reliability of Landslide Susceptibility Models Using Random Forest in Heterogeneous Landslide-Occurrence Region of the Indian Himalayas

6

Somnath Bera, Balamurugan Guru, Thomas Oommen, Vaibhav Kumar Upadhyay, and G. V. Ramana

Abstract

The task of modelling landslide susceptibility becomes complex when landslides manifest unique behaviours influenced by distinct sets of conditioning factors within a specific region. A landslide susceptibility model that omits the landslide types can increase its bias. This paper generates a landslide inventory database encompassing all landslide types within a study area. Additionally, the inventory underwent a classification process that resulted in its division into five distinct categories: (i) the entirety of landslides, (ii) debris slides in their entirety, (iii) debris slides induced by natural slope conditions, (iv) debris slides induced by cut slopes, and (v) rockfall-slide events. The interrelationship between these landslide groups and conditioning factors is analysed through the relationship index (RI). We then evaluated and compared different landslide susceptibility models using a random forest (RF) algorithm. Each model is based on a different set of landslide inventory. The inventory map covers the Kalimpong region (a part of Eastern Himalaya). The investigation revealed notable distinctions between the conditioning factors and the responses exhibited by natural slope-induced debris slides, cut slope-induced debris slides, and rockfall-slide events. Further, susceptibility values based on these groups of landslides were found to be different from each other. One of the research's interesting outcomes is that although the RF susceptibility models are based on all landslides, despite having an accuracy level of more than 0.9, they still cannot reflect the susceptibility of cut slope-induced landslides and the group of rockfall and slide. Therefore, the predictive performance of models in terms of statistics is necessary for building a susceptibility model

S. Bera
Department of Geography, Central University of South Bihar, Gaya 824236, India

B. Guru (✉)
Department of Geology, School of Earth Sciences, Central University of Tamil Nadu, Thiruvarur, India
e-mail: gurubala@cutn.ac.in

T. Oommen
Department of Geology and Geological Engineering, University of Mississippi, Oxford, MS 38677, USA

V. K. Upadhyay
Department of Data Science and Engineering, IISER, Bhopal, India

G. V. Ramana
Department of Civil Engineering, Indian Institute of Technology Delhi, Hauz Khas, New Delhi 110016, India

S. Talukdar et al. (eds.), *Environmental Risk and Resilience in the Changing World*, GIScience and Geo-environmental Modelling, https://doi.org/10.1007/978-3-031-62442-1_6

but cannot be the only metric to justify these models' reliability. The study recommends that the landslide susceptibility model should consider the type of each landslide separately rather than holistic way.

Keywords

Landslide susceptibility · Inventory · Machine learning · Random forest · GIS · Eastern Himalaya

6.1 Introduction

Landslides impact human society continuously in the mountainous region of the world. The landslide susceptibility map of a region depicts the degree of possibility of future-landslide occurrences; therefore, it commonly serves as a decision-making tool for hazard mitigation and spatial planning in a hilly area. In the last 20 years, landslide susceptibility studies have increased spectacularly. The approaches of landslide susceptibility broadly include physically based models, statistical and machine learning models. Physically based numerical models applied for simulating landslide occurrences. As the modelling procedure requires detailed geophysical parameters, it is more suitable at site-specific cases, whereas it is time-consuming and expensive on at larger scale. Comparatively, statistical and machine learning methods coupled with geographical information systems (GIS) and remote sensing (RS) are less time-consuming for landslide susceptibility prediction in larger areas. Therefore, in early, numerous statistical methods like frequency ratio (Ramesh and Anbazhagan 2015; Mirdda et al. 2019; Zhang et al. 2020), information value (Melo et al. 2012; Banerjee et al. 2018; Farooq and Akram 2021), weights of evidence (Pourghasemi et al. 2013; Getachew and Meten 2021), logistic regression (Melo and Zêzere 2017; Raja et al. 2018) have been applied for landslide susceptibility modelling. However, the statistical methods are not ideal for handling nonlinear relations between landslides and influencing factors. Therefore, machine learning techniques have been adopted for improving the predicting performance of susceptibility models and resolving the nonlinear complexity in very recently. The classical machine learning techniques such as artificial neural network (Bragagnolo et al. 2020), Bayesian networks (Liu et al. 2017; Pham et al. 2018), Naïve Bayes (Nhu et al. 2020), support vector machines (Achour and Pourghasemi 2020), and decision trees (Pradhan 2013; Addison and Oommen 2019) have been frequently used for susceptibility modelling and explicit nonlinearity in literature. For improving the generalization of the classical machine learning technique and optimizating model's parameters Deep learning methods have been applied in several studies (Dou et al. 2020; Thi Ngo et al. 2020; Bera et al. 2021a, b). The hybrid models such as adaptive neuro-fuzzy inference systems (Paryani et al. 2021; Razavi-Termeh et al. 2021) have been developed by combining fuzzy logic and artificial neural network to improve membership function weights. The ensemble method, such as random forest (Zhang et al. 2017; Bera et al. 2023) and generalized additive model (Pourghasemi and Rahmati, 2018), combines several machine learning classifiers to generate an optimal predictive model. Although a plethora of research exists in the area of application of analytical techniques to predict landslides, the selection of suitable techniques is challenging as the accuracy of the susceptibility model varies with the adopted algorithms. Therefore, many studies compare different techniques to acquire optimum accuracy. Pham et al. (2018) studied comparative performance of Bayes network, decision tree, Naïve Bayes, and support vector machines in landslide susceptibility mapping. The study found that support vector machines show the highest prediction accuracy of the model. On the other, Tien Bui et al. (2016a, b) found multi-layer perceptron neural networks have higher accuracy (90.2%) than support vector machine. Youssef and Pourghasemi (2021) compared landslide susceptibility models generated by artificial neural network, support vector machine, random forest, quadratic discriminant analysis, linear discriminant analysis, multivariate adaptive regression spline, and Naive Bayes techniques. The study found RF achieves the highest accuracy (95%) of the remaining techniques. Similarly, de Oliveira et al. (2019) and Kim et al. (2018) found RF brought significant accuracy in landslide susceptibility models. Apart from selecting techniques, the optimal susceptibility modelling is also complex as the reliability of a model also can vary due to many causes (Galli et al. 2008; Harp et al. 2011). Treating different types of landslides (i.e. debris slide, rock slide, etc.) as a whole is one of the causes of the susceptibility model's biases (van Westen et al. 2006). In the changing mountainous environment, different types of landslide are common on the natural slope and along the cut slope. Similarly, the Himalayas experience frequent landslides like debris slide, rock slide, and rockfall on the natural slope as usual and along cut slope due to rapid road construction. However, fewer studies are intended to develop landslide susceptibility models with differencing these types of landslides along the natural slope and cut slope. In the background, the objective of the study is to evaluate the reliability of landslide susceptibility models. The proposed models were developed by machine learning methods and included natural slope and cut slope-induced landslides in a heterogeneous landslide-occurrence region.

The study was concentrated in the Kalimpong region located in the Indian Himalayas. In the paper's first step, the inventory was categorized into five groups A: all landslides, B: debris slides, C: natural slope-induced debris slides (NDS), D: cut slope-induced debris slides (CDS), and E: rockfall-rock slides (RFRS) and analysed the relation of these groups with conditioning factors. Further, various ML models were developed and compared to select the best. The chosen models were random forest, decision tree, artificial neural network, linear support vector machine, and Naïve Bayes. Based on the performance, RF was selected to develop all groups of landslides. In the final step, models were applied based on these five groups of landslides and compared to evaluate the existent variations.

6.2 Study Area

The study area, i.e. Kalimpong, is located in the East Himalayas of India and extends about 238.7 km^2. The region has a subtropical highland climate with warm summer and cold-dry winter. The average annual rainfall is 2200 mm (ICIMOD 2019), and almost 80% of the total rainfall occurs under the influence of southwest monsoon. The distribution of rainfall is highly controlled by the local topography of this region which is highly rugged. Teesta and Relli are the principal rivers of the area. Numerous small streams drain in this region, significantly impacting landslide occurrences (Bera et al. 2020). The geology of this region is complex. The structure of geology is characterized by fold and fault systems. Since Main Central Thrust (MCT) and many active faults are crossing the region, it is subjected to frequent seismic activities. The area comprises rocks of several geologic formations such as Kanchenjunga gneiss, Chungthang, Gorubathan, Rangit pebble slate. The lithology of the region is described in Fig. 6.1 (prepared based on the Geological Survey of India). Kanchenjunga gneiss includes rocks of banded migmatite, gneiss, and mica schist (BMGGMC). Chungthang formation covers quartzite, mica schist, gneiss, and calcgranulite (QMSGC). Gorubathan covers three groups of rocks such as chlorite sericite schist and quartzite (CSSQ), mylonitic granite gneiss (MGG), quartz arenite, black slate, cherty phyllite (QABLCP). Rangit pebble slate includes boulder slate, conglomerate, phyllite (BSCP). Due to the differential weathering rate of these rocks, soil depth varies across the region.

Different landslide types, including debris slide, rock slide, and rockfall, are found in this region (Fig. 6.2). Almost every year, landslides are occurred naturally due to favourable geo-environment and rainfall during the monsoon season. In many cases, the landslides also happened along the road triggered by cutting the slopes (Fig. 6.3). The region has experienced fatal landslide disasters numerous times in history, i.e. 1899, 1950, 1968, 1980, 1996, 2007, and 2015 (Bera et al. 2021a).

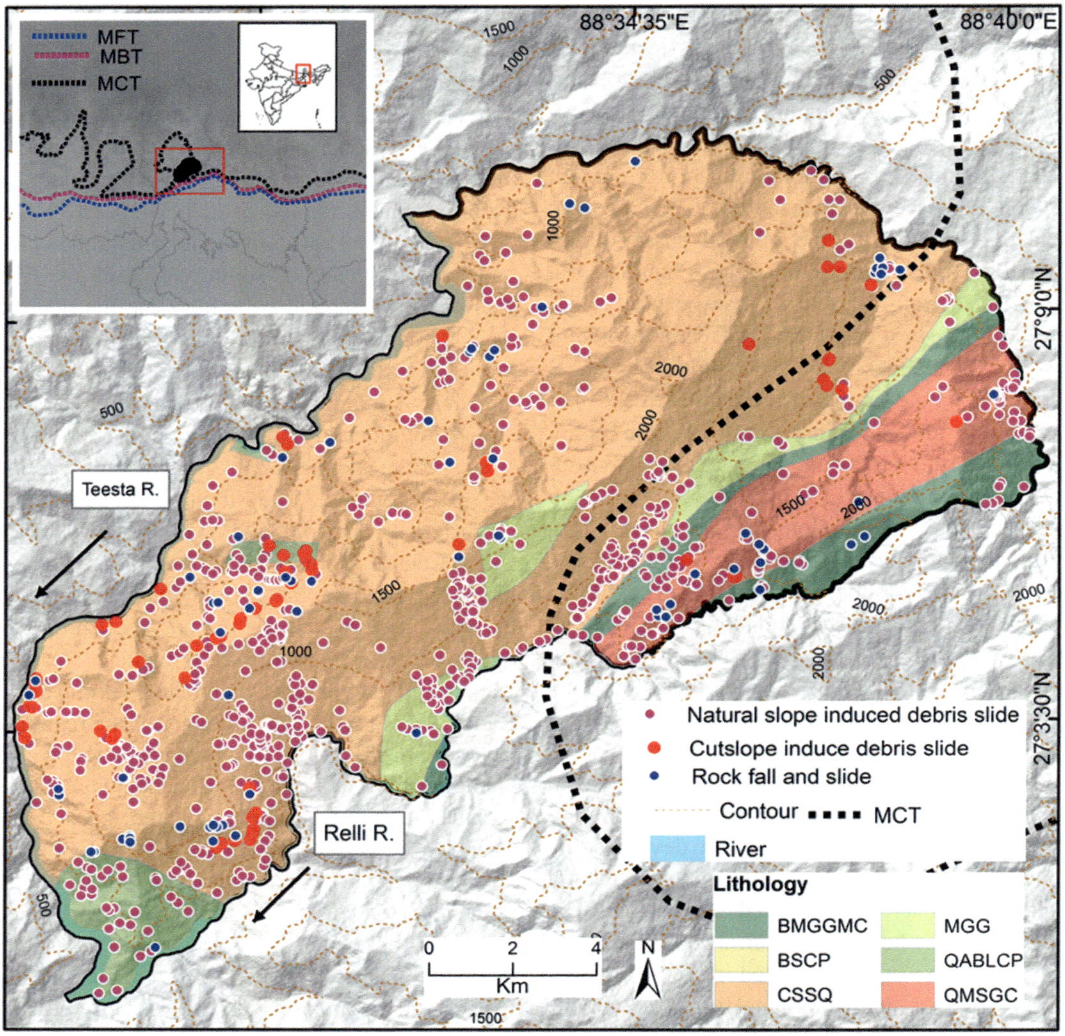

Fig. 6.1 Distribution of landslides and geophysical characteristics of the Kalimpong region

6.3 Materials and Methodology

6.3.1 Data Sources and Geospatial Database Generation

The process of methodology follows the steps: (i) data collection, categorization of landslide inventory and construction of the geospatial database, (ii) determination of the relation between landslides and conditioning factor, optimizing parameters for modelling susceptibility, (iii) generation of susceptibility models, (iv) model evaluation (Fig. 6.4).

The landslide inventory database was compiled using various data sources, including field surveys conducted between 2016 and 2018, multi-temporal satellite imagery from Google Earth (2002–2018), satellite images such as GeoEye-1 and WordView-2 (both with a resolution of 1.15 m, acquired in 2017), technical reports from the Geological Survey of India, previous literature references, and other relevant sources (e.g. Blogspot of Save The Hills). During

Fig. 6.2 Landslides of the study region: **a**, **b** NDS, **c**, **d** CDS, **e**, **f** RFRS. *Source* Field survey (2018)

field surveys, GPS was utilized to record the landslide locations, and attributes of the landslides were documented following the methodology described in Cruden and Varnes. All landslide occurrences were delineated as polygons using ArcGIS.

In this study, thirteen conditioning factors were considered, drawing from previous research and the specific geo-environmental conditions of the study area. These factors encompassed slope, aspect, curvature, topographic wetness index (TWI), stream power index (SPI), proximity to drainage networks, lithology, proximity to lineaments (linear geological features), soil depth, geomorphology, land use and land cover (LULC), normalized difference vegetation index (NDVI), and proximity to roads.

The factors such as slope, aspect, curvature, and TWI were derived from a digital elevation model (DEM) sourced from Advanced Land Observing Satellite (ALOS) data, with a resolution of 12.5 m. Soil depth information was obtained through field surveys at 223 locations, and kriging interpolation was employed to generate a soil depth map based on the sampled locations. The lithology geodatabase was created by digitizing the published map of the Geological Survey of India (GSI) (scale 1:200,000) within ArcGIS. Lineaments, drainage networks, road networks, LULC features, and geomorphological characteristics were digitized using satellite imagery (GeoEye-1, WorldView-2) and Google Earth images. Proximity maps for lineaments, drainage networks, and roads were generated by calculating the Euclidean distances. Landsat 8 satellite imagery with a resolution of 30 m was used for preparing the NDVI map. All thematic maps contained a pixel size of 12.5 m × 12.5 m.

Fig. 6.3 Changing landscape and activation of landslides before and after road construction in the study region

Five landslide groups, A, B, C, D, and E, were defined based on the inventory to assess the relationship between conditioning factors and landslides. Before modelling, the landslide points within each group (A, B, C, D, and E) were converted from polygons to points, where each point represented an area of 12.5m^2. The datasets for groups A, B, C, D, and E consisted of 40% landslide points (coded as "1") and 60% non-landslide points (coded as "0"). The landslide points in each group were 13,123 (A), 12,430 (B), 12,126 (C), 322 (D), and 666 (E), respectively. Subsequently, each dataset was spitted into training and testing data, with 80% of the landslide points assigned for training and 20% for validation. Training and testing data were selected through random sampling to ensure a representative distribution across the entire study region.

6.3.2 Relationship Analysis Between Landslides and Conditioning Factors

The role of conditioning factors on the spatial distribution of landslides is analysed using the relationship index (RI). Firstly, the frequency ratio (Fr) is computed to measure the association between past landslides and conditioning factors. It is calculated as the ratio of the percentage of landslide pixels (LPi) in the ith class of a conditioning factor to the percentage of total pixels

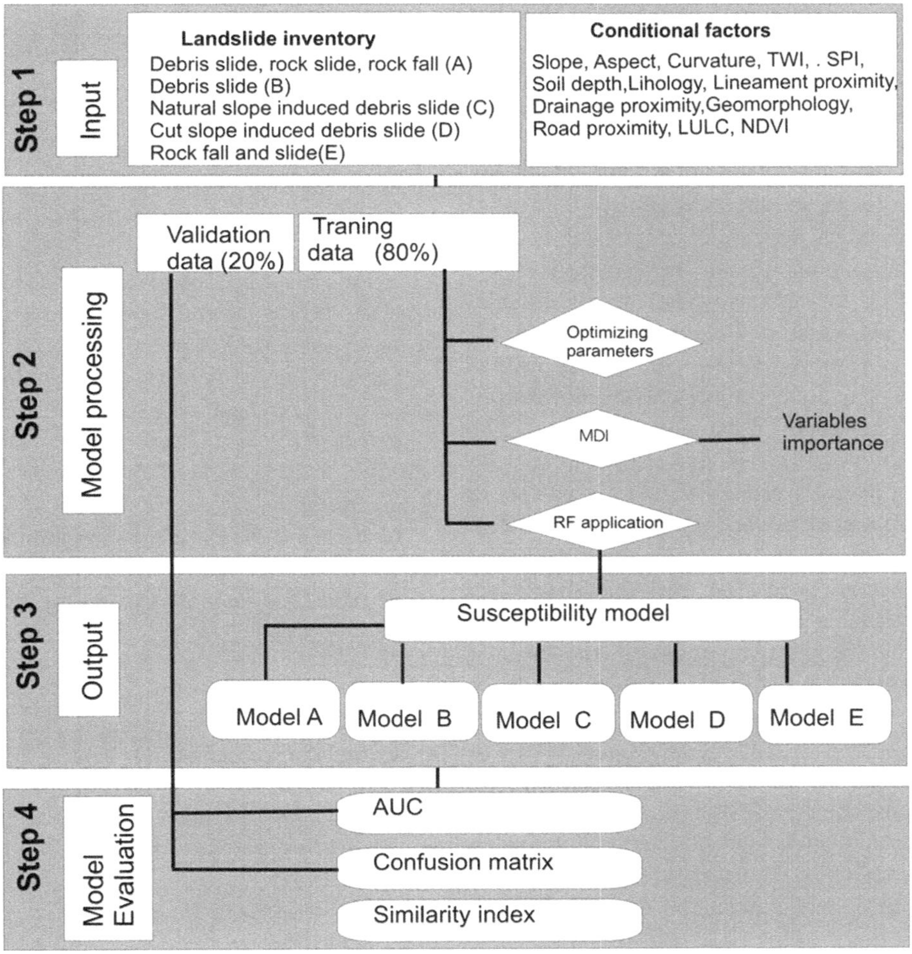

Fig. 6.4 Methodological flowchart

(CPi) in the same category (Eq. 6.1). The Fr is then normalized, and the RI is determined using Eq. 6.2.

$$\mathrm{Fr} = L_{\mathrm{Pi}} / C_{\mathrm{pi}} \tag{6.1}$$

$$\mathrm{RI} = \frac{(X_{\mathrm{FY}} - \mathrm{Min}_{\mathrm{FY}})}{(\mathrm{Max}_{\mathrm{FY}} - \mathrm{Min}_{\mathrm{FY}})} \tag{6.2}$$

Here, L_{Pi} represents the percentage of landslide pixels in the *i*th class of a conditioning factor, C_{pi} represents the percentage of total pixels in the *i*th class of a conditioning factor. X_{Fr} denotes the frequency ratio of class '*X*' within a factor, $\mathrm{Min}_{\mathrm{Fr}}$ represents the minimum value of frequency ratio within the factor's domain, and $\mathrm{Max}_{\mathrm{Fr}}$ represents the maximum value of frequency ratio within the factor's domain.

6.3.3 Random Forest (RF) for Landslide Susceptibility Modelling

Random forest (RF) is an ensemble machine learning technique for developing landslide susceptibility models. RF is particularly suited for modelling complex systems and their interdependencies due to its non-parametric nature. It can handle discrete and continuous datasets effectively, making it a popular method in landslide susceptibility modelling. RF has demonstrated higher predictive accuracy than other methods such as decision trees, Naïve Bayes, linear support vector machine, and artificial neural network (see Table 6.4). Consequently, RF was chosen as the modelling approach for landslide susceptibility.

RF constructs decision trees by using bagging and subsets of training data. The resulting model is then applied to predict susceptibility on an independent test dataset. To develop accurate RF models, two key parameters, ntree (the number of trees) and mtry (the number of variables considered at each node), need to be optimized. The number of trees (ntree) was varied from 1 to 10,500 at intervals of 500, while mtry was set to the square root of the total number of independent variables. This process generated 105,001 different RF prediction models for each landslide type. The correlation coefficient was used to quantify the relationship between the response variable and the independent variables. The prediction error of the models was evaluated using the "out-of-bag" (OOB) scores (Fig. 6.11). The Mean Decrease in Impurity (MDI) metric was employed to estimate the importance values of the variables. Next, MDI is calculated for classification by measuring the total decrease in node impurity averaged over all trees in the ensemble model.

6.3.3.1 Accuracy Assessment

The accuracy of the models was evaluated by adopting various metrics, including overall accuracy, error rate, precision, recall, F-measure, ROC area (success rate), and ROC area (prediction rate).

$$\text{Recall} = \text{TP}/(\text{TP} + \text{FN}) \tag{6.3}$$

$$\text{Precision} = \frac{(\text{TP})}{(\text{TP} + \text{FP})} \tag{6.4}$$

$$\text{Overall Accuracy} = \frac{(\text{TP} + \text{TN})}{(\text{TP} + \text{TN} + \text{FP} + \text{FN})} \tag{6.5}$$

$$\text{Error Rate} = (1 - \text{Overall Accuracy}) \tag{6.6}$$

$$F\text{-measure} = 2 * \text{precision} * \text{recall}/(\text{precision} + \text{recall}) \tag{6.7}$$

TP represents true positives, TN denotes true negatives, FN represents false negatives, and FP denotes false positives in the confusion matrix.

The ROC curve is derived by plotting sensitivity in Y-axis against 1-specificity in X-axis. A perfect model would have a ROC value of 1, while a value below 0.5 indicates a non-informative model.

To compare the different susceptibility models, the average deviation (D) between the pixels of the models is calculated using Eq. 6.8.

$$D = \left(S_i - S_j\right)/N \tag{6.8}$$

Here, D represents the average deviation, S_i and S_j denote the susceptibility values of a pixel between models i and j at the exact location, and N represents the total pixel number in a model (in all models, it is equal).

6.4 Results

The landslide inventory used in this study consists of a total of 1072 landslides recorded between 1968 and 2018. However, due to insufficient spatial information for landslide events prior to 2002 (such as those in 1968 and 1996), these older landslides were excluded from the analysis. This resulted in a dataset comprising 905 landslides, ensuring a more unbiased analysis of their relationship with conditioning factors and susceptibility models in the subsequent sections.

Table 6.1 Frequency and area of landslides of different groups

Group of landslides	Total number	Total area (m^2)	Mean area (m^2)	Standard deviation
All debris slide, rockfall, and slide (A)	905	2,066,546.9	2283.4	10,433.1
All debris slides (B)	847	1,968,199.3	2323.7	10,729.4
Natural slope-induced debris slide (C)	781	1,919,375.8	2457.5	11,153
Cutslope induce debris slide (D)	66	48,823.5	739.7	1646.4
Rockfall and slide (E)	58	98,347.5	1695.6	4111.4

The inventory includes three main types of landslides: NDS, CDS, and RFRS (see Fig. 6.1). Table 6.1 presents the statistics for these landslide types. Among them, NDS are the most prevalent, accounting for 781 out of the total inventory. The number of CDS and RFRS is 66 and 58, respectively. In terms of relative size, NDS exhibit a significant dominance compared to other types. The average area of NDS is estimated at 2457.59 m^2, while the mean area of active CDS and the group of RFRS are 739.75 m^2 and 1695.65 m^2, respectively.

6.4.1 Relation Between Landslides and Conditioning Factors

Figure 6.5 illustrates the relationship between influential conditioning factors and the distribution of landslides. This relationship analysis is conducted for the five landslide groups (A, B, C, D, and E). The relationship index (RI) values reveal a positive correlation between all landslide types and slopes (Fig. 6.5a). The concentration of debris slides is highest within a slope range of 34–42°, while very steep slopes of 42–75° exhibit fewer debris slides due to limited debris availability. Conversely, rockfall and rock slides dominate the occurrence within slopes of 42–75°.

Regarding aspect, the density of landslides is higher in the east, south, southeast, and southwest directions, while CDS exhibit higher density in the north and northeast aspects (Fig. 6.5b). This heterogeneous distribution suggests that NDS are strongly influenced by monsoonal rainfall, particularly impacting the south, southwest, and southeast aspects. On the other hand, the CDS in this region are relatively less influenced by aspect. The density of all landslide types is higher in areas with concave curvature (Fig. 6.5c).

The RI values show a positive correlation between overall landslide density, debris slide density, and NDS density with the topographic wetness index (TWI) values (Fig. 6.5d). Similarly, the RI values for the classes of stream power index (SPI) indicate that the response of NDS differs from that of CDS, and RFRS (Fig. 6.5e). The RI values sharply decrease with increasing distance in the case of all landslide groups for drainage proximity (Fig. 6.5f), indicating that drainage strongly influences all types of landslides.

In terms of lithology, the RI values for all landslide types are higher in the QABLCP group, which covers approximately 1% of the total area (Fig. 6.5g). This area is characterized by stream erosion and steeper slopes, which are favourable conditions for landslide initiation. The occurrence of landslides within this small area causes a higher RI value. However, due to the 1:200,000 scale of the lithology map, it is challenging to justify the precise role of lithology in landslide occurrence.

Regarding lineament proximity, the RI values exhibit a positive and linear trend for the "A" landslide group, indicating that lineaments strongly influence overall landslides. However, the trend of RI values is not consistent across all landslide groups. For groups "B" and "C," the RI values consistently decrease with increasing distance from lineament. In the case of CDS and RFRS, although the landslides are concentrated within a 300 m stretch, the trend of RI values is

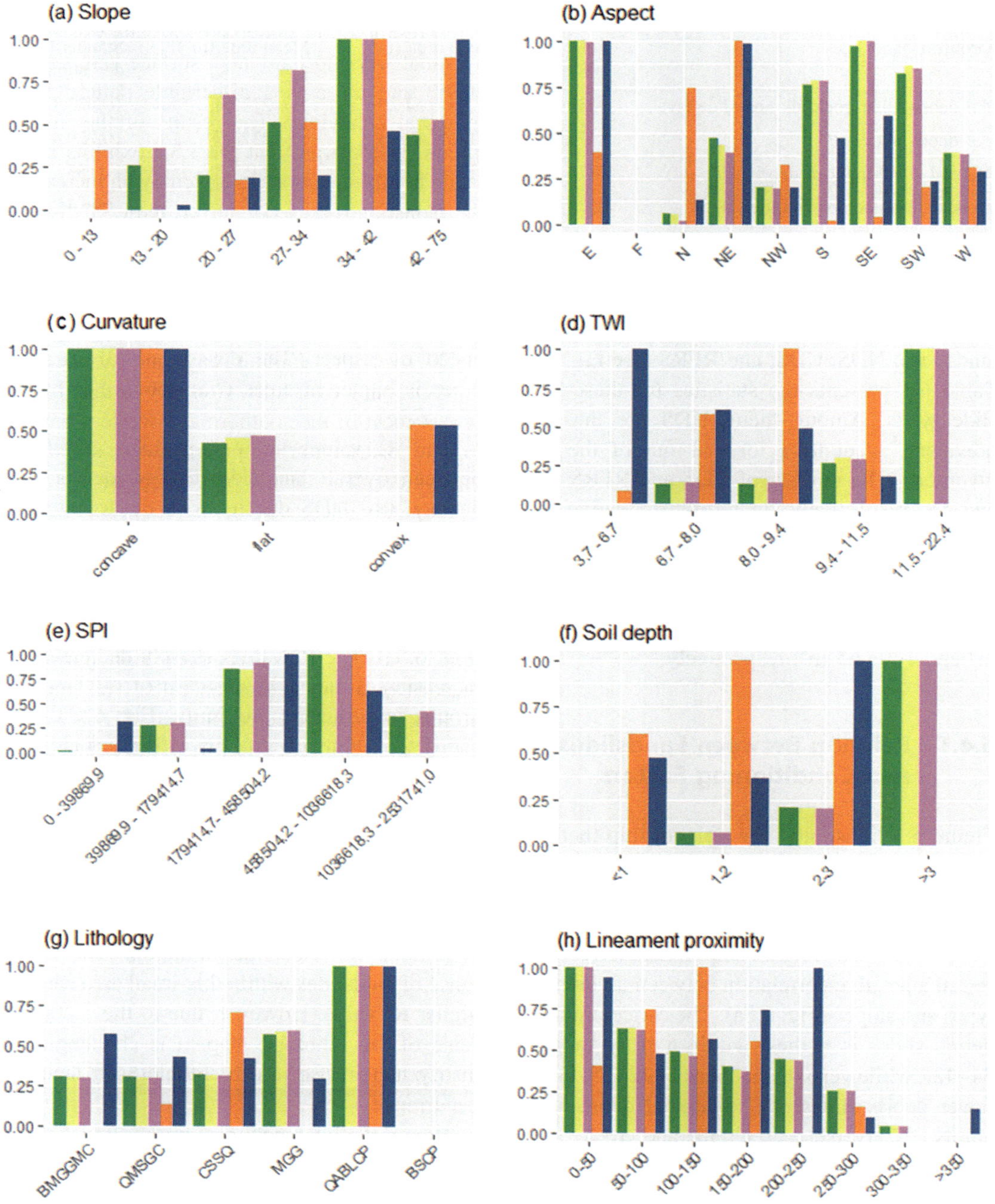

Fig. 6.5 Relationship index (RI) values between various groups of landslides and conditioning factors. *Note* The unit of slope is degrees, while lineament proximity, drainage proximity, road proximity, and soil depth are measured in metres. The land cover categories are as follows: WB—Water body, BL—Barren land, DF—Deciduous forest, SF—Scrub forest, GL—Grassland, B—Built-up area, A—Agricultural land, C—Cinchona plantation. HT—Highly dissected terrain, MT—Moderately dissected terrain, LT—Low dissected terrain, RT—River terrace, DH—Denudation hill, VF—Valley fill, TR—Toe removal, WB—Water body, Su—Summit

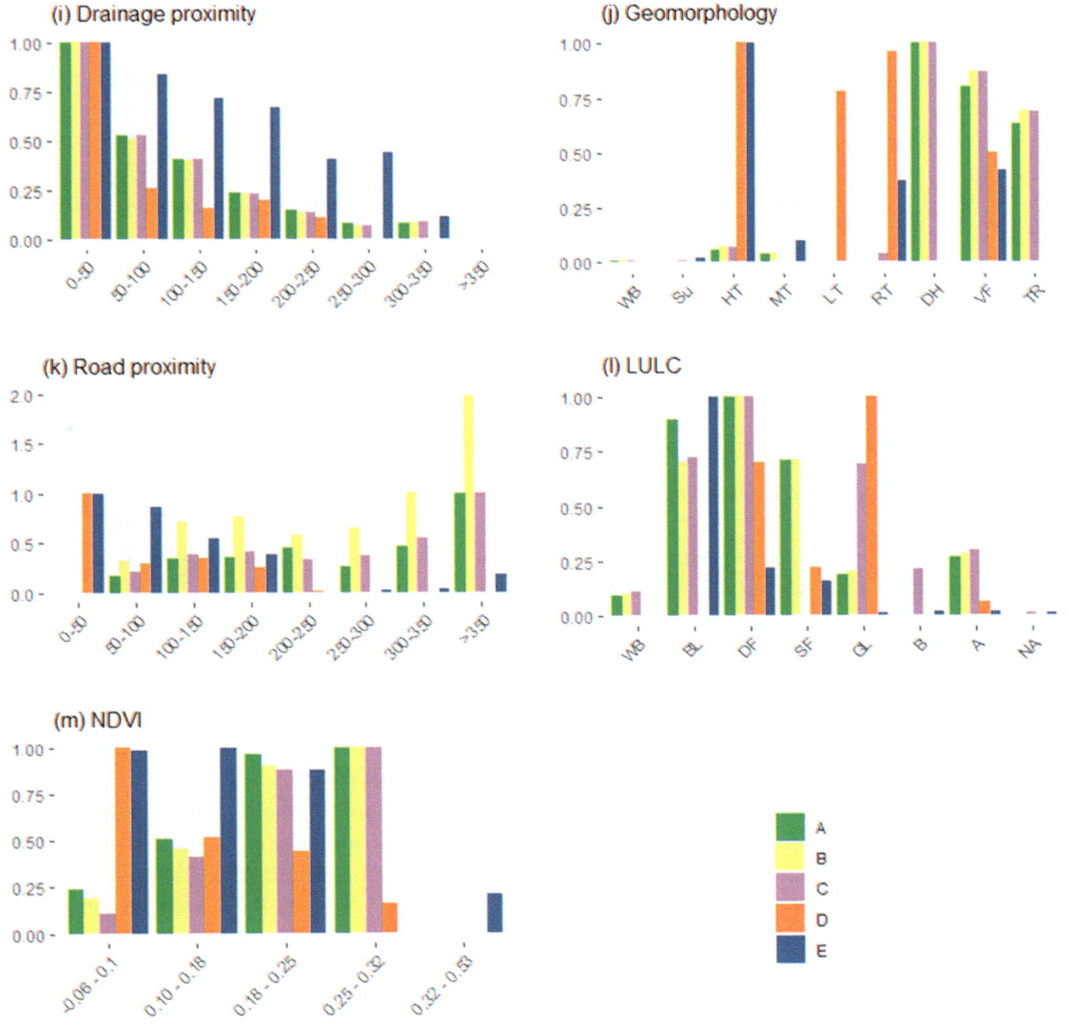

Fig. 6.5 (continued)

not linear (Fig. 6.5h). This suggests that NDS are more influenced by lineament proximity compared to CDS and RFRS in this region.

The behaviour of different landslide groups exhibits distinct patterns across various geomorphological categories (Fig. 6.5i). The RI values indicate that NDS are more densely distributed in denudation hills, areas with toe removal, and valley fills. Conversely, CDS and RFRS occurrences are more prevalent in highly dissected terrains. Summit locations at the ridge tops tend to have fewer landslide incidents. Regarding soil depth, the RI values for overall landslides, all debris slides, and NDS generally increase with greater soil depth (Fig. 6.5j). However, the density of CDS and RFRS is lowest in areas with soil depths greater than 3 m. The concentration of NDS is highest within the NDVI range of 0.18 to 0.32 (Fig. 6.5k). In contrast, RFRS density is elevated between the NDVI range of −0.06 to 0.18, which can be attributed to barren rock areas and limited forest cover. The RI values for different LULC features exhibit variations among the different landslide groups. The density of overall landslides, all debris slides, and NDS is higher in deciduous forest regions (Fig. 6.5l). On the other hand, RFRS density is greater in barren lands due to the presence of steep slopes and exposed rocks. The RI values for road proximity differ significantly between the classes of NDS and CDS (Fig. 6.5m). Higher RI values for CDS, RFRS within shorter road distances indicate that the occurrence of such landslides is strongly influenced by road-cutting activities. Conversely, a higher RI value for NDS suggests that factors other than road proximity play a more dominant role in initiating these types of landslides. As the majority of landslides are initiated along natural slope, the RI value for overall landslides depicts an inverse relationship between road proximity and landslide occurrences.

The association of RI values among the inventory groups shows notable differences (Fig. 6.6). There is a high association of RI values between groups A, B, and C (R^2 greater than 0.8). However, the association between groups C, D, and E is considerably low (R^2 not exceeding 0.3). Since 93% of landslides are debris slides, and 85% are NDS in group A, there is a strong association among these groups. On the other hand, the mechanisms and responses to conditioning factors of NDS (C), CDS (D), and RFRS (E) differ significantly. Therefore, the association of RI values among these groups of landslides is not significant. The number of CDS (D) and RFRS (E) is significantly lower in the overall landslide domain (A). Consequently, determining the impact of conditioning factors on past CDS and RFRS using all landslide categories is challenging to justify.

6.4.2 Landslide Susceptibility Models

The present study developed five distinct susceptibility models, namely Model A, Model B, Model C, Model D, and Model E. These models were constructed using a random forest algorithm with the best combination of conditioning factors (Fig. 6.7). Analysis revealed that as the number of trees (ntree) increased, the R-square value showed a significant increase until reaching 500 trees. After this point, minimal variations were observed with further increases in ntree values. Based on these findings, ntree values of 500 were selected for Models B, C, and D, while Model A utilized 5500 trees.

To assess the relative predictive ability of each conditioning factor in the models, the normalized importance value for each factor was calculated (Fig. 6.8). The results indicated that the relative importance of conditioning factors was relatively consistent among Models A, B, and C. Soil depth emerged as the most influential factor in these three models. However, the trend of relative importance differed for Models D (CDS) and E (rockfall and slide). In the case of cut slope-induced landslide susceptibility (D), road proximity exhibited the highest predictive ability, while for rockfall and slide susceptibility, slope was the most important factor. Curvature consistently demonstrated the least importance across all five models.

6.4.2.1 Landslide Susceptibility Zonation

The susceptibility maps were generated for classifying the probability value. The probability value ranges from 0 to 0.25, 0.25 to 0.50, 0.50 to 0.75, and 0.75 to 1.00 have been considered as "low," "moderate," "high," and "very high" category of susceptibility, respectively (Fig. 6.9). The very high susceptible zones of Models A, B, C, D, and E cover 3.59%, 3.68%, 3.89%, 30.19%, and 0.32% of the study region. Although the coverage area of very high susceptible zones is relatively low, past landslides are concentrated higher within these classes. Therefore, the density of landslide is high in the very high susceptible zone. The result also shows that landslide density increases with increasing susceptibility value in almost all models (Table 6.2). However, its consistency is varied among the models. The landslide density value is relatively less consistent for model D, i.e. cut slope-induced debris slide compared to the Models A, B, C, and E.

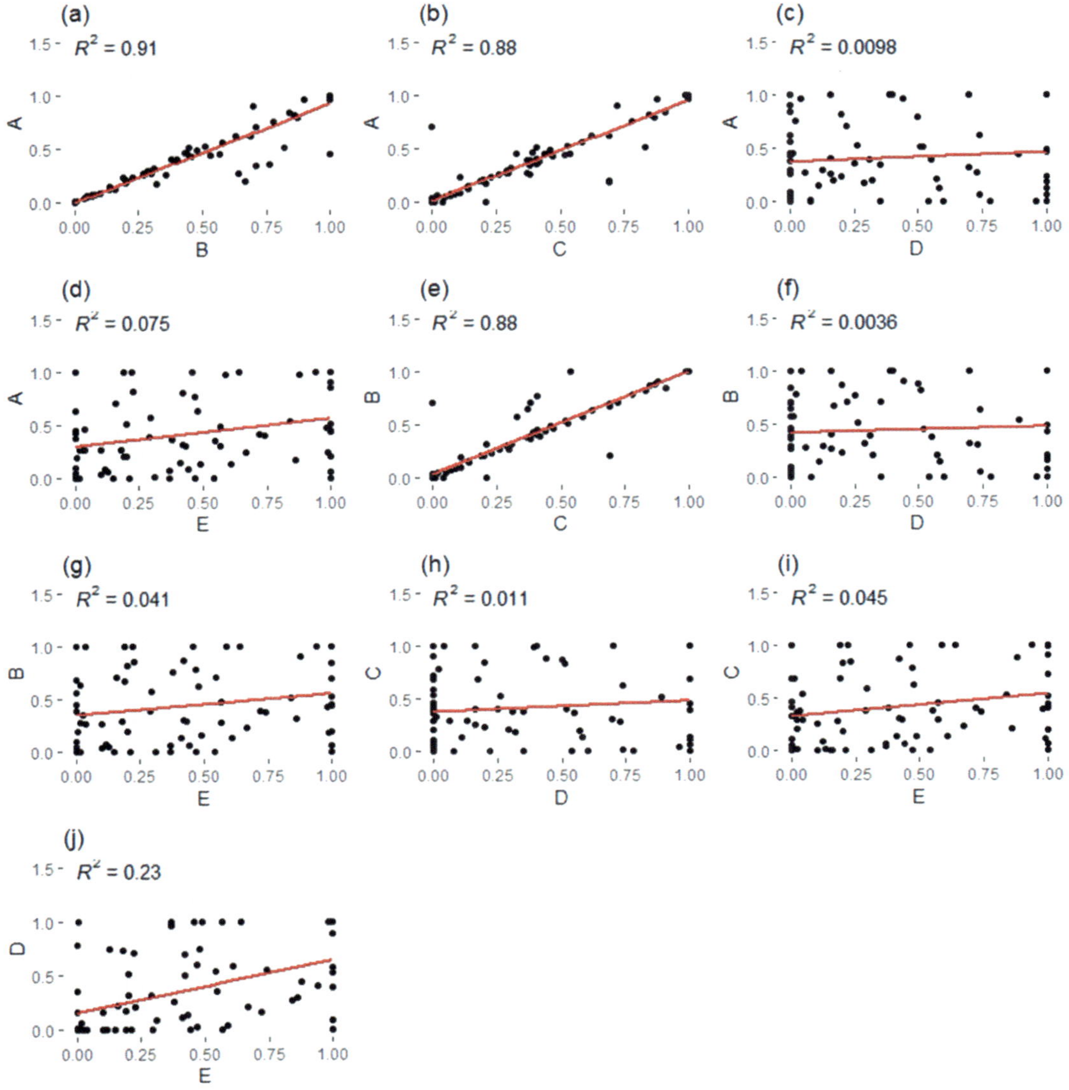

Fig. 6.6 Correlation between RI values among the different groups of landslides

6.4.2.2 Accuracy Assessment

The overall accuracy, ROC area, and F-measure are more than 0.9, and the error rate is less than 0.1 in all models except model D (Table 6.3). Since the overall accuracy of model D is 0.451, it is not applicable in the study region. It reflects the conditioning factors, although suitable for NDS and RFRS, these are not suitable for CDS susceptibility modelling.

6.4.2.3 The Similarity Index Among the Models

The similarity between the susceptibility models was evaluated using the average deviation (Fig. 6.10). The average deviation between susceptibility models serves as a measure of differentiation. A higher average deviation implies greater differentiation, while a lower value indicates higher similarity among the

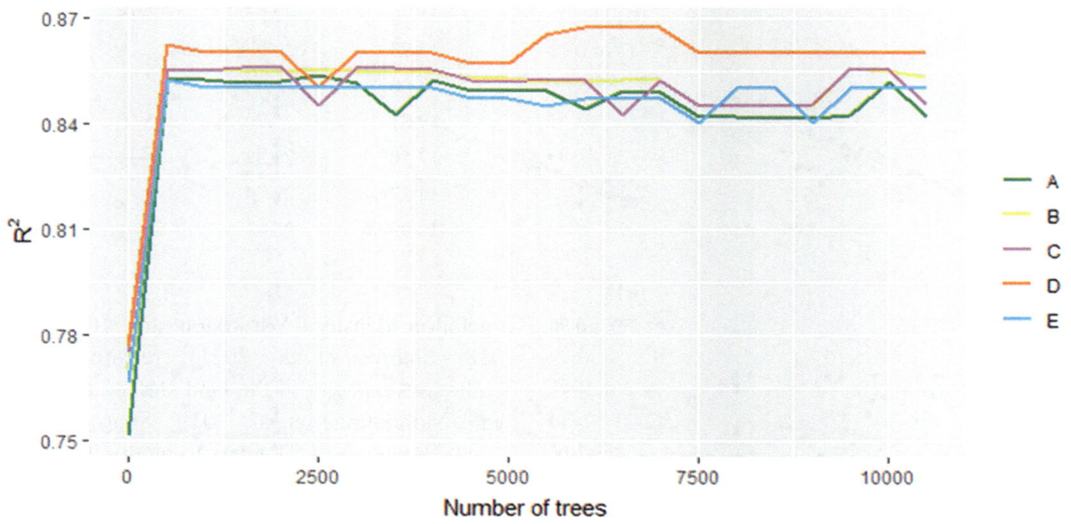

Fig. 6.7 *R*-square of models (A, B, C, D, E) based on various combination trees

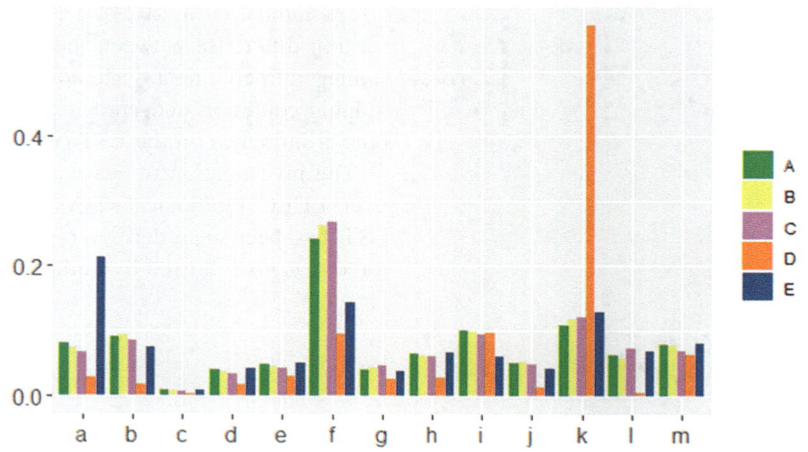

Fig. 6.8 Normalized importance value of variables on the base of predicting ability of the susceptibility models (A, B, C, D, E). (Here, a. Slope, b. Aspect, c. Curvature, d. TWI, e. SPI, f. Soil depth, g. Lithology, h. Lineament proximity, i. Drainage proximity,j. Geomorphology, k. Road proximity, l. LULC, m. NDVI)

models. The analysis revealed that the average deviation was low when we compared Models A, B, and C to each other. Conversely, the deviation was significantly higher when comparing Models A, B, and C with models D and E.

6.5 Discussion

The present study assessed the susceptibility models based on different categories of landslides. The objective of categorization is to

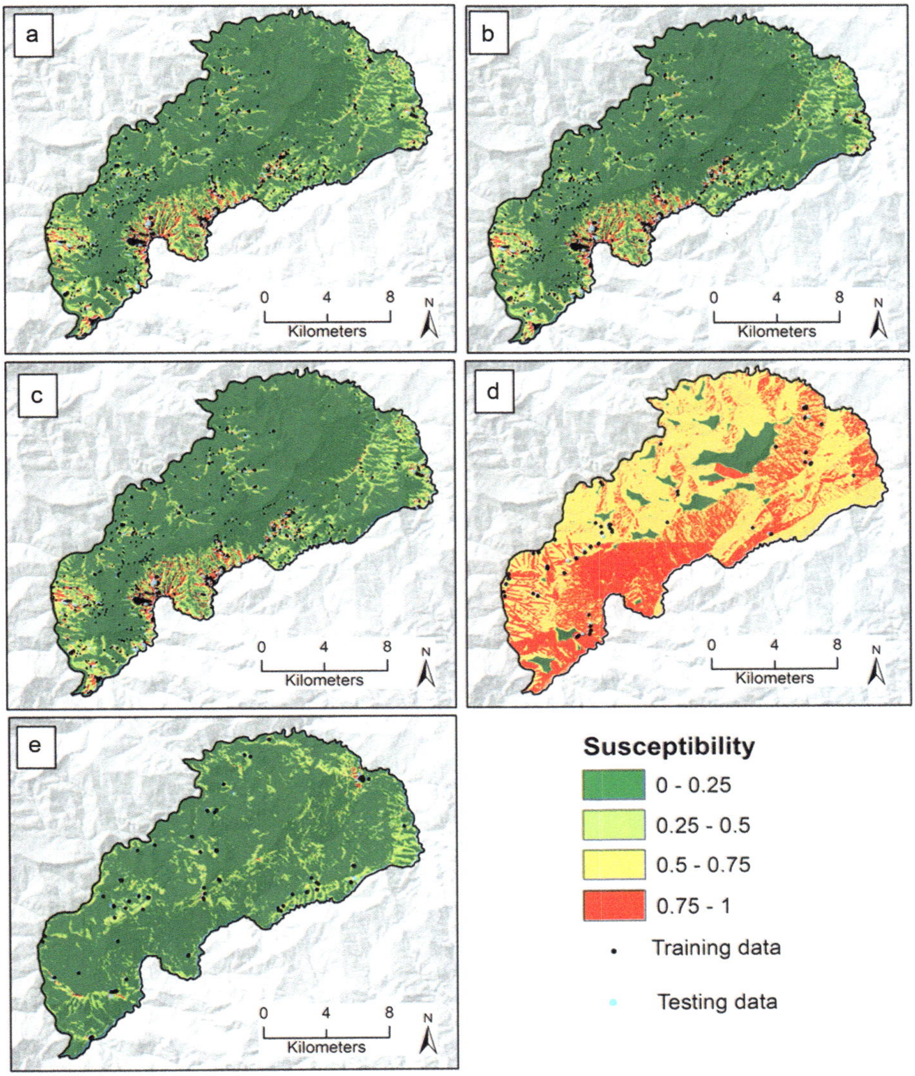

Fig. 6.9 Landslide susceptibility models using random forest: **a** all debris slide, rockfall and slide, **b** debris slide, **c** NDS, **d** CDS, and **e** RFRS

evaluate the reliability and effectiveness of the models. The findings reveal that Models A, B, and C exhibit high similarity, while models D and E differ significantly. In this section, we discuss the reasons behind spatial variations of susceptibility values, potential biases in models, and the importance of inventory categorization.

6.5.1 Spatial Variation of Susceptibility Values and the Role of Conditioning Factors

The susceptibility map demonstrates significant variation in susceptibility index values among NDS, CDS, and RFRS occurrences.

Table 6.2 Statistics of susceptibility classes of A, B, C, D, and E models

Models	Susceptibility class	Pixels of susceptibility classes	Percentage of pixels in susceptibility classes	Pixels of landslides	Percentage of landslides	Density of landslides
A	Low	1,132,280	74.11	229	1.75	0.02
	Moderate	245,485	16.07	1221	9.30	0.58
	High	95,250	6.23	2960	22.56	3.62
	Very high	54,915	3.59	8713	66.39	18.47
B	Low	1,160,138	75.93	236	1.90	0.03
	Moderate	216,761	14.19	1243	10.00	0.70
	High	94,798	6.20	2692	21.66	3.49
	Very high	56,233	3.68	8259	66.44	18.05
C	Low	1,168,062	76.45	268	2.21	0.03
	Moderate	205,474	13.45	1002	8.25	0.61
	High	95,027	6.22	2340	19.28	3.10
	Very high	59,367	3.89	8529	70.26	18.08
D	Low	97,516	6.38	15	4.67	0.73
	Moderate	7260	0.48	0	0.00	0.00
	High	961,918	62.96	205	63.86	1.01
	Very high	461,236	30.19	101	31.46	1.04
E	Low	1,274,707	83.43	35	5.26	0.06
	Moderate	208,915	13.67	35	5.26	0.38
	High	39,343	2.57	144	21.62	8.40
	Very high	4965	0.32	452	67.87	208.86

Table 6.3 Accuracy assessment of models

	Model A	Model B	Model C	Model D	Model E
Overall accuracy	0.930	0.924	0.939	0.451	0.968
Error rate	0.070	0.076	0.061	0.549	0.032
Precision	0.940	0.948	0.948	0.426	0.943
Recall	0.874	0.854	0.892	1.000	0.943
F-measure	0.906	0.898	0.919	0.598	0.943
ROC area (success rate)	0.985	0.985	0.987	0.534	0.968
ROC area (prediction rate)	0.985	0.983	0.988	0.486	0.967

The susceptibility to NDS is concentrated in the lower part of the region, where higher soil depth and stream/river erosion contribute to increased susceptibility. The upper and summit areas, characterized by flatter terrain and less stream erosion, have lower soil depth due to reduced deposition of eroded material. As NDS are predominantly concentrated in areas with higher soil depth, the predictive capability of this model is high.

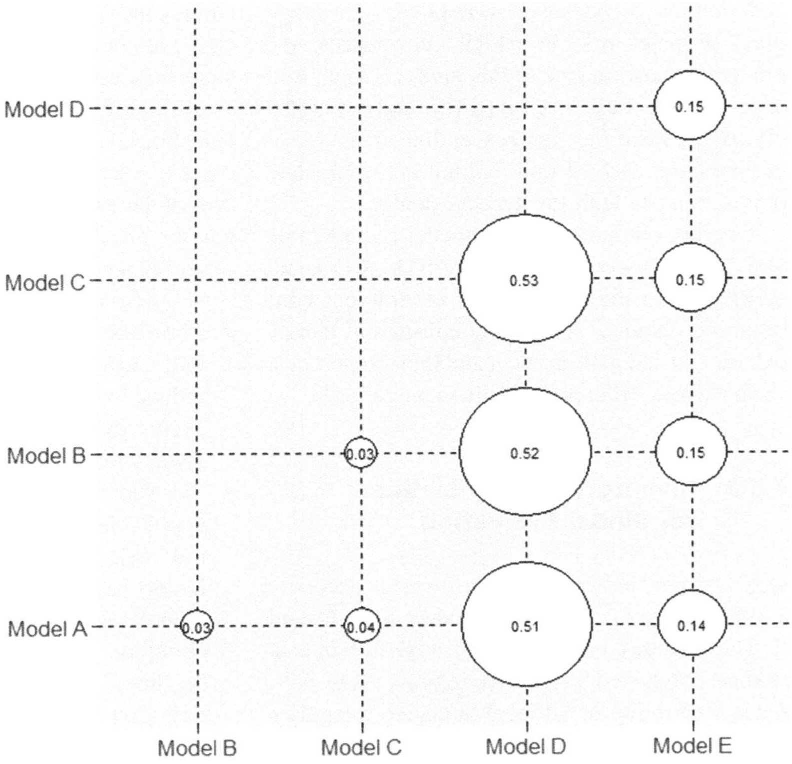

Fig. 6.10 Pairwise average deviation between pixels of the susceptibility models, illustrating the variations and distinctions between them

The susceptibility pattern of CDS differs significantly from that of NDS. While factors like slope, aspect, and soil depth are correlated with NDS, they do not exert a significant influence on CDS. Other factors show weak linear relationships, except for road and drainage proximity. This finding contrasts with previous susceptibility studies conducted along Himalayan roads, where similar conditioning factors were considered. The discrepancy may be attributed to variations in road types, such as highways, newly constructed roads, and other metalled roads, as previous studies predominantly focused on highways.

The spatial prediction of CDS exhibits high uncertainty due to the types of roadways and the timing of slope cutting, which impact the initiation of debris slides in the study region. Field observations indicate a higher frequency of debris slides along newly constructed roads, particularly when cutting activities occur just before the monsoon season. Consequently, areas with low susceptibility to NDS may become prone to debris slides after slope-cutting activities for road construction. Since road construction practices evolve over time, the susceptibility of the region also varies. Therefore, susceptibility is not static and requires updating in response to changing scenarios.

The uncertainty in the susceptibility of CDS may also arise from data quality. These slides are concentrated in the vicinity of roads, where local factors can influence their occurrence. High-resolution datasets are necessary to capture these local factors effectively.

Rockfall and rock slide events occur less frequently compared to debris slides. The susceptibility map indicates that a significant portion of the study area exhibits low susceptibility to RFRS. Highly susceptible areas for these types of landslides are predominantly found in steep slopes characterized by exposed rock surfaces and shallow or no soil cover. Given that slope plays a major role in RFRS occurrences, the predictive performance of this model is high. Soil depth has less influence on RFRS, and the model effectively identifies non-susceptible areas due to the predominance of non-rockfall and rock slide points in areas with greater soil depth.

Overall, our study highlights the considerable variation in susceptibility among NDS, CDS, and RFRS due to the combination of different conditioning factors. Not all conditioning factors exhibit consistent relevance, and their importance varies across different landslide categories.

6.5.2 Inventory Categorization for Model Evaluation

Our findings indicate that the overall accuracy and ROC area exceed 0.9 in Models A, B, C, and E. These values are distinctively higher than past studies conducted in similar regions. However, the susceptibility of different landslide categories varies significantly across the region. The similarity in susceptibility values is higher only among Models A, B, and C. This discrepancy arises because debris slide is dominated in Model A (approximately 85%). The limited occurrence of RFRS in the dataset is not adequately reflected in the susceptibility map when using all landslide types. Consequently, even though a model may exhibit excellent accuracy, it remains unreliable for evaluating the model. Hence, the categorization and integration of inventory data necessitates for achieving reliability in landslide susceptibility mapping within a heterogeneous environment.

6.6 Conclusion

Building landslide susceptibility is complex when a region experiences heterogonous types of landslides. In this context, omitting the heterogenic characteristics of landslide might cause biases in a susceptibility model. To overcome the biases of a susceptibility model, understanding the landslide inventory is an essential step. Considering its significance, the paper prepared landslide inventory encompassing multiple types, and then, the study conducted the relationship analysis between conditioning factors and different groups of landslides. Lastly, the study compared different types of susceptibility models using the different sets of landslides and provided further insights into the biases of the models. The key findings of the study are summarized as follows:

- The conditioning factors associated with natural slope-induced debris slides, cut slope-induced debris slides, and rockfall-slide events exhibit distinct responses that are not always linear.
- The same set of conditioning factors are not applicable to different types of landslides. At the regional scale, the present conditioning factors are not sufficient to assign cut slope-induced landslide susceptibility. In future, the cut slope-induced landslide susceptibility can be improved, altering conditioning factors based on the local environment.
- The similarity between susceptibility models based on all landslides and models based on debris slides is high due to debris slide dominance (about 85%). At the same time, the difference is high in the case of rockfall and slide-based model as well as the cut slope-based model. Therefore, the study concludes that the susceptibility models based on all landslides, despite having very good accuracy value, it is not applicable for the susceptibility of the rockfall and slide as well as cut slope-induced debris slides. In other words, the landslide model developed based on all data is biased by the dominant landslide type in the inventory.

Based on the findings, we suggest that the reliability of spatial prediction of landslides should not be justified solely based on improving predictive performance; qualitative evaluation based on the local environment is also needed to gain insights into the model's uncertainties. Since a susceptibility model depends on landslides' characterization, it should be developed differently for different categories of landslides.

In future, an integrated approach considering the importance of diversity of landslide inventory might be more reliable to develop a landslide susceptibility model in this kind of region.

Acknowledgements The research was supported by Junior Research Fellowship of the University Grant Commission (India) (UGC-Ref. No.: 3181) and the Digital Globe Foundation, USA.

Appendix

See Fig. 6.11 and Table 6.4.

References

Achour Y, Pourghasemi HR (2020) How do machine learning techniques help in increasing accuracy of landslide susceptibility maps? Geosci Front 11(3):871–883. https://doi.org/10.1016/j.gsf.2019.10.001

Addison P, Oommen T (2019) Post-fire debris flow modeling analyses: case study of the post-Thomas Fire event in California. Nat Hazards 100:329–343. https://doi.org/10.1007/s11069-019-03814-x

Banerjee P, Ghose MK, Pradhan R (2018) Analytic hierarchy process and information value method-based landslide susceptibility mapping and vehicle vulnerability assessment along a highway in Sikkim Himalaya. Arab J Geosci 11:1–18. https://doi.org/10.1007/s12517-018-3488-4

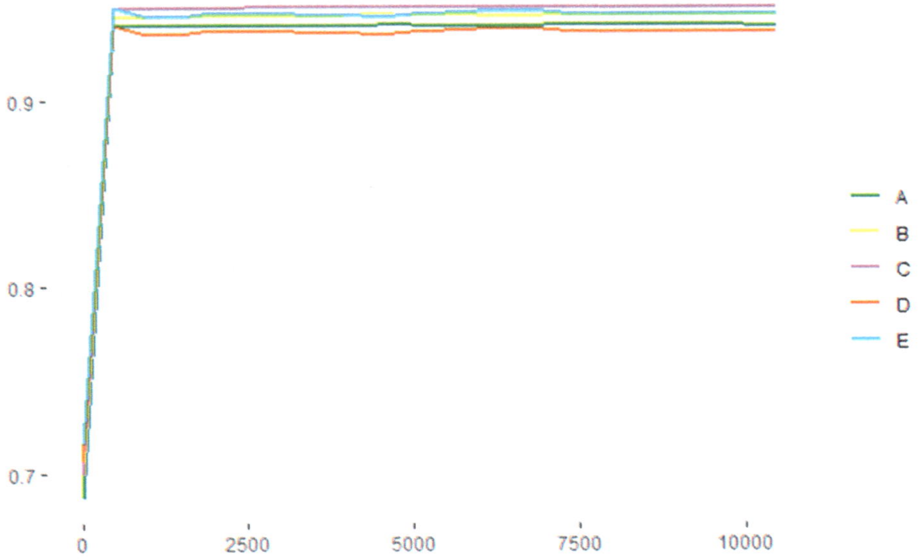

Fig. 6.11 Out-of-bag (OOB) error of the models (*X*-axis shows OOB error value and *Y*-axis shows the number of trees). (A: all landslides, B: debris slides, C: natural slope-induced debris slides, D: cut slope-induced debris slides, E: rockfall-rock slides)

Table 6.4 Training and validation accuracy of various models [based on group all landslides (A)]

Sl. no	Model	Training accuracy	Validation accuracy
1	Decision tree	79.34	73.15
2	Naïve Bayes	76.77	74.91
3	Linear support vector machine	72.33	73.18
4	Artificial neural network	83.33	82.45

Bera S, Guru B, Chatterjee R, Shaw R (2020) Geographic variation of resilience to landslide hazard: a household-based comparative studies in Kalimpong hilly region, India. Int J Dis Risk Red 46:101456

Bera S, Melo R, Guru B (2021a) Assessment of exposed elements in a changing built environment by using an integrated model of debris flow initiation and runout (Kalimpong region, Himalaya). Bull Eng Geol Environ 80:7131–7152. https://doi.org/10.1007/s10064-021-02352-w

Bera S, Upadhyay VK, Guru B, Oommen T (2021b) Landslide inventory and susceptibility models considering the landslide typology using deep learning: Himalayas, India. Nat Hazards 108:1257–1289. https://doi.org/10.1007/s11069-021-04731-8

Bera S, Gnyawali K, Dahal K et al (2023) Assessment of shelter location-allocation for multi-hazard emergency evacuation. Int J Disaster Risk Reduct 84:103435. https://doi.org/10.1016/j.ijdrr.2022.103435

Bragagnolo L, da Silva RV, Grzybowski JMV (2020) Landslide susceptibility mapping with r. landslide: a free open-source GIS-integrated tool based on artificial neural networks. Environ Model Softw 123: 104565. https://doi.org/10.1016/j.envsoft.2019.104565

de Oliveira GG, Ruiz LFC, Guasselli LA, Haetinger C (2019) Random forest and artificial neural networks in landslide susceptibility modeling: a case study of the Fão River Basin, Southern Brazil. Nat Haz 99(2):1049–1073. https://doi.org/10.1007/s11069-019-03795-x

Dou J, Yunus AP, Merghadi A et al (2020) Different sampling strategies for predicting landslide susceptibilities are deemed less consequential with deep learning. Sci Total Environ 720:137320. https://doi.org/10.1016/j.scitotenv.2020.137320

Farooq S, Akram MS (2021) Landslide susceptibility mapping using information value method in Jhelum Valley of the Himalayas. Arab J Geosci 14(10):824. https://doi.org/10.1007/s12517-021-07147-7

Galli M, Ardizzone F, Cardinali M et al (2008) Comparing landslide inventory maps. Geomorphology 94:268–289. https://doi.org/10.1016/j.geomorph.2006.09.023

Getachew N, Meten M (2021) Weights of evidence modeling for landslide susceptibility mapping of Kabi-Gebro locality, Gundomeskel area, Central Ethiopia. In: Geoenvironmental disasters. https://doi.org/10.1186/s40677-021-00177-z

Harp EL, Keefer DK, Sato HP, Yagi H (2011) Landslide inventories: the essential part of seismic landslide hazard analyses. Eng Geol 122:9–21. https://doi.org/10.1016/j.enggeo.2010.06.013

International Centre For Integrated Mountain Development (ICIMOD) (2019) Conserving springs as climate change adaptation action: lessons from Chibo–Pashyor Watershed, Teesta River Basin, Kalimpong, West Bengal, India

Kim JC, Lee S, Jung HS, Lee S (2018) Landslide susceptibility mapping using random forest and boosted tree models in Pyeong-Chang, Korea. Geocarto Int 33(9):1000–1015. https://doi.org/10.1080/10106049.2017.1323964

Liu B, Siu YL, Mitchell G (2017) A quantitative model for estimating risk from multiple interacting natural hazards: an application to northeast Zhejiang, China. Stoch Environ Res Risk Assess 31:1319–1340. https://doi.org/10.1007/s00477-016-1250-6

Melo R, Vieira G, Caselli A, Ramos M (2012) Susceptibility modelling of hummocky terrain distribution using the information value method (Deception Island, Antarctic Peninsula). Geomorphology 155:88–95. https://doi.org/10.1016/j.geomorph.2011.12.027

Melo R, Zêzere JL (2017) Modeling debris flow initiation and run-out in recently burned areas using data-driven methods. Nat Hazards 88:1373–1407. https://doi.org/10.1007/s11069-017-2921-4

Mirdda HA, Bera S, Siddiqui MA, Singh B (2019) Analysis of bi-variate statistical and multi-criteria decision-making models in landslide susceptibility mapping in lower Mandakini Valley, India. GeoJournal 85:681–701. https://doi.org/10.1007/s10708-019-09991-3

Nhu VH, Shirzadi A, Shahabi H et al (2020) Shallow landslide susceptibility mapping: a comparison between logistic model tree, logistic regression, naïve bayes tree, artificial neural network, and support vector machine algorithms. Int J Environ Res Public Health. https://doi.org/10.3390/ijerph17082749

Paryani S, Neshat A, Pradhan B (2021) Spatial landslide susceptibility mapping using integrating an adaptive neuro-fuzzy inference system (ANFIS) with two multi-criteria decision-making approaches. Theor Appl Climatol 146:489–509. https://doi.org/10.1007/s00704-021-03695-w

Pham BT, Prakash I, Khosravi K, et al (2018) A comparison of Support Vector Machines and Bayesian algorithms for landslide susceptibility modelling. Geocarto Int 0:000. https://doi.org/10.1080/10106049.2018.1489422

Pourghasemi HR, Rahmati O (2018) Prediction of the landslide susceptibility: Which algorithm, which precision? CATENA. https://doi.org/10.1016/j.catena.2017.11.022

Pourghasemi HR, Pradhan B, Gokceoglu C et al (2013) Application of weights-of-evidence and certainty factor models and their comparison in landslide susceptibility mapping at Haraz watershed, Iran. Arab J Geosci 6:2351–2365

Pradhan B (2013) A comparative study on the predictive ability of the decision tree, support vector machine and neuro-fuzzy models in landslide susceptibility mapping using GIS. Comput Geosci. https://doi.org/10.1016/j.cageo.2012.08.023

Raja NB, Çiçek I, Türkoğlu N et al (2018) Correction to: Landslide susceptibility mapping of the Sera River Basin using logistic regression model. Nat Hazards. https://doi.org/10.1007/s11069-017-3145-3

Ramesh V, Anbazhagan S (2015) Landslide susceptibility mapping along Kolli hills Ghat road section (India) using frequency ratio, relative effect and fuzzy logic models. Environ Earth Sci 73:8009–8021. https://doi.org/10.1007/s12665-014-3954-6

Razavi-Termeh SV, Shirani K, Pasandi M (2021) Mapping of landslide susceptibility using the combination of neuro-fuzzy inference system (ANFIS), ant colony (ANFIS-ACOR), and differential evolution (ANFIS-DE) models. Bull Eng Geol Environ. https://doi.org/10.1007/s10064-020-02048-7

Thi Ngo PT, Panahi M, Khosravi K et al (2020) Evaluation of deep learning algorithms for national scale landslide susceptibility mapping of Iran. Geosci Front. https://doi.org/10.1016/j.gsf.2020.06.013

Tien Bui D, Tuan TA, Klempe H et al (2016a) Spatial prediction models for shallow landslide hazards: a comparative assessment of the efficacy of support vector machines, artificial neural networks, kernel logistic regression, and logistic model tree. Landslides 13:361–378. https://doi.org/10.1007/s10346-015-0557-6

Tien Bui D, Tuan TA, Klempe H et al (2016b) Spatial prediction models for shallow landslide hazards: a comparative assessment of the efficacy of support vector machines, artificial neural networks, kernel logistic regression, and logistic model tree. Landslides. https://doi.org/10.1007/s10346-015-0557-6

van Westen CJ, van Asch TWJ, Soeters R (2006) Landslide hazard and risk zonation—Why is it still so difficult? Bull Eng Geol Environ. https://doi.org/10.1007/s10064-005-0023-0

Youssef AM, Pourghasemi HR (2021) Landslide susceptibility mapping using machine learning algorithms and comparison of their performance at Abha Basin, Asir Region, Saudi Arabia. Geosci Front 12(2):639–655. https://doi.org/10.1016/j.gsf.2020.05.010

Zhang YX, Lan HX, Li LP et al (2020) Optimizing the frequency ratio method for landslide susceptibility assessment: a case study of the Caiyuan Basin in the southeast mountainous area of China. J Mountain Sci 17(2):340–357. https://doi.org/10.1007/s11629-019-5702-6

Zhang K, Wu X, Niu R, et al (2017) The assessment of landslide susceptibility mapping using random forest and decision tree methods in the Three Gorges Reservoir area, China. Environ Earth Sci 76:1–20. https://doi.org/10.1007/s12665-017-6731-5

7 Integrating Explainable Artificial Intelligence and Machine Learning for Understanding Water Pollution and Its Management

Swapan Talukdar, Shahfahad, Ishita Afreen Ahmed, Mohd. Rihan, and Atiqur Rahman

Abstract

Water quality assessment is vital for water resource management and environmental sustainability. This study estimated the water quality index (WQI) for Loktak Lake, Manipur, India, evaluated predictive models for WQI, and used explainable artificial intelligence (XAI) techniques for water pollution management. Entropy theory was employed to determine WQI, with weights assigned based on parameter importance. WQI values ranged from 80.59 to 100.00, indicating varying water quality levels. The study classified samples into good (24%), moderately polluted (25%), and severely polluted (26%) categories. Predictive models were implemented using the AdaBoost and random forest algorithms, with grid search-based hyper-parameter tuning. The random forest model outperformed AdaBoost, showing high accuracy in predicting water quality classes. Explainable AI techniques, SHAP, and LIME were used to gain insights into parameter impacts on WQI prediction. BOD, COD, and turbidity negatively influenced WQI, while temperature and dissolved oxygen had positive effects. Site-specific analysis using LIME revealed parameter behavior at specific locations, guiding targeted management strategies. The study recommended strategies to improve water quality, including controlling and reducing BOD, COD, and turbidity, enhancing dissolved oxygen levels, monitoring temperature and pH, addressing nitrates and TDS, and implementing site-specific measures. Policy implications involve adopting the identified strategies by water resource managers and policymakers. Regular monitoring, adaptive management, and interdisciplinary collaboration are crucial for sustainable water resource management and improved water quality. Considering local regulations and stakeholder involvement will contribute to successful implementation and positive environmental outcomes. Therefore, this study provided valuable insights into water quality assessment and management. The predictive models exhibited strong performance, and XAI techniques facilitated a better understanding of influential parameters. Implementing the recommended strategies can lead to effective water pollution control and resource conservation.

S. Talukdar
Department of Geography, Asutosh College, University of Calcutta, Kolkata, India

Shahfahad · I. A. Ahmed · Mohd.Rihan · A. Rahman (✉)
Department of Geography, Faculty of Natural Sciences, Jamia Millia Islamia, New Delhi 110025, India
e-mail: arahman2@jmi.ac.in

I. A. Ahmed
Department of Civil Engineering, Indian Institute of Technology, New Delhi 110016, India

S. Talukdar et al. (eds.), *Environmental Risk and Resilience in the Changing World*, GIScience and Geo-environmental Modelling, https://doi.org/10.1007/978-3-031-62442-1_7

Keywords

Water quality index (WQI) · Predictive models · Explainable artificial intelligence (XAI) · Pollution management · Management strategies

7.1 Introduction

Water pollution is a critical environmental issue that endangers the health and sustainability of ecosystems around the world (Ighalo and Adeniyi 2020; Verhoeven et al. 2006). As industrialization, urbanization, and agricultural practices continue to expand, numerous contaminants enter water bodies, poisoning freshwater supplies and damaging the delicate balance of aquatic ecosystems (Bashir et al. 2020). Toxic chemicals, heavy metals, pathogens, and excess nutrients are among the contaminants emitted by industrial discharges, sewage, agricultural runoff, and poor waste management (Fan et al. 2022; Li et al. 2021). Water pollution has far-reaching implications, affecting not only aquatic organisms but also human health, because contaminated water sources can contribute to the spread of waterborne diseases (Conant Jr et al. 2019, Sall et al. 2020). Moreover, the polluted water can have long-term effects on ecosystems, changing food chains, diminishing biodiversity, and degrading aquatic habitat health (Karaouzas et al. 2021; Yamamuro et al. 2019). As a result, knowing the causes, consequences, and potential mitigation techniques for water contamination is critical to ensuring the integrity and availability of clean water resources for current and future generations.

To evaluate and analyze the quality of water, a number of techniques have been developed, ranging from conventional, labor-intensive techniques to advanced artificial intelligence (AI)-based methods. These techniques include measuring temperature, pH, dissolved oxygen, turbidity, nutritional levels, heavy metal concentrations, and the presence of pathogens, among other physical, chemical, and biological characteristics (Bhateria and Jain 2016). Traditional methods include manual sampling and laboratory analysis, while modern techniques employ real-time monitoring systems, remote sensing, and AI (Ahmed et al. 2020; Barzegar et al. 2020; Kuklina et al. 2013). With the increasing availability of innovative technologies, the accurate and timely assessment of water quality is becoming more feasible, enabling efficient decision-making for water resource management and pollution prevention (Cosgrove and Loucks 2015).

Real-time monitoring and early identification of pollution incidents could be made possible by machine learning algorithms and AI approaches, which have the capacity to analyze enormous volumes of complicated water quality data, discover trends, and generate reliable predictions (Zhou 2020). These technologies can also help with the integration of diverse data sources, including sensor networks and remote sensing, to provide a thorough understanding of the dynamics of water quality (Bhardwaj et al. 2022). The machine learning and AI techniques have been applied for assessing urban expansion (Shahfahad et al. 2022; Talukdar et al. 2024), flood and landslide hazard mapping (Shen et al. 2022; Talukdar et al. 2020), draught monitoring (Kikon and Deka 2022), soil erosion mapping (Alqadhi et al. 2023), forest fire susceptibility mapping (Shahfahad et al. 2022), etc. Explainable AI (XAI), which provides transparent and comprehensible insights into difficult decision-making processes, has become a potent tool in a number of fields in recent years (Kim et al. 2023; Naikoo et al. 2023; Pradhan et al. 2023).

The incorporation of XAI techniques in the context of water quality analysis has enormous potential to increase our understanding of water quality and deliver useful information for better water resource management. One of the most significant advantages of employing XAI in water quality analysis is its ability to interpret the intricate relationships between a variety of elements impacting water quality (Park et al. 2022). For decision-makers and stakeholders in water resource management, the explainability component of XAI also offers an unprecedented level of confidence and acceptability. Therefore, the objectives are:

1. To predict WQI using AdaBoost and RF model.
2. To employ XAI techniques, specifically SHAP and LIME modeling, for assessing the influence of water quality parameters on the prediction of WQI.
3. To develop site-specific management strategies based on the results obtained from the SHAP and LIME analyses.

The novelty of this work lies in the application of explainable artificial intelligence (XAI) techniques, namely SHAP and LIME modeling, to investigate the impact of various water quality parameters on WQI prediction. By utilizing these techniques, the study aims to provide a deeper understanding of the specific contribution of each parameter to the overall WQI, thereby enhancing the interpretability and transparency of the prediction model.

It is hypothesized that the SHAP values, representing the influence of each water quality parameter on the WQI prediction, will reveal significant associations between parameter values and the overall water quality. Additionally, it is anticipated that the site-specific analysis conducted through LIME modeling will uncover the relative importance of different parameters in distinct locations, enabling the development of tailored management strategies for addressing water pollution concerns.

7.2 Materials and Methods

7.2.1 Study Area

Loktak Lake, located in the northeastern state of Manipur, India, is the largest freshwater lake in the region (Fig. 7.1). It holds immense ecological and socio-economic significance, serving as a crucial habitat for numerous species of flora and fauna, as well as supporting the livelihoods of local communities. However, in recent years, Loktak Lake has been facing severe pollution challenges. The major contributing factors to its pollution are anthropogenic activities, including the discharge of untreated domestic and industrial waste, agricultural runoff, and the unregulated use of fertilizers and pesticides in the surrounding areas. As a result, the lake has experienced a significant decline in water quality. Increased levels of nutrients such as nitrogen and phosphorus have led to eutrophication, causing excessive growth of algae and other aquatic plants. This excessive growth not only depletes oxygen levels but also hampers the penetration of sunlight, disrupting the lake's ecological balance. Furthermore, the pollution has had adverse effects on the lake's biodiversity, leading to the loss of certain species and disturbance in the food chain. The deteriorating water quality of Loktak Lake poses a threat to the overall ecosystem and the livelihoods of local communities who depend on its resources. Urgent measures are required to mitigate pollution sources, improve waste management practices, and promote sustainable agricultural practices to restore and preserve the ecological integrity of this vital freshwater ecosystem.

7.2.2 Water Samples Collection and Selection of the Water Quality Parameters

Water samples from 30 different sites in Loktak Lake were collected during the pre-monsoon season at a depth of 20–30 cm, following BIS guidelines. The sampling sites were strategically chosen to represent various lake characteristics and critical areas. Composite sampling was employed to gather water from multiple locations. Using a portable multiple water analyzer (model HI98129, Hanna India), essential physical parameters like pH, EC, temperature, turbidity, and total dissolved solids (TDS) were immediately measured.

The collected water samples were stored in pre-treated polyethylene bottles at 4 °C and later analyzed in the laboratory. Prior to analysis, 0.45-μm filters were used to filter the samples, and the American Public Health Association (APHA) standard technique was followed to assess parameters such as dissolved oxygen (DO), biochemical oxygen demand (BOD), chemical

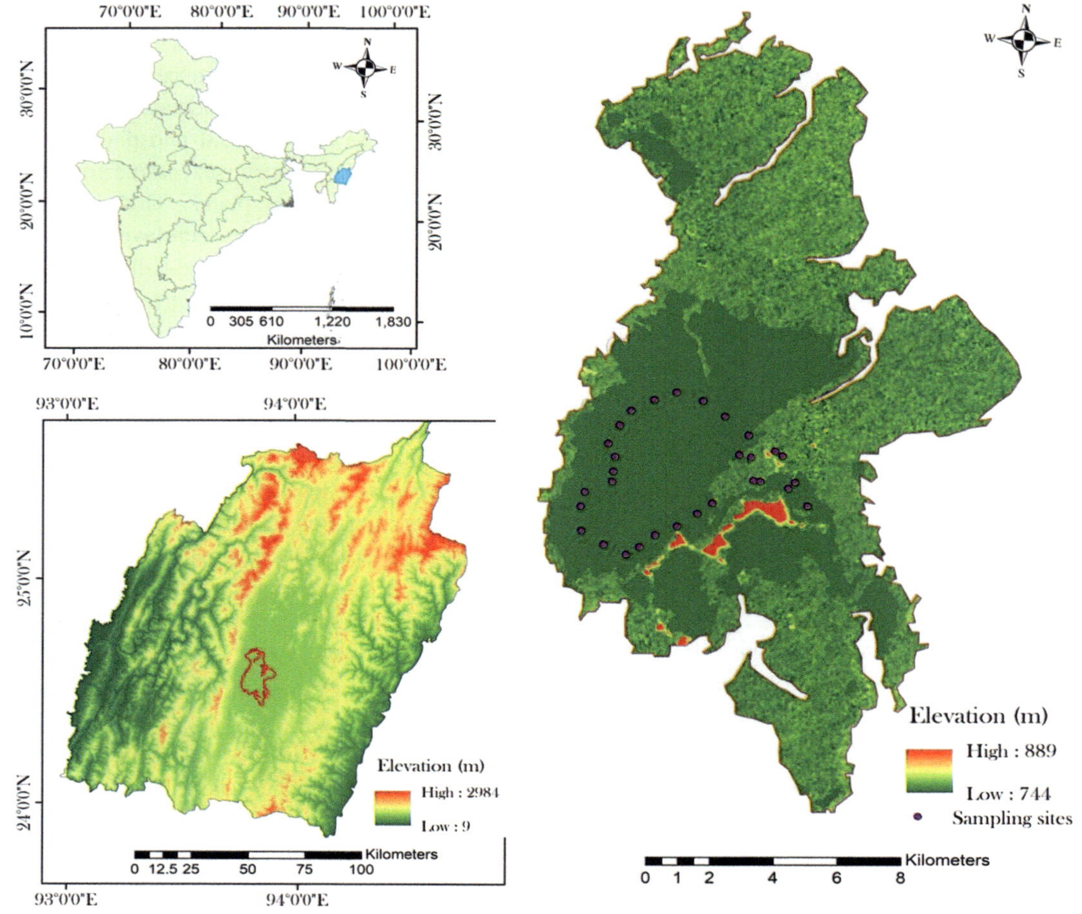

Fig. 7.1 Location of the study area with sampling points

oxygen demand (COD), and nitrate (NO3−). For DO, BOD, COD, and NO3-analysis, methods such as Winkler acid modification titrimetric, bottle incubation, open reflux, and UV spectrophotometric were used, respectively.

7.2.3 WQI Estimation Using Entropy Theory

Entropy theory is a widely used approach in water quality assessment for estimating the water quality index (WQI). This method is based on the principle of information entropy, which measures the disorder or randomness within a system. The underlying concept is that water quality parameters with higher variability or dispersion are indicative of poorer water quality. The estimation process begins by selecting a set of key water quality parameters that are representative of the overall water quality status. These parameters typically include variables such as pH, dissolved oxygen (DO), biochemical oxygen demand (BOD), total suspended solids (TSS), and nutrient concentrations. Next, the observed values of these selected parameters are normalized using appropriate transformation techniques to bring them into a comparable range. Normalization ensures that each parameter contributes proportionally to the overall WQI calculation, regardless of its original unit or scale. Once the parameters are normalized, their entropy values are calculated using the Shannon entropy formula. The Shannon entropy quantifies the uncertainty or information content of a random variable. In the context of water quality assessment, it represents the degree of disorder or variability within the system. The entropy values for each parameter are then weighted based on their relative importance in influencing water quality. The weighting factors are determined using expert judgment, stakeholder preferences, or statistical analysis techniques. These weights reflect the significance of each parameter in contributing to the overall water quality status. Finally, the weighted entropy values are aggregated using an appropriate aggregation method, such as a weighted sum or a weighted average. The resulting aggregated entropy value represents the WQI, which provides a single numerical value to characterize the overall water quality condition. A higher WQI value indicates poorer water quality, as it reflects a higher level of disorder or variability in the selected parameters. Conversely, a lower WQI value suggests better water quality, indicating a lower degree of uncertainty or variability within the system. The WQI estimation using entropy theory provides a quantitative and comprehensive assessment of water quality by considering multiple parameters simultaneously. It offers a systematic approach to evaluate and compare water quality across different locations and time periods, aiding in the identification of priority areas for water management and pollution control efforts.

7.2.4 WQI Prediction Using Optimized AdaBoost and RF Models

7.2.4.1 AdaBoost Model

The AdaBoost algorithm is an ensemble learning method that combines multiple weak learners to create a strong predictive model. In the context of WQI prediction, the AdaBoost algorithm can be employed to estimate the relationship between water quality parameters and the corresponding WQI values. The key idea behind AdaBoost is to iteratively train a series of weak classifiers or regression models, with each subsequent model focusing on correctly predicting the instances that were misclassified by the previous models. The misclassified instances are assigned higher weights, which ensure that subsequent models pay more attention to them. By iteratively combining these weak models, AdaBoost constructs a strong ensemble model that achieves better predictive performance. The hyper-parameters of the AdaBoost algorithm, such as the number of weak learners and the learning rate, can be optimized using a grid search approach to find the best combination of parameters that maximizes the WQI prediction accuracy.

7.2.4.2 Random Forest (RF) Model

The random forest algorithm is another ensemble learning technique widely used for predictive modeling, including WQI prediction. RF combines multiple decision trees to create a robust and accurate prediction model. In the context of WQI prediction, RF can handle both classification and regression tasks effectively. RF constructs a multitude of decision trees by bootstrapping the training dataset and selecting random subsets of predictor variables for each tree. During the training process, each tree is grown to its maximum depth without pruning, resulting in a collection of uncorrelated trees. The final prediction is obtained by averaging or taking the majority vote of the predictions made by individual trees. RF provides several advantages, including handling high-dimensional datasets, automatic feature selection, and robustness against overfitting. To optimize the RF model's performance for WQI prediction, a grid search approach can be utilized to tune hyper-parameters such as the number of trees, maximum tree depth, and the number of variables considered at each split.

By employing grid search-based AdaBoost and RF models for WQI prediction, it is possible to leverage the strengths of ensemble learning techniques to achieve accurate and reliable predictions. The grid search technique allows for systematic exploration of various hyper-parameter combinations, enabling the selection of optimal parameter values that maximize the predictive performance of these models.

7.2.5 Validation of the Model

Confusion matrix analysis is a technique used to evaluate the performance of a classification model by comparing the predicted labels with the true labels of a dataset. From the provided values of precision (pre), recall (rec), specificity (spe), F1-score (f1), and geometric mean (geo), we can derive the following theories:

Precision (pre): Precision is a metric that measures the accuracy of positive predictions made by the model. It is calculated as the ratio of true positive (TP) predictions to the sum of true positive and false positive (FP) predictions. A precision value of 0.95 indicates that out of all the positive predictions made by the model, 95% were correct. In other words, the model has a high level of precision, suggesting that it is effective in correctly identifying positive instances.

Recall (rec): Recall, also known as sensitivity or true positive rate, measures the model's ability to correctly identify positive instances out of all the actual positive instances. It is calculated as the ratio of true positive predictions to the sum of true positive and false negative (FN) predictions. A recall value of 0.94 indicates that the model can capture 94% of the positive instances present in the dataset. It suggests that the model has a high level of sensitivity in detecting positive instances.

Specificity (spe): Specificity measures the model's ability to correctly identify negative instances out of all the actual negative instances. It is calculated as the ratio of true negative (TN) predictions to the sum of true negative and false positive predictions. A specificity value of 0.98 indicates that the model has a high level of accuracy in classifying negative instances. It suggests that the model can effectively distinguish and correctly classify negative instances.

***F*1-score (*f*1)**: The *F*1-score is the harmonic mean of precision and recall. It provides a balanced measure of the model's performance by considering both precision and recall simultaneously. An *F*1-score of 0.94 suggests that the model has a good balance between precision and recall, indicating that it can achieve both high precision and high recall simultaneously.

Geometric mean (geo): The geometric mean is another metric used to evaluate classification models' performance, particularly when dealing with imbalanced datasets. It is calculated as the square root of the product of sensitivity (recall) and specificity. A geometric mean value is not provided in the given information, so we cannot infer its specific value or interpret its significance.

7.2.6 Implementation of XAI

Explainable artificial intelligence (XAI) refers to the methods and techniques employed to make machine learning models more transparent and interpretable. It aims to provide insights into how models make predictions or decisions, thereby enhancing their trustworthiness and facilitating human understanding. In the context of water quality assessment, XAI techniques can be applied to gain insights into the factors influencing the WQI predictions.

7.2.6.1 SHAP Analysis

Shapley additive explanations (SHAP) analysis is an XAI technique that quantifies the contribution of each feature or parameter to the prediction outcome. It utilizes game theory concepts to assign a value to each feature based on its impact on the prediction. In the context of water quality assessment, SHAP analysis can be applied to determine the specific impact of each parameter (such as pH, TDS, EC, temperature, turbidity, DO, BOD, COD, nitrates) on the WQI prediction. The SHAP values obtained provide an indication of the magnitude and direction of influence that each parameter has on the WQI prediction.

7.2.6.2 Site-Specific Analysis Using LIME

Local Interpretable Model-agnostic Explanations (LIME) is another XAI technique used to explain the predictions of machine learning models. It provides interpretable explanations for individual predictions by approximating the model's behavior locally around a specific data instance. In the context of water quality assessment, site-specific analysis using LIME can be employed to understand the factors influencing the WQI prediction for a particular site or location. By generating explanations specific to that site, LIME enables the identification of important features or parameters that contribute to the WQI prediction at that specific location.

Overall, the implementation of XAI techniques like SHAP analysis and site-specific analysis using LIME can enhance the interpretability of the water quality assessment models. These techniques enable researchers and stakeholders to gain insights into the factors driving the WQI predictions and provide transparency in the decision-making process. Such interpretability is crucial for effective decision-making, policy formulation, and resource allocation in water management and pollution control.

7.3 Results

7.3.1 WQI Estimation

The water quality index (WQI) was determined using the entropy theory method, which assigns weights to different parameters based on their importance in assessing water quality. The weights assigned to each parameter are as follows: pH (0.92), TDS (0.88), DO (0.87), BOD (0.89), COD (0.86), EC (0.89), turbidity (0.91), temperature (0.91), and nitrate (0.87). These weights reflect the significance of each parameter in evaluating water quality.

The calculated WQI values ranging from 80.59 to 100.00 represent the diverse quality levels observed across the study area. These values indicate variations in water quality, with some locations exhibiting lower WQI values, suggesting better water quality, while others have higher values, indicating potential pollution or degradation.

By employing the classification system proposed by Uddin et al. (2022), the obtained WQI values can be categorized into different classes. In this study, 24% of the samples were classified as having good water quality. However, it is noteworthy that moderately polluted and severely polluted conditions were found in 25 and 26% of the samples, respectively. These findings highlight the presence of areas within the study region that require attention and potential remediation measures to address water pollution and ensure improved water quality (Fig. 7.2).

7.3.2 WQI Assessment

In this study, the WQI assessment has been in done two ways: first is to predict the WQI and second is to validate the predicted models.

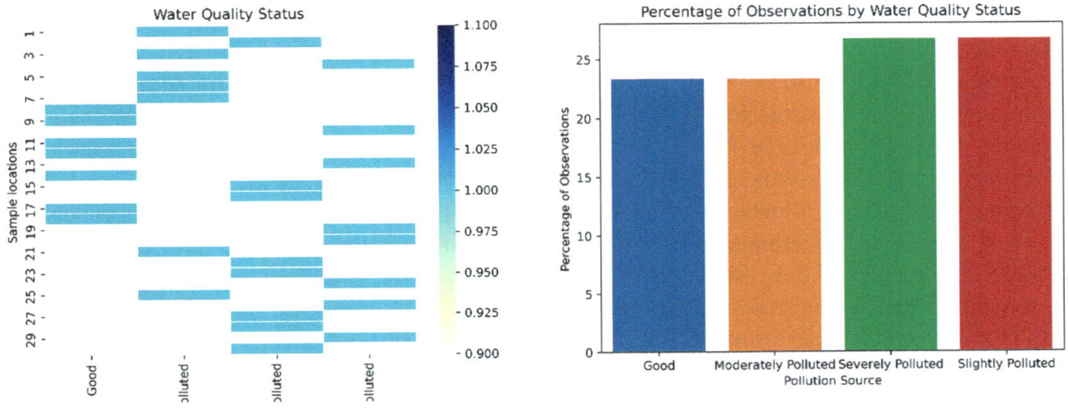

Fig. 7.2 Categorization of WQI for all samples in Loktak Lake

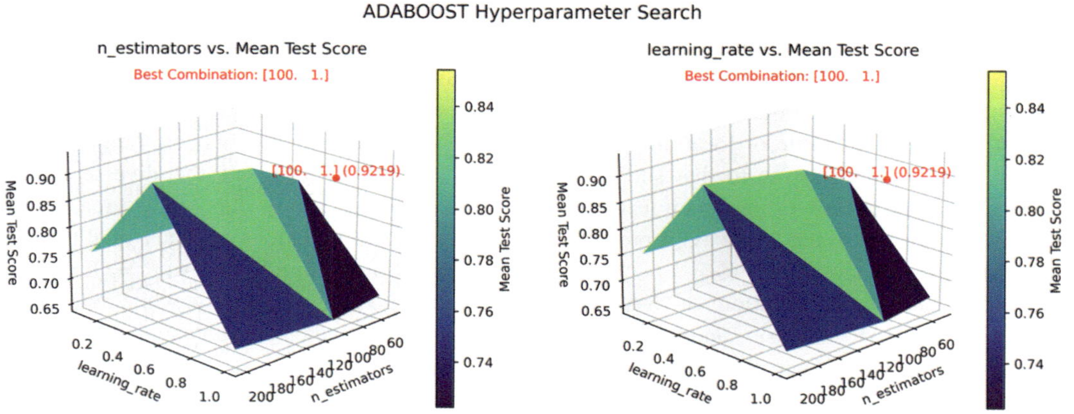

Fig. 7.3 Hyper-parameter search for AdaBoost algorithm for finding optimal parameters

7.3.2.1 Implementation of the Models

The WQI was predicted using the grid search-based AdaBoost and random forest (RF) algorithms implemented in the Python programming language. The hyper-parameters for AdaBoost and RF were tuned through a grid search process to identify the optimal combination of parameter values. For AdaBoost, the grid search involved varying the number of estimators (50, 100, 200) and learning rates (0.1, 0.5, 1.0). Similarly, for RF, the grid search explored different values for the number of estimators (50, 100, 200) and maximum depth (3, 5, 7).

The grid search results provided insights into the performance of the models with different hyper-parameter settings. For AdaBoost, the mean test scores and corresponding best parameter combinations were as follows (Fig. 7.3): 0.7063 (learning_rate: 0.1, n_estimators: 50), 0.7469 (learning_rate: 0.1, n_estimators: 100), 0.7516 (learning_rate: 0.1, n_estimators: 200), 0.8438 (learning_rate: 0.5, n_estimators: 50), 0.8938 (learning_rate: 0.5, n_estimators: 100), 0.9219 (learning_rate: 0.5, n_estimators: 200), 0.6687 (learning_rate: 1.0, n_estimators: 50), 0.6547 (learning_rate: 1.0, n_estimators: 100), and 0.6687 (learning_rate: 1.0, n_estimators: 200).

Regarding RF, the mean test scores and corresponding best parameter combinations were as follows (Fig. 7.4): 0.9062 (max_depth: 3, n_estimators: 50), 0.9141 (max_depth: 3, n_estimators: 100), 0.9078 (max_depth: 3, n_estimators: 200), 0.9578 (max_depth: 5, n_estimators: 50), 0.9609 (max_depth: 5, n_estimators: 100), 0.9641 (max_depth: 5, n_estimators: 200), 0.9891 (max_depth: 7, n_estimators: 50), 0.9875 (max_depth: 7, n_estimators: 100), and 0.9891 (max_depth: 7, n_estimators: 200).

Based on the performance of the models, the best AdaBoost model was selected using the hyper-parameters: learning_rate of 0.5 and n_estimators of 200. Similarly, the best RF model was identified with the hyper-parameters: max_depth of 7 and n_estimators of 200. These models exhibited high accuracy in predicting the WQI. Therefore, the grid search process facilitated the selection of optimal hyper-parameter settings for the AdaBoost and RF models. These models demonstrate promising accuracy and can be utilized for water quality prediction based on the identified hyper-parameter settings.

7.3.2.2 Validation of the Models

The predicted models were evaluated using various indices, including the confusion matrix and other performance metrics. For the AdaBoost model, the confusion matrix analysis yielded the following results: precision (pre) of 0.95, recall (rec) of 0.94, specificity (spe) of 0.98, *F*1-score (*f*1) of 0.94, geometric mean (geo) of 0.96, and index of balanced accuracy (iba) of 0.92. These indices were computed for each class (0, 1, 2, and 3) and averaged over the entire dataset, which consisted of 160 samples (Fig. 7.5).

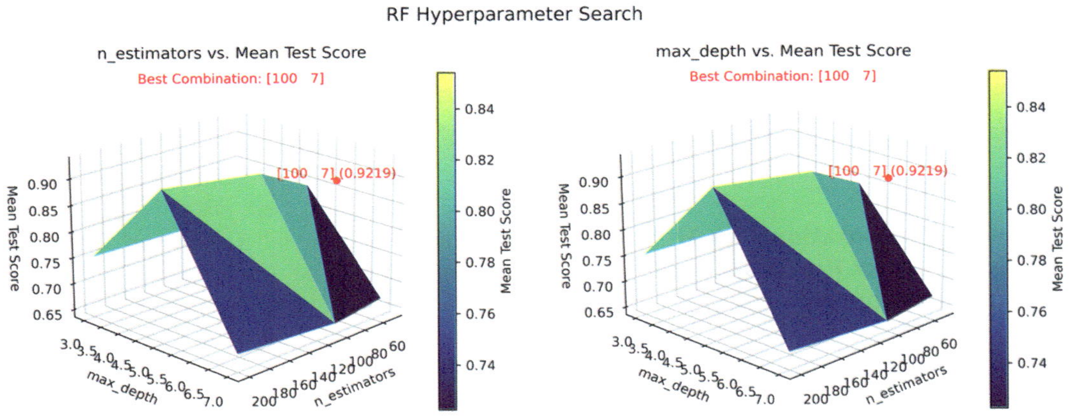

Fig. 7.4 Hyper-parameter search for RF algorithm for finding optimal parameters

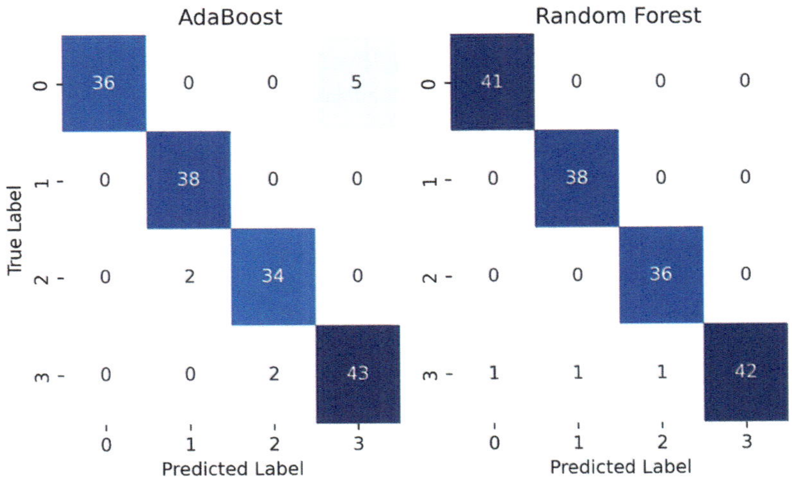

Fig. 7.5 Confusion matrix for accuracy assessment for AdaBoost and random forest models

Similarly, for the RF model, the classification report provided the following results: precision (pre) of 0.98, recall (rec) of 0.98, specificity (spe) of 0.99, $F1$-score ($f1$) of 0.98, geometric mean (geo) of 0.99, and index of balanced accuracy (iba) of 0.97. These indices were computed for each class and averaged over the entire dataset.

The obtained results indicate the performance of the predicted models in accurately classifying the water quality samples. Both the AdaBoost and RF models demonstrated high precision, recall, and $F1$-scores, indicating their ability to correctly identify samples belonging to different water quality classes. The specificity scores indicate the models' capability to accurately classify non-target samples.

The high geometric mean and index of balanced accuracy values suggest that both models performed well in maintaining a balance between sensitivity and specificity. Overall, the models exhibited strong performance and achieved accurate predictions for the water quality classes, as evident from the high values of the evaluated indices.

These results provide valuable insights into the effectiveness of the AdaBoost and RF models in predicting and classifying water quality, indicating their potential utility in real-world applications such as water resource management and environmental monitoring. Also, it can be stated that RF model outperformed AdaBoost model, which can be used for further analysis.

7.3.3 Implementation of XAI for Water Pollution Management

We employed XAI techniques, specifically SHAP and LIME modeling, for management purposes in water quality analysis. These techniques provide insights into the impact of different parameters on the prediction of the WQI using the RF model. Additionally, we conducted site-specific analysis using the LIME package to understand parameter behavior at different sampling sites. Based on these analyses, we can derive management strategies to improve water quality.

7.3.3.1 SHAP Analysis

The SHAP summary plot generated from the RF model revealed the impact of various parameters on the WQI prediction (Fig. 7.6). The plot represents high and low parameter values using different shades of red and blue colors, respectively. High parameter values indicate a strong impact on the WQI prediction, while low values suggest a weaker influence.

Based on the SHAP summary plot, the following observations were made:

Parameters with high values, such as biological oxygen demand (BOD), chemical oxygen demand (COD), and turbidity, had a significant impact on the WQI prediction. Elevated levels of these parameters were associated with poorer water quality. Conversely, high values of temperature and turbidity had a negative influence on the WQI prediction, suggesting that higher temperatures and turbidity levels tend to degrade water quality. These SHAP values indicate the magnitude and direction of influence that each parameter has on the WQI prediction. Positive SHAP values indicate that an increase in the parameter's value positively impacts the WQI, while negative values indicate a negative impact. Larger absolute SHAP values indicate a higher influence of the parameter on the WQI prediction. Analyzing the specific parameter SHAP values, we can infer the following:

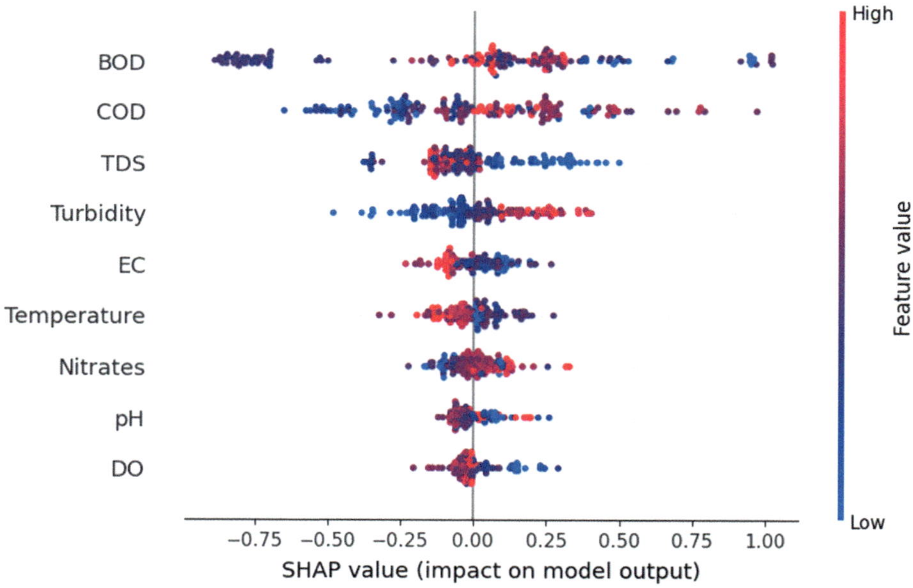

Fig. 7.6 SHAP summary plot for estimating the influence of parameters for water quality

- pH: Higher pH values have a negative impact on the WQI prediction.
- TDS: Higher TDS levels negatively influence the WQI prediction.
- EC: Higher electrical conductivity has a positive impact on the WQI prediction.
- Temperature: Lower temperatures are associated with a higher WQI prediction.
- Turbidity: Increased turbidity levels negatively affect the WQI prediction.
- DO: Higher dissolved oxygen levels have a positive impact on the WQI prediction.
- BOD: Higher biochemical oxygen demand values negatively influence the WQI prediction.
- COD: Increased chemical oxygen demand has a positive impact on the WQI prediction.
- Nitrates: Elevated nitrate levels negatively affect the WQI prediction.

Understanding these SHAP values helps us identify the parameters that have the most significant influence on the WQI and water pollution. By managing and controlling the parameters accordingly, such as reducing pollution sources, improving wastewater treatment, and implementing appropriate mitigation measures, we can positively impact water quality and improve the WQI.

7.3.3.2 Site-Specific Analysis Using LIME

To understand how different parameters behave at specific sites, we employed the LIME package for site-specific analysis (Fig. 7.7). We selected six sampling sites from each category and determined the feature importance for each site using LIME. The feature importance values indicate the contribution of each parameter to the prediction at a given site. Here are the results for the site-specific analysis:

Sample 1:

- The most important features for this site were BOD, nitrates, and EC. Higher BOD and EC values, along with lower nitrates levels, were associated with a higher predicted WQI.

Sample 2:

- The significant features for this site were turbidity, BOD, and nitrates. Elevated turbidity and BOD values contributed to a higher predicted WQI.

Sample 5:

- The key features at this site were COD, turbidity, and nitrates. Higher COD and turbidity values, along with lower nitrates levels, were linked to a higher predicted WQI.

Sample 9:

- The most influential features for this site were COD, BOD, and EC. Lower COD levels and higher BOD and EC values were associated with a higher predicted WQI.

Sample 10:

- The significant features for this site were COD, turbidity, and BOD. Elevated COD and turbidity values, along with lower BOD levels, contributed to a higher predicted WQI.

Sample 16:

- The most important features at this site were BOD, turbidity, and nitrates. Lower BOD and turbidity values, along with higher nitrates levels, were linked to a lower predicted WQI.

7.3.4 Management Strategies

Based on the SHAP analysis and site-specific analysis using LIME, the following management strategies can be derived:

1. Control and reduce BOD, COD, and turbidity: These parameters had a significant impact on the WQI prediction and were associated with degraded water quality. Implementing measures to control and reduce organic and chemical pollution sources, improving

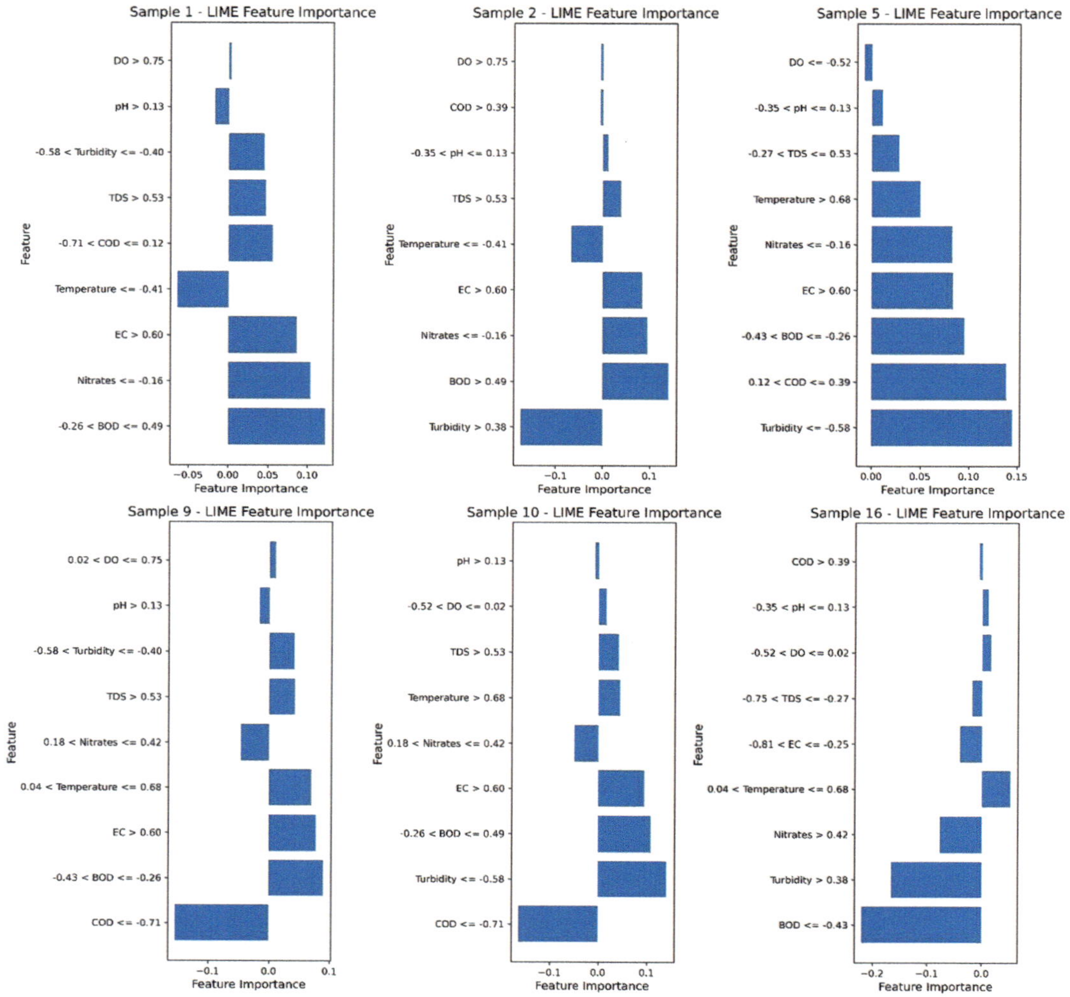

Fig. 7.7 Site-specific LIME-based feature influence for water quality degradation

wastewater treatment processes, and implementing sediment control practices can help mitigate the negative impact of these parameters.

2. Enhance dissolved oxygen (DO) levels: DO was found to have a positive influence on the WQI prediction. Increasing DO levels can be achieved through measures such as increasing water circulation, reducing nutrient inputs, and implementing aeration techniques.
3. Monitor and manage temperature and pH: Higher temperatures and deviations from optimal pH values negatively impacted the WQI prediction. Regular monitoring of temperature and pH, along with appropriate management strategies, such as shading to reduce temperature fluctuations and managing pH through buffering or treatment methods, can help maintain favorable water quality conditions.
4. Address nitrates and TDS: Nitrates and TDS were identified as important parameters in the site-specific analysis. Implementing agricultural best management practices, promoting proper fertilizer usage, and adopting appropriate land management techniques can help

reduce nitrate levels. Additionally, managing water sources and implementing suitable treatment methods can help control and maintain TDS levels within acceptable limits.

5. Implement site-specific management measures: The site-specific analysis using LIME highlighted the variation in feature importance across different sampling sites. Therefore, implementing site-specific management measures based on the identified influential parameters can help address specific water quality concerns at each location.

It is essential to note that these strategies are based on the provided analysis results. Considering additional factors, local regulations, and stakeholder involvement is crucial for effective water quality management. Regular monitoring, adaptive management approaches, and interdisciplinary collaboration will aid in achieving sustainable water resource management and preserving water quality.

7.4 Conclusion

The research aimed to assess water quality using the WQI and predict WQI values using the AdaBoost and RF algorithms. It also utilized XAI techniques, SHAP and LIME, for water pollution management. The WQI values indicated variations in water quality across the study area, with some locations showing good water quality and others indicating potential pollution or degradation. The AdaBoost and RF models demonstrated high accuracy in predicting WQI, with RF outperforming AdaBoost. SHAP analysis revealed the impact of various parameters on WQI prediction, while LIME provided site-specific insights into parameter behavior. The research introduced the application of entropy theory-based WQI estimation and XAI techniques, SHAP and LIME, for water quality analysis and management. It also explored the use of AdaBoost and RF algorithms for WQI prediction, providing optimal hyperparameter settings for these models.

Some limitations of the research include the use of specific algorithms for WQI prediction, which might not account for all possible factors affecting water quality. Additionally, the analysis was conducted based on specific water quality parameters, and other relevant factors may have been excluded. The study's scope might be limited to the study area, and generalizing the findings to other regions requires caution.

The research highlights the importance of managing and controlling water quality parameters such as BOD, COD, turbidity, DO, nitrates, TDS, temperature, and pH to improve overall water quality. It emphasizes the need for targeted measures in areas with severe pollution or degradation, such as improving wastewater treatment, reducing pollution sources, and implementing sediment control practices. Policymakers should consider site-specific management measures based on parameter behavior at different locations for effective water quality management.

The research opens up avenues for further investigation in water quality assessment and prediction using advanced algorithms and incorporating additional parameters. Exploring the application of XAI techniques in other environmental and ecological studies can provide valuable insights for decision-making. Integrating real-time data collection and adaptive management approaches would enhance the effectiveness of water quality management strategies. Moreover, conducting similar studies in different regions and comparing the results would contribute to a better understanding of water quality dynamics on a broader scale.

References

Ahmed U, Mumtaz R, Anwar H, Mumtaz, S, Qamar AM (2020) Water quality monitoring: from conventional to emerging technologies. Water Supply 20(1): 28–45

Alqadhi S, Mallick J, Talukdar S, Alkahtani M (2023) An artificial intelligence-based assessment of soil erosion probability indices and contributing factors in the Abha-Khamis watershed, Saudi Arabia. Front Ecol Evol 11:1189184

Barzegar R, Aalami MT, Adamowski J (2020) Short-term water quality variable prediction using a hybrid CNN–LSTM deep learning model. Stoch Env Res Risk Assess 34(2):415–433

Bashir I, Lone FA, Bhat RA, Mir SA, Dar ZA, Dar SA (2020) Concerns and threats of contamination on aquatic ecosystems. In: Bioremediation and biotechnology: sustainable approaches to pollution degradation, pp 1–26

Bhardwaj A, Dagar V, Khan MO, Aggarwal A, Alvarado R, Kumar M, Irfan M, Proshad R (2022) Smart IoT and machine learning-based framework for water quality assessment and device component monitoring. Environ Sci Pollut Res 29(30):46018–46036

Bhateria R, Jain D (2016) Water quality assessment of lake water: a review. Sustain Water Res Manag 2:161–173

Conant B Jr, Robinson CE, Hinton MJ, Russell HA (2019) A framework for conceptualizing groundwater-surface water interactions and identifying potential impacts on water quality, water quantity, and ecosystems. J Hydrol 574:609–627

Cosgrove WJ, Loucks DP (2015) Water management: current and future challenges and research directions. Water Resour Res 51(6):4823–4839

Fan C, Liu Y, Liu C, Zhao W, Hao N, Guo W, Yuan J, Zhao J (2022) Water quality characteristics, sources, and assessment of surface water in an industrial mining city, southwest of China. Environ Monit Assess 194(4):259

Ighalo JO, Adeniyi AG (2020) A comprehensive review of water quality monitoring and assessment in Nigeria. Chemosphere 260:127569

Karaouzas I, Kapetanaki N, Mentzafou A, Kanellopoulos TD, Skoulikidis N (2021) Heavy metal contamination status in Greek surface waters: a review with application and evaluation of pollution indices. Chemosphere 263:128192

Kikon A, Deka PC (2022) Artificial intelligence application in drought assessment, monitoring and forecasting: a review. Stoch Env Res Risk Assess 36(5):1197–1214

Kim M, Kim D, Jin D, Kim G (2023) Application of explainable artificial intelligence (XAI) in urban growth modeling: a case study of Seoul metropolitan area, Korea. Land 12(2):420

Kuklina I, Kouba A, Kozák P (2013) Real-time monitoring of water quality using fish and crayfish as bio-indicators: a review. Environ Monit Assess 185:5043–5053

Li P, Karunanidhi D, Subramani T, Srinivasamoorthy K (2021) Sources and consequences of groundwater contamination. Arch Environ Contam Toxicol 80:1–10

Naikoo MW, Talukdar S, Ishtiaq M, Rahman A (2023) Modelling built-up land expansion probability using the integrated fuzzy logic and coupling coordination degree model. J Environ Manage 325:116441

Park J, Lee WH, Kim KT, Park CY, Lee S, Heo TY (2022) Interpretation of ensemble learning to predict water quality using explainable artificial intelligence. Sci Total Environ 832:155070

Pradhan B, Lee S, Dikshit A, Kim H (2023) Spatial flood susceptibility mapping using an explainable artificial intelligence (XAI) model. Geosci Front 14(6):101625

Sall ML, Diaw AKD, Gningue-Sall D, Efremova Aaron S, Aaron JJ (2020) Toxic heavy metals: impact on the environment and human health, and treatment with conducting organic polymers, a review. Environ Sci Pollut Res 27:29927–29942

Shahfahad T, Das S, Naikoo T, Rihan MWM, Rahman A (2022) Forest fire susceptibility mapping by integrating remote sensing and machine learning algorithms. Adv Remote Sens Forest Monitor 179–195

Shen H, Huang F, Fan X, Shahabi H, Shirzadi A, Wang D, Peng C, Zhao X, Chen W (2022) Improving the performance of artificial intelligence models using the rotation forest technique for landslide susceptibility mapping. Int J Environ Sci Technol 1–16

Talukdar S, Ghose B, Shahfahad SR, Mahato S, Pham QB, Linh NTT, Costache R, Avand M (2020) Flood susceptibility modeling in Teesta River basin, Bangladesh using novel ensembles of bagging algorithms. Stochastic Environ Res Risk Assess 34:2277–2300

Talukdar S, Naikoo MW, Rahman A (2024) Urban expansion and vegetation dynamics: the role of protected areas in preventing vegetation loss in a growing mega city. Habitat Int 150:103129

Uddin MG, Nash S, Diganta MTM, Rahman A, Olbert AI (2022) Robust machine learning algorithms for predicting coastal water quality index. J Environ Manage 321:115923

Verhoeven JT, Arheimer B, Yin C, Hefting MM (2006) Regional and global concerns over wetlands and water quality. Trends Ecol Evol 21(2):96–103

Yamamuro M, Komuro T, Kamiya H, Kato T, Hasegawa H, Kameda Y (2019) Neonicotinoids disrupt aquatic food webs and decrease fishery yields. Science 366(6465):620–623

Zhou Y (2020) Real-time probabilistic forecasting of river water quality under data missing situation: deep learning plus post-processing techniques. J Hydrol 589:125164

Characterizing the Pollution Sources and Ecological Water Quality of the Tropical Rivers, India: A Comparative Study of the Bhagirathi-Hooghly, Jalangi, and Mathabhanga-Churni Rivers

8

Biplab Sarkar, Aznarul Islam, Swapan Talukdar, Abdur Rahman, and Kaustubh Lahiri

Abstract

The present study has been carried out on the water quality assessment of the three selected rivers with 10 sampling sites: Bhagirathi-Hooghly, Jalangi, and Mathabhanga-Churni located in the tropical region of India. The quality has been assessed based on the concentration of 14 parameters, two general water quality indices and two other ecological water quality indices. The results from samples (N = 120) indicate few valuable insights on the spatial and seasonality of the water quality. Particularly, the Jalangi and Mathabhanga-Churni Rivers are highly vulnerable due to very low dissolve oxygen (<3 mg l^{-1}) and high organic pollution (Fig. 8.5). WQI represent Bhagirathi-Hooghly (WQI range:63.71 for BH1-68.94 for BH5) river as poor and Jalangi and Mathabhanga-Churni as poor to very poor water quality (WQI range:73.355 for JK-90.64 for MC1). Similarly, the OPI indicates high organic pollution in the upper stretch of the Mathabhanga-Churni and moderate pollution for Bhagirathi-Hooghly. The HCA and discriminant analysis have also confirmed the significant difference of water quality between Bhagirathi-Hooghly and Mathabhanga-Churni. The causes of this phenomena have also been discussed from the natural and human induced point and non-point sources of pollution. The discharge variation is the crucial natural factor while discharge of wastewater from industries, urban and agricultural runoff are the major causes of water pollution.

Keywords

Organic pollution · Eutrophication · Point and non-point sources of pollution · Tropical rivers

B. Sarkar (✉) · A. Islam
Department of Geography, Aliah University, 17 Gorachand Road, Kolkata, West Bengal 700014, India
e-mail: biplab25sarkar@gmail.com

S. Talukdar
Department of Geography, Asutosh College, University of Calcutta, Kolkata, India

A. Rahman · K. Lahiri
Department of Geography, Dr. Kanailal Bhattacharyya College, Howrah, West Bengal, India

8.1 Introduction

Assessing water quality is an integral part of water resource management because water deviating from standard quality threatens human health (Janoska and Gruszecka-Kosowska 2020; Jolly et al. 2023) and the aquatic ecosystem (Cui et al. 2021; Haque et al. 2023; Na et al. 2023).

S. Talukdar et al. (eds.), *Environmental Risk and Resilience in the Changing World*, GIScience and Geo-environmental Modelling, https://doi.org/10.1007/978-3-031-62442-1_8

Pollution of rivers has become a sensitive issue for last few decades as the water quality of most of the rivers in the world is scaling down. As a result, pressure on surface water resources is increasing acutely. Thus, researchers need to examine water quality and detect the possible pollution sources at different spatial and temporal scales (Boyacioglu and Boyacioglu 2008; Mirzaei Aminiyan et el. 2018). Generally, the study of water quality aims to provide authentic information on causes, consequences (Mirzaei Aminiyan et al. 2018; Koukal et al. 2004), and spatio-temporal variation (Shammi et al. 2017) of water quality to planners enabling them to compose recommendations in the decision-making process for future action to achieve sustainable water resource management. Researchers have developed several water quality indices (WQIs) using physic-chemical and biological properties of water to suit their indices more accurately in different regions (Ahmed et al. 2017; Alves et al. 2018; Medeiros et al. 2017; Bora and Goswami 2017a, b). Quality of surface water is controlled by both the natural (climatic variability, vegetation cover, and soil erosion) and anthropogenic processes (urban sewerage, industrial effluents, and agrochemicals) prevailing in the river basin (Ding et al. 2016; Dwivedi et al. 2018; Kibena et al. 2014). A significant variation is commonly found in the concentration of pollutants in rivers due to seasonal variations in precipitation, surface and sub-surface, etc. (Sarkar and Islam 2020). Likewise, escalated human activities are continuously modifying the basin's hydrogeomorphologic character. The high rate of urbanization makes cities significant contributors to water pollution wherefrom municipal wastes (Sarkar et al. 2021), and domestic and industrial effluents are discharged continuously into the river (Sharma et al. 2015). Moreover, rainfall runoff as stormwater contaminated by dust, soil sediments, toxic metals, etc. impinges on the river. Similarly, with the introduction of pesticides and fertilizers, animal manure induces nutritional imbalances in water mixing so many nutrients through agricultural runoff that causes organic pollution in water. In many scholastic works the conventional multivariate statistical techniques such as cluster analysis (CA) (Alves et al. 2018), discriminant analysis (DA) (Duan et al. 2016), and principal component analysis (PCA) (Ahmed et al. 2017; Ebqa'ai and Ibrahim 2017; Ogwueleka 2015) are widely embraced to assess water quality and identify the positional sources of water pollution. Research on river water pollution source identification has made significant progress in recent years, but several research gaps still exist. One key research gap is the integration of multiple pollutant sources. Many studies focus on identifying individual sources of pollution, but rivers are often impacted by a combination of point sources and non-point sources. Temporal and spatial variability in river water quality is another gap, as current sampling efforts may not capture the full extent of this variability. Additionally, there is a need for better data integration and modelling approaches to effectively analyse diverse datasets. Source attribution, stakeholder engagement, and policy implementation are also areas that require further research. Addressing these research gaps will enhance our understanding of river water pollution sources and contribute to more effective pollution control strategies.

8.2 Database and Methodology

8.2.1 Study Area

This study was administered on three rivers namely the Bhagirathi-Hooghly, Jalangi, and Mathabhanga-Churni of tectonically active Bengal delta (Allison 1997; Allison and Kepple 2001; Goodberd 2003; Kuhel et al. 2005; Sarkar and Islan 2019, 2021; Kuehl et al. 1997; Rudra, 2014). The Bhagirathi-Hooghly river runs for about 500 miles taking off on the right bank of the Ganga River and falls into the Bay of Bengal (Guchhait et al. 2016). About 350 miles from its off-take to Nabadwip is a non-tidal channel known as Bhagirathi River while the rest part, i.e. 150 miles from Nabadwip to its confluence is a tidal channel known as Hooghly River (Islam and Guchhait 2017). The left bank tributaries of the river Bhagirathi are Jalangi and

Mathabhanga-Churni. The Jalangi River originates from the Ganga River at the location and runs for about 200 miles and impinges on the Bhagirathi River at Nabadwip (Das 2015). Similarly, the Mathabhanga-Churni River takes off from the same river Ganga at the location and runs for about 200 miles it meets with the Bhagirathi River at Chakdah (Sarkar et al. 2021). The rivers of the Bengal Delta are to a great extent controlled by neotectonics. Until the twelfth century, the lion's share of the discharge of the Ganga River used to pass through the Bhagirathi River other than the Padma River (Rudra 2018). But during the sixteenth century continuous shifting of the Ganga towards the Padma, due to the tilting of the Bengal delta towards the east during the late Holocene period, made the Padma the main water course of the Ganga River leaving the Bhagirathi river as a secondary river (Chapman and Rudra 2015). Moreover, many of its southwestward-flowing distributaries have become beheaded, which once significantly contributed to high discharge into the Bay of Bengal (Parua 2010). Thus, these rivers have undergone steady deterioration and though the estimation of the actual rate and the extent of this progressive deterioration is not possible due to lack of available data, few scholars attempted to establish the phenomena from different perspectives. The deterioration is also the result of climate change reported in so many scholastic works. These deteriorated rivers are highly vulnerable to water pollution from many points and non-point sources of pollution. (Fig. 8.1; Table 8.1).

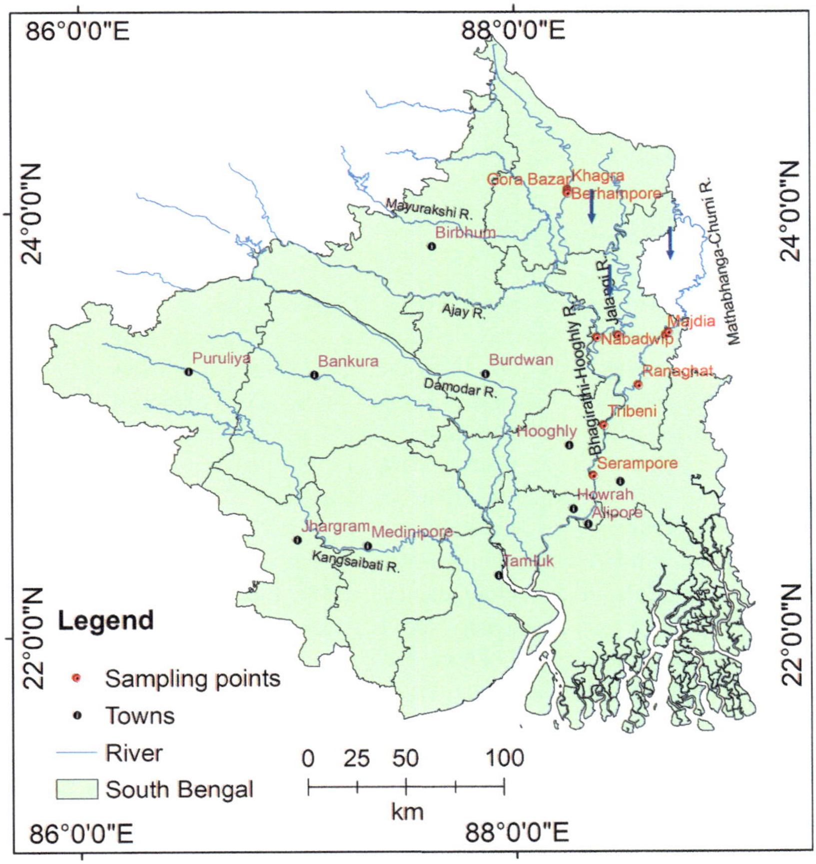

Fig. 8.1 Location of the study area

Table 8.1 Location of the water sampling points and the description of the pollution-generating factors

Sites	Geographical location	Description of sampling sites
1	24° 06′ 26′′ N, 88° 15′ 29′′ E	Agriculture, the principal land use; right bank of the river is receiving pollutants from agricultural runoff and domestic wastes while the left bank is from urban runoff, industrial wastes from bell metal industry, and many small-scale industry
2	24° 05′ 55′′ N, 88° 16′ 04′′ E	Administrative headquarter of Murshidabad district; water quality is influenced by municipal and domestic wastes; sewage from untreated waste water of small-scale industries including; runoff from agriculture
3	24° 05′ 31′′ N, 88° 15′ 03′′ E	Right bank of the river is receiving pollutants from agricultural runoff and domestic wastes while the left bank is from urban runoff
4	23° 24′ 42′′ N, 88° 22′ 51′′ E	Confluence points of the Bhagirathi and the Jalangi river; agricultural runoff, brickfield industry, municipal and domestic wastes are main sources of pollution
5	22° 59′ 15′′ N, 88° 24′ 08′′ E	Thermal power plant, jute mill, tissues factory, burning ghat (funeral site), rayon mill
6	22° 45′ 09′′ N, 88° 21′ 07′′ E	Battery factory, paper and pulp mill
7	23° 24′ 49′′ N, 88° 28′ 12′′ E	Agricultural runoff, domestic sewage, and discharge of urban wastewater are prominent in this location and continuously effecting the river health.
8	23° 28′ 14′′ N, 88° 44′ 43′′ E	Agricultural runoff, domestic sewage, sugar factory located in Bangladesh
9	23° 24′ 17′′ N, 88° 42′ 37′′ E	Mathabhanga river bifurcates into channels Churni and Icchamathi; predominant agricultural activity, printing and dyeing industry are found
10	23° 10′ 40′′ N, 88° 33′ 40′′ E	Dyeing industry, textile industry, brickfield industry

8.2.2 Database

From the selected three rivers a total of 120 water samples have been considered (72 from the Bhagirathi-Hooghly River, 12 from the Jalangi River, and 24 from the Mathabhanga-Churni River). For this study, ten sites were selected as monitoring stations among them Khagra (site 1), Berhampore (site 2), Gorabazar (site 3), Nabadwip (site4), Tribeni (site 5), Serampore (site 6) from Bhagirathi-Hooghly while Krishnanagar (site 7) from Jalangi and Gobindapur (site 8), Majhdia (site 9), and Ranaghat (site 10) from Mathabhanga-Churni River. From each monitoring station monthly water samples have been considered representing the whole year. Additionally, all of the water samples have been divided into three groups to reflect the pre-monsoon season (from February to May), monsoon season (from June to September), and post-monsoon season (from October to January) in order to examine the seasonality of the river water quality. Finally, the parameters used for the present study are biological oxygen demand (BOD), total dissolved solids (TDS), sulphate total hardness (TH), total suspended solids, (TSS) pH, electrical conductivity (EC), total alkalinity (TA), chloride, nitrate, phosphate, and dissolved oxygen (DO).

8.2.3 Methods of Data Analysis

8.2.3.1 General Water Quality Assessment

(a) **Weighted Arithmetic Water Quality Index (WAWQI)**

The WAWQI is used to assess and summarize the overall water quality of a given water body based on multiple water quality parameters (Brown et al. 1972). It computes the final numerical value determining the overall quality of the water, taking into account the relative importance or weight of different parameters. Common parameters included in this index are BOD, Chloride, DO, Sulphate, Total alkalinity, TDS, pH, Hardness, EC, and TSS (Khatri et al. 2020). The calculation of the WAWQI involves several steps (Eq. 8.1–8.4)

$$\text{WAWQI} = \sum Q_i W_i / \sum W_i \tag{8.1}$$

$$Q_i = \left(\frac{Q_{\text{actual}} - Q_{\text{ideal}}}{Q_{\text{standard}} - Q_{\text{ideal}}} \right) \times 100 \tag{8.2}$$

$$W_i = K \sum \frac{1}{S_{\text{standard}}} \tag{8.3}$$

$$K = \frac{1}{\frac{1}{s1} + \frac{1}{s2} + \ldots + \frac{1}{sn}} \tag{8.4}$$

Q_i represents parameter's quality calculated by Eq. 8.2, where Q_{actual} is parameter's actual concentration, Q_{ideal} is the ideal concentration [$Q_{\text{ideal}} = 0$, except pH (considered as 7) and DO (to be 14.6 mg l^{-1})], Q_{standard} is the parameter's standard permissible concentration. Wi (Eq. 8.3) is the parameter's unit weight (Table 8.2) derived with the help of '*K*' (Eq. 8.4) is the constant of proportionality and inversely proportional to recommended standard ($S_{\textit{standard}}$). (WAWQI) value can be categorized into five classes based on the overall water quality assessment. 0–25 (excellent), 26–50 (good), both suitable for drinking, domestic, agricultural (irrigation), and industrial uses, 51–75 (poor), and 76–100 (very poor), only suitable for agriculture and industrial uses. Moreover, the value > 100 represents very poor water quality and is unsuitable for all uses (Hagage et al. 2022).

(b) **Ved Prakash Index (VPI)**

Ved Prakash developed this water quality index to evaluate river water quality, particularly for rivers in the Indian territory (Abbasi and Abbasi 2012). With weights modified from the NSFWQI, it is primarily based on the CPCB water quality guidelines for various types of usage. The DO, BOD, pH, and faecal coliforms variables were chosen for this index. The sub-index formulae (Table 8.3) have been considered to quantify the values of the sub-indexes needed for this index (Bhutiani et al. 2016). Finally, based on the values of subindices, the final index has been derived. Therefore, the index is applied to measure the water quality of the selected sites of the rivers. The index value 63–100 indicates good to excellent, 50–63 as medium to good, 38–50 bad, and < 38 as bad to very bad water quality (Abbasi and Abbasi 2012).

Table 8.2 Wi of the selected parameters used for computing WQI (Bora and Goswami 2017a, b)

Parameters	Q_{standard} of ICMR/BIS	*Wn*
pH	6.5–8.5	0.215
EC	300	0.0061
TDS	500	0.00366
TSS	500	0.00366
Total hardness	300	0.0061
Chloride	250	0.00732
DO	5	0.366
BOD	5	0.366
Sulphate	150	0.0122
Total alkalinity	120	0.01525
		$\sum Wn = 1.001$

Table 8.3 Sub-index function (Ii) for computing VPI (Abbasi and Abbasi 2012)

Parameters	Range applicable (x)	Sub-index function (Ii)
DO (% saturation)	0–40%	IDO = 0.18 + 0.66x
	40–100%	IDO = −13.5 + 1.17x
	100–140%	IDO = 163.34 − 0.62x
BOD (mg l^{-1})	0–10	IBOD = 96.67 − 7.0x
	10–30	IBOD = 38.9 − 1.23x
pH	2–5	IPH = 16.1 + 7.35x
	5–7.3	IPH = −47.61 + 20.09x
	7.3–10	IPH = 316.96 − 29.85x
	10–12	IPH = 96.17 − 8.0x
Faecal Coliform	10^3–10^5	Icoli = 42.33 − 7.75 log10 x
	> 10^5	Icoli = 2

8.2.3.2 Ecological Water Quality Assessment

(a) **Organic Pollution Index (OPI)**

The OPI is a method for calculating the level of organic contamination in water bodies, particularly in rivers, lakes, and other aquatic ecosystems. It enables a quantitative evaluation of the presence and consequences of organic pollutants in the water. The OPI takes into account a variety of attributes that are related to organic pollution, such as total organic carbon (TOC), BOD, COD, and other important indicators of organic contamination. In Eq. 8.5, it is specified what formula has been utilized (Sarkar and Islam 2020).

$$\mathrm{OPI} = \frac{\mathrm{BOD}}{\mathrm{BODs}} + \frac{\mathrm{COD}}{\mathrm{CODs}} + \frac{\mathrm{Nitrate}}{\mathrm{Nitrates}} + \frac{\mathrm{Phosphate}}{\mathrm{Phosphates}} \tag{8.5}$$

The denominators are the standard permissible concentrations. On the OPI scale, 0 indicates excellent, 1 indicates good, 1 indicates the start of contamination, 2 indicates considerable contamination, 3 indicates very contaminated, and 4 indicates extremely high contamination.

(b) **Eutrophication Index**

The Eutrophication index takes into account multiple parameters that indicate the level of eutrophication, such as nutrient concentrations, chlorophyll-a concentration (an indicator of algal biomass), and water clarity (measured by Secchi depth). Therefore, the level of eutrophication has been computed using Eq. 8.6 (Sarkar and Islam 2020).

$$\mathrm{EI} = \frac{\mathrm{COD} \times \mathrm{DIN} \times \mathrm{DIP}}{4500} \times 10^6 \tag{8.6}$$

where the dissolved inorganic N (DIN) represents the concentration of NO_{3-} N and NO_{2-} N and dissolved inorganic phosphate (PO_4^{3-}) denotes DIP. The EI has a critical threshold at which EI1 denotes the absence of eutrophication in water and EI > 1 denotes its presence.

8.2.3.3 Multivariate Statistical Techniques

(a) **Cluster Analysis (CA)**

Cluster analysis is used to categorize objects into meaningful groups that include related but different objects (Alves et al. 2018). To cluster similar objects into higher clusters, a common technique known as hierarchical clustering has been considered. In this clustering process, Euclidean distance is utilized for comparing the similarity of samples. The process is continued until all samples form a single cluster, which is represented by a Dendogram or tree diagram (Ogwueleka 2015).

Clustering the sampling sites using discriminant analysis (DA) provides a useful framework for understanding spatial patterns of river water pollution (Duan et al. 2016). It helps identify areas that share similar water quality profiles, which can assist in targeted monitoring efforts, prioritizing pollution control measures, and implementing site-specific management strategies. Therefore DA has been used to identify similarities and differences in water quality parameters among different sites of the selected rivers and classify them into distinct clusters.

(b) **Principal Component Analysis (PCA)**

In the field of multivariate statistical analysis, PCA is an innovative technique that is widely employed to minimize the dimension of enormous sets of data with minimum original information lost (Ahmed et al. 2017). In a technical sense, it may be stated that this method extracts the eigenvalues and eigenvectors from the covariance matrix of the original variables. The original correlated data are multiplied by the eigenvectors in this process to produce uncorrelated (orthogonal) variables, often known as PCs. The PCs are produced by linearly combining eigenvectors, a set of coefficients (loadings or weightings), and the original variables. Additionally, eigenvalues are essential in illustrating the significance of PCs Ebqa'ai and Ibrahim (2017). The most important PCs are those with the greatest eigenvalues. Equation 8.7 is a mathematical equation that represents the PCA method.

$$Z_{ij} = \alpha_{i1}X_{1j} + \alpha_{i2}X_{2j} + \alpha_{i3}X_{3j} + \ldots \alpha_{im}X_{mj} \quad (8.7)$$

In the Eq. 8.7, X stands for the measured value of the variables, Z for the component score, for the component loading, i for the component number, j for the sample number, and m for the total number of variables.

8.3 Results and Discussion

8.3.1 Water Quality Parameters and Their Spatio-Temporality

The statistical variations of the selected water quality parameters of all the selected monitoring stations of the three rivers are presented in Fig. 8.2a–j. Regarding pH, the samples exhibits the normal range and remain within the desirable limit as per the guideline of the BIS. The average EC are 624 µs/cm for the Bhagirathi River, 381.14 µs/cm for the Jalangi River, and 551.75 for the Mathabhanga-Churni River. Therefore, the average EC have crossed the BIS reference limit (300 µs/cm) with the highest observed in the Bhagirathi River. The same is noticed for the value of TDS. The average values of TDS are found as 624 mg l^{-1} for the Bhagirathi River, 381.14 mg l^{-1} for the Jalangi River, and 551.75 mgl^{-1} for the Mathabhanga-Churni River. Therefore, like EC all the TDS are not in the desirable limit (300 µs/cm by BIS). Besides, BOD, DO, and TA reveal the contrasting character.

DO has been the most concerned parameter for the Janalgi and Mathabhanga-Churni Rivers. Even critically low concentrations (1–3 mg l^{-1}) are frequently observed for the Mathabhanga-Churni River which during pre-monsoon season induced by industrial pollution. However, the concentration of DO in the Bhagirathi River is relatively high for all time where the average of it is recorded as > 5 mg l^{-1}. Regarding BOD, all the average values have been recorded as high for the Janalgi and Mathabhanga-Churni Rivers and relatively low for the Bhagirathi River. Moreover, TA, TSS, TH, chloride, and sulphate exhibit high concentration and not desirable for general uses.

Three unique groupings, or clusters, are shown by the cluster analysis carried out on the observed basic water quality parameters. BOD, DO, pH, sulphate, chloride, and TSS are all present in

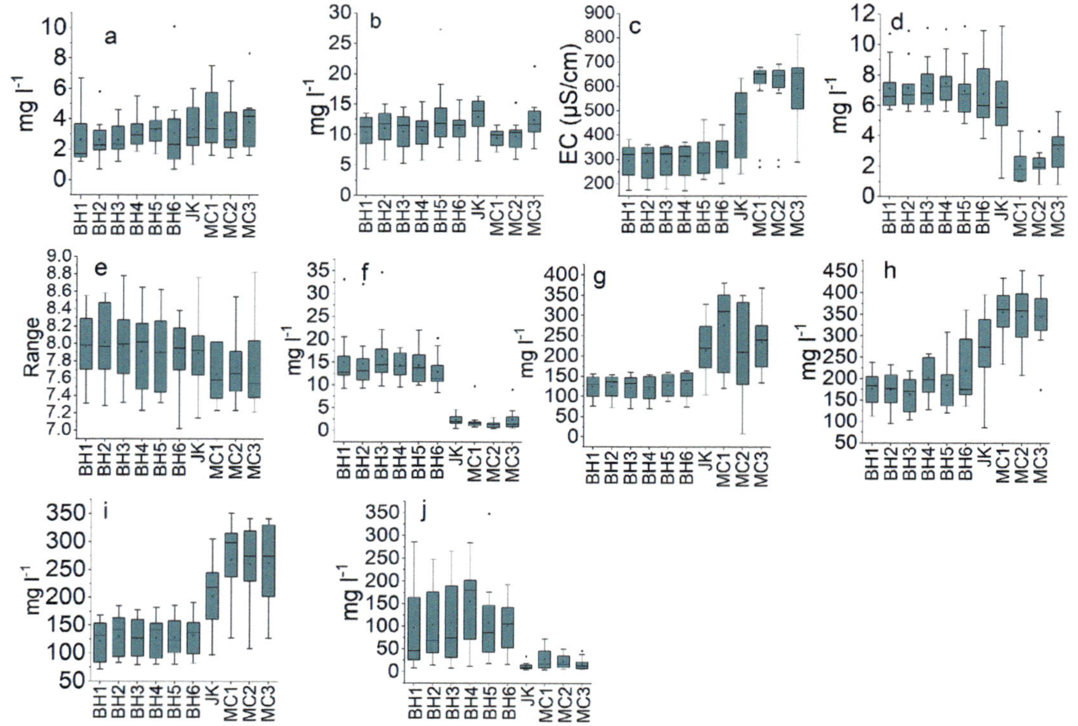

Fig. 8.2 Box plots of the selected water quality parameters **a** BOD, **b** chloride, **c** EC, **d** DO, **e** pH, **f** sulphate, **g** total alkalinity, **h** TDS, **i** total hardness, **j** TSS

Cluster 1 (Fig. 8.3a). The interconnected relationship between them shows nutrient enrichment and organic water contamination. Total alkalinity, total hardness, and TDS are all found in Cluster 2, showing the presence of alkaline materials in water, particularly bicarbonate, carbonate, and hydroxide ions. Cluster analysis is also applied to gauge the similarity of sampling sites. The sampling locations that make up a given cluster share similarities concerning the parameters being studied. All of the water quality parameters that have been studied for the 10 sampling locations have been divided into three main groups. Cluster 1 grouped all the sampling sites selected for the Bhagirathi-Hooghly River (Fig. 8.3b). Besides, the sampling points located in the Mathabhanga-Churni River form a different cluster. It is interesting to note that the clusters are formed river specific. This is because the nature of the rivers are different in themes of discharge, i.e. the Bhagirathi river is dam regulates and experiences considerable discharge throughout the year whereas Jalangi and Mathabhanga-Churni is a non-perennial river and considerable discharge is available only for the monsoon season. Additionally, the statistical comparison of the selected water quality parameters among the three rivers reveals few observations. Specifically, no significant variation was found in BOD ($p = 0.49$), chloride ($p = 0.33$), and pH ($p = 0.44$) in all monitoring stations but the variation was significant in EC ($p = 4.31\text{E-}19$), DO ($p = 5.7\text{E-}19$), sulphate ($p = 1.18\text{E-}25$), TA ($p = 3.35\text{E-}11$), TDS ($p = 2.12\text{E-}18$), TH ($p = 7.8\text{E-}18$), and TSS ($p = 4.58\text{E-}08$).

8.3.2 WAWQI, VPI, and Their Spatio-Temporally

WAWQI calculated for the selected monitoring stations of the Bhagirathi-Hooghly, Jalangi, and Mathabhanga-Churni Rivers exhibit poor to very poor water quality. Specifically, the Bhagirathi-

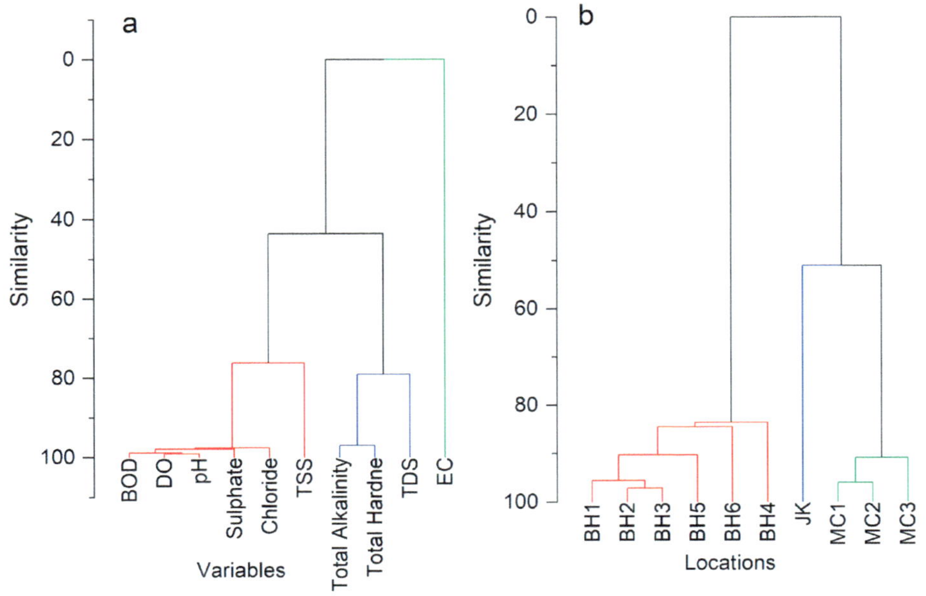

Fig. 8.3 Dendrogram representing the hierarchical clusters of **a** water quality parameters, and **b** sampling sites

Hooghli River is representing poor water quality and indicates its suitability for irrigation and industrial. However, it is also pertinent to mention here that the monitoring stations located at the upper stretch of the river are good and with increasing the distance downstream the quality of water is going to decrease. The Jalangi River is also exhibiting the same water quality. However, the worst water quality has been reported for the Mathabhanga-Churni River. Additionally, a contrasting pattern has been observed for this river where the upper stretch is presenting poor water quality and the lower stretch is better.

Regarding the QWI, no water sample has been recorded in the range 0–25 indicating excellent and 26–50 indicating good water quality. The range for WQI is recorded from 57.13 to 73.79 and all the six sampling sites exhibit the 51–75 category indicating poor water quality and suitable for irrigation and industrial purpose

Table 8.4 Water quality of the selected sites according to WAWQI and VPI

Rivers	Locations	Values of WAWQI	Water quality status	Possible uses	Values of VPI	Water quality status
Bhagirathi-Hooghli	Gorabazar	63.71	Poor	Irrigation and industrial	63.28	Good
	Khagra	64.30			61.60	Moderate to good
	Berhampore	64.78			61.66	
	Nabadwip	66.11			63.62	Good
	Tribeni	68.94			61.65	Moderate to good
	Serampore	67.99			60.35	
Jalangi river	Krishnanagar	73.55			56.73	
Mathabhanga-Churni	Gobindapur	90.64	Very poor	Irrigation	38.84	Bad
	Majhdia	86.63			39.43	
	Ranaghat	86.24			43.51	

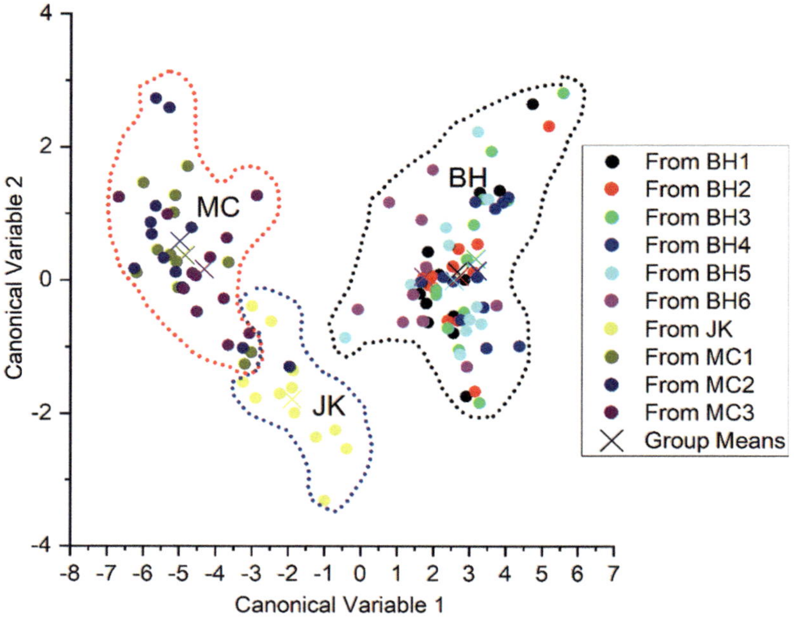

Fig. 8.4 Discriminant analysis on the sampling sites based on the water quality water on two canonical axes representing WAWQI and their sampling sites

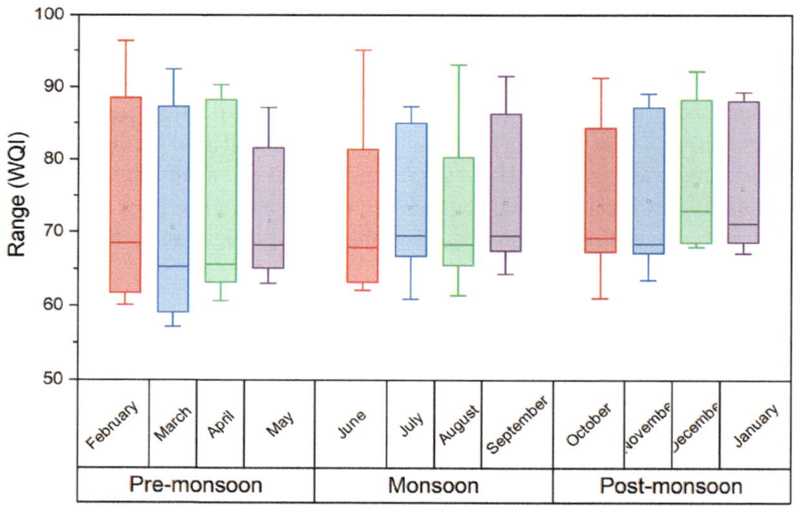

Fig. 8.5 Seasonal variation of WAWQI of the selected rivers

(Table 8.4). For the Jalangi River, the average WQI is found as 75.55 indicating very poor water quality and 41% of water quality are falling in this category. However, the rest of the samples exhibit the 51–75 category indicating poor water quality. Regarding the Mathabhanga-Churni River, the range for WQI is recorded from 81.33 to 96.36 and all the values come under the 57–100 category indicating very poor water quality (Fig. 8.4).

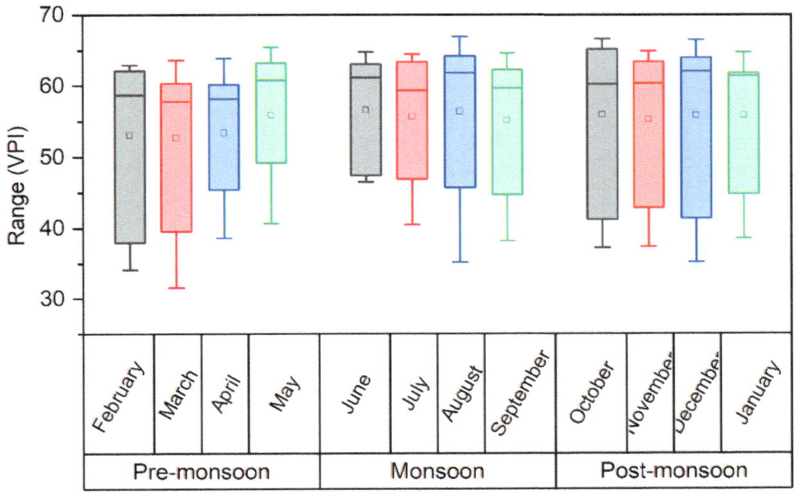

Fig. 8.6 Seasonal variation of VPI of the selected rivers

Regarding VPI, the range for the Bhagirathi River is recorded from 55.48–66.85 therefore no samples are portraying bad and very bad water quality. Additionally, 56.94% of water samples are falling in the category of medium and the rest of the water samples are portraying good water quality. For the Jalangi River, the range is observed as: 48.23–62.88 where 25% of samples are bad, and the rest of the samples are medium water quality. Regarding the Mathabhanga-Churni River 27% of water samples fall in the < 38 categories indicating very bad water quality and the rest of the samples are representing very bad water quality (Table 8.4). Discriminant analysis has also been applied to WQI to gauge the similarity of sampling sites which portrays the similarity with the clusters formed based on the water quality parameters. The locations chosen for sampling along the Bhagirathi-Hooghly River are clustered together in Cluster 1. The sample sites along the Mathabhanga-Churni River also constitute a distinct cluster. It's noteworthy to observe that the clusters are generated in particular rivers due to the discharge variations. Moreover, seasonality has also been observed for both indices (Figs. 8.5 and 8.6). Where the pre-monsoon season is observed better in comparison with the other seasons.

8.3.3 Ecological Water Quality and Their Spatio-Temporality

The average OPI 2.61, 2.98, and 3.41 are recorded for the Bhagirathi-Hooghly, Jalangi, and Mathabhanga-Churni Rivers, respectively. Regarding the Bhagirathi-Hooghly River OPI ranges from 2.32 (Gorabazar) to 2.89 (Nabadwip) and hence indicating somewhat polluted by organisms. Even, no sample has been recorded to represent waster as excellent or extremely polluted. In the Bhagirathi-Hooghly River 4.80, 54.57, and 42.53 percent are recognized to be good (class 1), beginning to be contaminated (class 2), lightly polluted (class 3), and polluted (class 4), respectively. Besides, a correlation has been drawn between the OIP and its factors where COD and BOD are highly correlated (for example COD: r = 0.92, BOD: r = 0.98 at Khagra; COD: r = 0.84, BOD: r = 0.95 at Serampore).

Regarding the Jalangi River, 4.80, 30.57, and 62.53 percent are found in class 1, 2, 3, and 4, respectively. The correlations exhibit that OPI is highly correlated with COD (0.92) and BOD (0.98) at Krishnanagar. Regarding the Mathabhanga-Churni River, among the total samples 4.80, 44.57, 32.53, 10.34, 7.22 percent

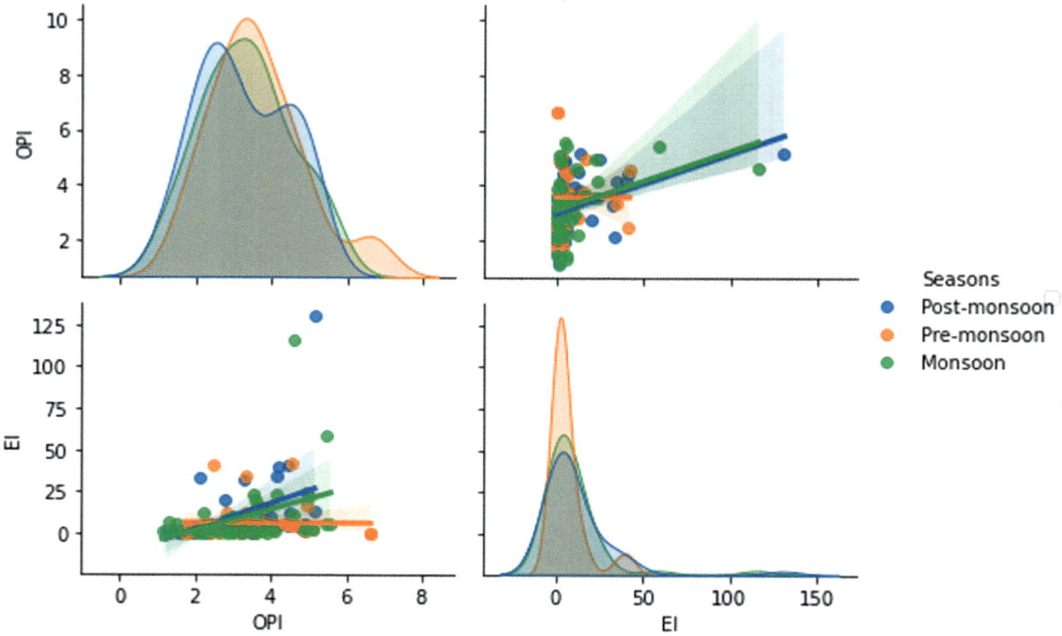

Fig. 8.7 Correlation between OPI and EI of different seasons

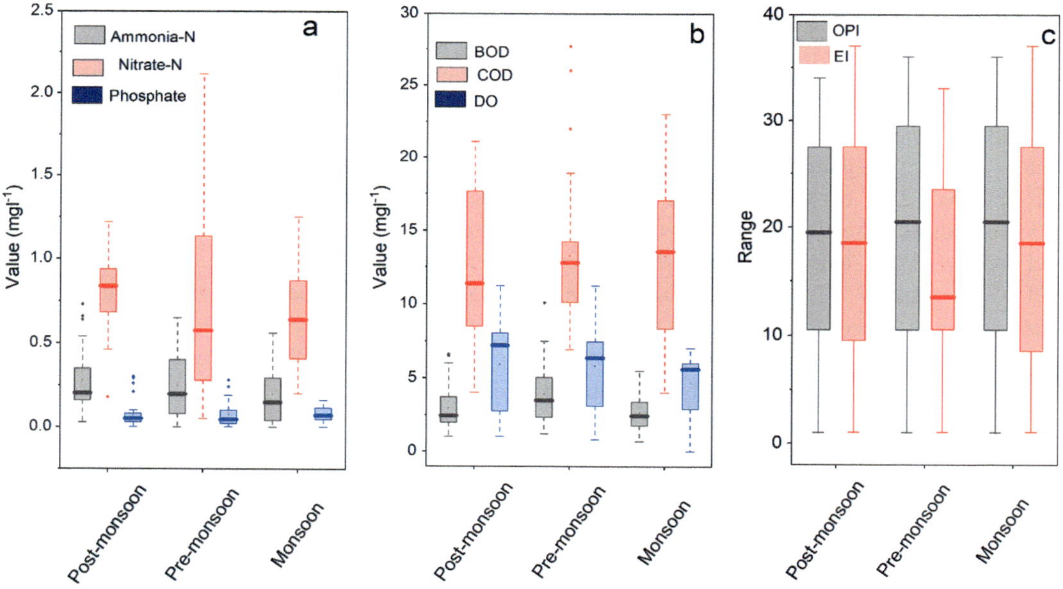

Fig. 8.8 Seasonal variation of parameters involving ecological water quality and related indices. **a** Ammonia-N, Nitrate-N, Phosphate, **b** BOD, COD, DO, **c** OPI, EI

are found in in class 1, 2, 3, and 4, respectively. The EI for all rivers exceeds the critical value 1 representing eutrophied condition. A strong correlation has also been observed between OPI and EI (Fig. 8.7). Additionally, the seasonality of the ecological water quality parameters and their related indices have also been observed for all the rivers. Among the three seasons pre-monsoon season portrays relatively poor water quality for ecological health (Fig. 8.8a–c).

8.3.4 Identifying the Factors of Pollution Through PCA

PCA algorithm has been used to both data sets. One data set represents the Bhagirathi River, and the other combines the Jalangi and Mathabhanga-Churni Rivers. Regarding PCA of the Bhagirathi River, two principal components (PC) explained the variance 63.04% of the original data. The PC loading plot exhibits high positive loadings on BOD, TDS and chloride, in PC1 and TA and EC in PC2. Contrastingly, high negative loading was observed for pH and Sulphate in PC1 and DO and TSS in PC2 (Table 8.5). Similarly, for the Jalangi and Mathabhanga-Churni River, PC1 and PC2 explained 64.31% variance. The loading plot portrays that DO, chloride, EC have high positive loadings in PC1, and TDS, TSS, and total hardness in PC2. As a result, the high positive loadings of TDS, EC, TSS, and chloride correspond to the active role of agricultural and urban runoff. Besides, the low negative loading of DO is associated with nutrient pollution from agricultural sources as well as organic pollution from sewage and industrial wastewater that depletes oxygen (Fig. 8.9).

8.3.5 Natural and Anthropogenic Control of the Pollution

The drainage system of the Bengal Delta is significantly impacted by the three chosen rivers. However, they do exhibit a range of pollution characteristics as a result of the substantial variations in their hydro-morphological characteristics and human interventions in the river basin. The potential causes inducing river pollution have been outlined in the following sections.

The variation in water pollution in the selected river is due to their different discharge pattern. The Bhagirathi River is controlled since the Farakka Barrage was installed in the Ganga River in 1975 (Islam and Guchhait 2017; Rahman and Rahaman 2018). Since then, the river is maintained with considerable discharge to support ecological health and navigability, even in the dry period. Moreover, during the monsoon period, the river is fed by its eastern (Janangi and Mathabhanga-Churni) western tributaries like Ajay and Damodar (Guchhait et al. 2016). Therefore, the massive discharge induces the

Table 8.5 PCA component plots for the Bhagirathi-Hooghly River and the Jalangi and Mathabhanga-Churni Rivers

Parameters	Components for Bhagirathi-Hooghly		Components for Jalangi and Mathabhanga-Churni	
	1	2	1	2
BOD	0.907	0.191	0.581	0.814
Chloride	0.285	0.395	0.940	−0.342
EC	0.541	0.812	0.994	0.111
DO	−0.013	−0.984	0.936	−0.352
pH	−0.961	0.008	0.039	−0.999
Sulphate	−0.773	−0.486	0.867	0.498
TA	0.296	0.955	−0.929	0.370
TDS	0.829	0.323	−0.413	0.911
Total hardness	0.371	0.541	−0.035	0.999
TSS	0.611	−0.774	−0.457	0.890

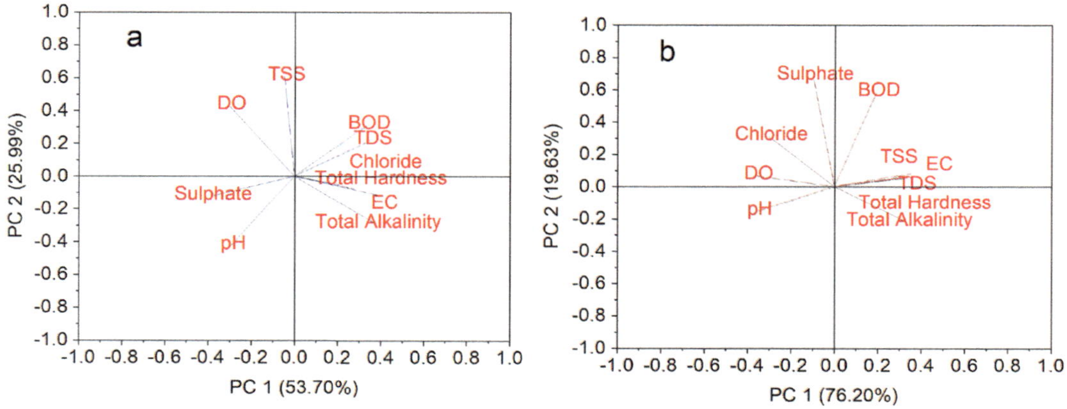

Fig. 8.9 PCA loading plot **a** for the samples located on the Bhagirathi-Hooghly River, **b** the Jalangi and Mathabhang-Churni Rivers

auto-purification capacity. Besides, the Jalangi and the Mathabhanga-Churni River are moribund rivers. Both the river is being fed by the Ganga River only in the rainy season and they remain disconnected from the Ganga River in the remaining months. Therefore, the rivers face an acute problem of water flow in the dry season, and the rivers loss their auto-purification capacity leaving them polluted rivers (Chakrabarty and Das2006; Sarkar and Islam 2021). The phenomenon that the improved river water is found during the monsoon season and it is being deteriorated when the discharge is critical during the dry months (Sarkar et al. 2021).

Like the natural causes, anthropogenic interventions within the river basin are also important for controlling the quality of the river water and ecosystem. The anthropogenic sources of pollution can be investigated from point and non-point sources of pollution. Regarding the point source of pollution, Industrial discharge, and Municipal Sewage are playing an important role for all three rivers while agricultural runoff and urban runoff are important non-point sources of pollution.

The Bhagirathi is highly affected by the industrial activities along its banks releasing untreated or poorly treated wastewater, laden with heavy metals, toxic chemicals, and organic compounds (Mitra et al. 2018; Samanta et al. 2017). This industrial pollution not only contaminates the river but also poses a threat to the diverse aquatic life and fragile ecosystems that rely on its waters. On the left bank 34 of the 54 drains are identified adding 1179 MLD of wastewater to the Ganga River, while 20 are placed on the right bank adding 600 MLD (Fig. 8.10) (CPCB 2013). The drains on the left bank may be divided into 10 zones, their combined daily BOD contribution is 74624 kg. Which contributes (74%) 55,443 kg per day from the Kolkata area. There are two cities after Kolkata: Kamarhati (6310 kg/day, or 8%) and Khardaha (2330 kg/day, or 3%). Other zones' contributions are little and distributed quite equally. The drains on the right bank may be divided into six zones, including Hoogli, Serampur, Rishra, Bally, Howrah, and Uluberia, contributing a total of 22,514 kg/day of BOD. Significant correlations between the discharge of wastewater and their contribution to BOD have also been observed for both banks (Fig. 8.11). Most important drains (draining BOD of more than 1000 kg per day) include the Hasting Ghat Road Canal, Najer Ganj Khal, Bagh Khal, Chhatra Canal, etc. Domestic sewage, another significant contributor, introduces pathogens, nutrients, and various contaminants into the river, endangering both human health and the well-being of aquatic organisms. Improper solid waste disposal compounds the issue, as plastic waste and other debris accumulate along the river banks, degrading the water quality and harming

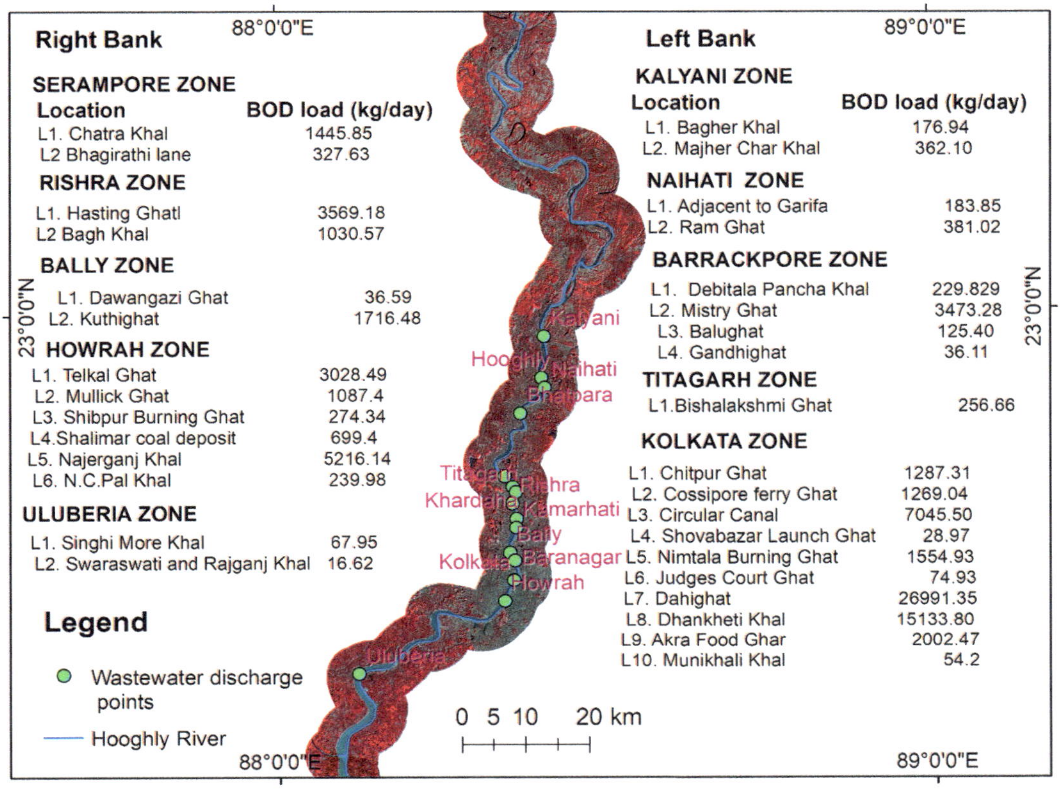

Fig. 8.10 Point source of pollution in the selected stretch (from Kalyani to Uluberia) of the Hooghly River

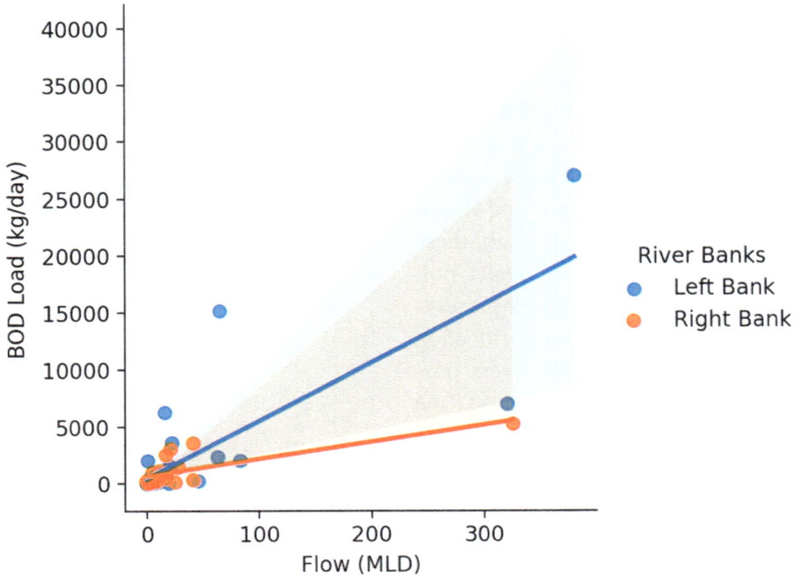

Fig. 8.11 Correlation between the discharge of wastewater and their contribution to BOD load measured in (kg/day)

the delicate aquatic ecosystems (Kanuri et al. 2020). However, like the Ganga River, the Bhagirathi River possesses a natural auto-purification capacity that helps mitigate the effects of pollution (Meher et al. 2015). Through various physical, chemical, and biological processes, the river can restore its water quality to some extent. Dilution plays a role as the river flow disperses and dilutes pollutants, reducing their concentration. Sedimentation allows heavier particles and some pollutants to settle down. Sunlight contributes to the breakdown of pollutants through photolysis, while biological degradation by microorganisms and algae helps break down organic pollutants. Additionally, the river's movement and oxygenation support the survival of aquatic organisms and aid in the decomposition of organic matter (Dutta and Choudhury 2021). However, it is important to note that the auto-purification capacity of the Bhagirathi River has limitations. To ensure its long-term health, it is essential to address pollution sources and implement effective pollution control measures.

On the other hand, untreated sewage and wastewater from urban areas, including residential and commercial sources, are discharged into the Churni River, contributing to pollution (Panigrahi and Bakshi 2014). The investigation identifies 9 pollution generating canalsin the Ranaghat Municipality (Fig. 8.12, Table 8.6) (Bakshi et al. 2016). Sreenathpur (2.799 MLD) and Basko Khal (3.638 MLD) are the two drains that contribute the most to the sewage water's high BOD (48 mg l^{-1} for Sreenathpur and Basko Khal, respectively) (Table 8.5). As a result, these outfalls have led to higher levels of water pollution. The pollution of the Churni River presents a challenge to its natural auto-purification capacity (Sarkar et al. 2021). While rivers possess inherent mechanisms to self-cleanse, the pollution levels in the Churni River often exceed their natural capacity for restoration. The high influx of industrial discharges, untreated sewage, and agricultural runoff overwhelms the river's ability to dilute and disperse pollutants (Sanyal et al. 2015). The excess nutrient load, toxic chemicals, and heavy metals in the water hinder

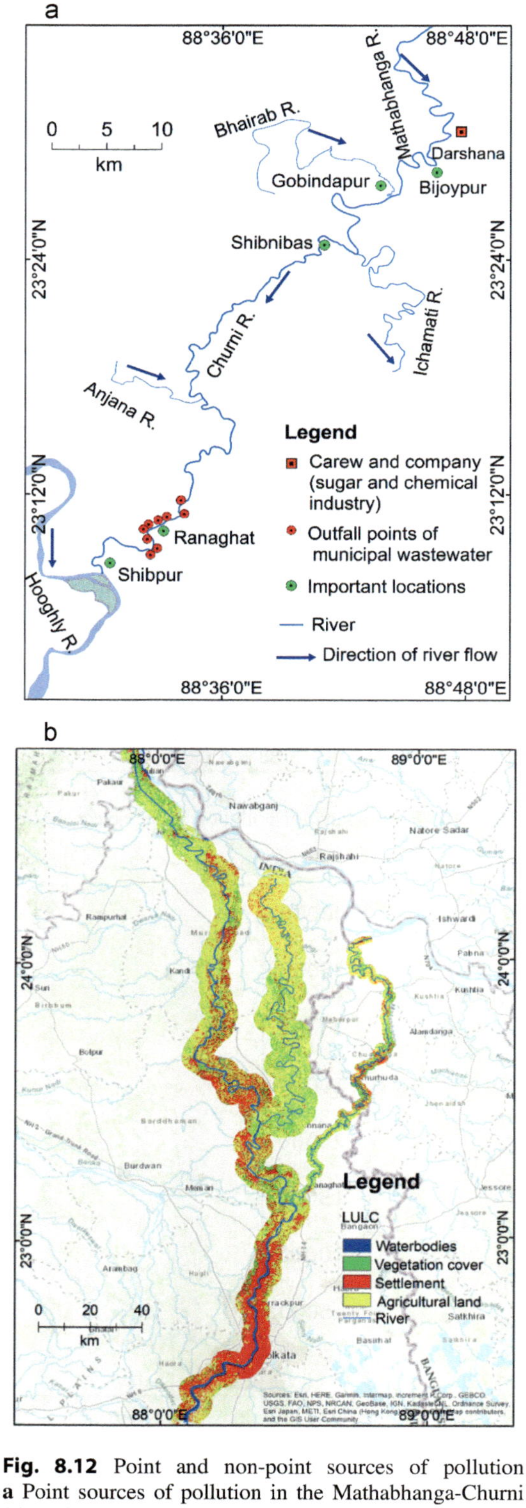

Fig. 8.12 Point and non-point sources of pollution **a** Point sources of pollution in the Mathabhanga-Churni River. **b** Impact of LULC on the river water quality

the effectiveness of sedimentation, photolysis, and biological degradation processes (Panigrahi and Pattnaik 2021). Additionally, the accumulation of solid waste along the riverbanks further hampers the self-purification ability. As a result, the Churni River struggles to maintain water quality and ecological balance (Das and Chakrabarty 2007). To combat this problem, it is crucial to address pollution sources and implement effective pollution control measures to reduce the burden on the river's auto-purification capacity and restore the health of the Churni River.

Nonpoint sources of pollution in the Bhagirathi-Hooghly River in West Bengal, India, can arise from various activities. Agricultural runoff contributes to pollution through the release of pesticides, fertilizers, and sediment into the river. Urban areas also play a role as stormwater runoff carries pollutants like oils, heavy metals, and debris from roads and rooftops into the river. Deforestation and improper land management practices accelerate soil erosion, leading to sedimentation in the river. Industrial activities near the river can introduce pollutants through runoff from storage areas and construction sites. Open waste disposal and inadequate landfill management contribute to pollution when leachate reaches the river. The high nutrient content from agricultural runoff triggers eutrophication, causing oxygen depletion and disrupting the delicate balance of the ecosystem.

8.4 Conclusion

The present research on river water quality assessment has provided important insights into understanding the water quality, factors of pollution, and health of river ecosystems. The study has shown that various parameters such as dissolved oxygen levels, BOD, COD, and nutrient concentrations (nitrate, and phosphate), play a crucial role in assessing river water quality. The discharge variability as a natural factor plays a significant role in changing the pollution concentration. Besides, the impacts of human activities, such as industrial discharges, agricultural runoff, and

Table 8.6 Outfall points of the municipal wastewater in the Mathabhanga-Churni River and pollution of water

Landmark of drain/outfall	Locational coordinate		Dimension (mm)		Water depth (mm)	Velocity (M/Sec)	Current flow	Projected flow (2035)	BOD (mg l^{-1})	Faecal coliforms (MPN/100 mL)
	North	East	Width	Depth			MLD	MLD		
Chaitanya Ghat	23°10'58.94"	88°33'58.08"	2500	1750	247	0.068	0.518	0.682	59.5	300,000
Radharani Seba Sadan	23°10'48.49"	88°33'50.48"	1000	1650	15	0.4	0.333	0.497	47.82	1,300,000
Nadia Samaj Pally	23°10'48.45"	88°33'50.49"	600	600	15	0.429	1.296	1.46	54.24	210,000
Chhoto Najar	23°10'44.41"	88°33'43.77"	1200	1350	25	0.5	0.091	0.255	71.3	190,000
Churipara	23°10'39.04"	88°33'33.10"	350	600	15	0.2	0.822	0.986	70	160,000
Thanapara–I	23°10'38.83"	88°33'17.64"	900	1500	141	0.075	0.187	0.351	92	220,000
Thanapara–II	23°10'36.21"	88°33'15.64"	450	1200	12	0.4	0.456	0.62	58	140,000
Sreenathpur	23°10'25.03"	88°33'12.07"	600	1350	22	0.4	2.799	2.963	48	900,000
Basko Khal	23°11'2.63"	88°33'54.96"	1800	1500	36	0.5	3.638	3.802	68	230,000

Source CPCB, 2020

urbanization, on river ecosystems have been highlighted. The discharge of wastewater from the industries and urban centres located near the river has contaminated the river. However, the scenario is cruel for the lower part of the Hooghly River and the upper stretch of the Mathabhanga-Churni River. The findings emphasize the importance of implementing effective management strategies, including pollution control measures, restoration efforts, and the adoption of sustainable practices, to protect and restore the ecological integrity of rivers. Continued research in this field is essential for developing targeted and science-based approaches to preserve the quality of river water and ensure the sustainability of these vital freshwater resources.

References

Abbasi T, Abbasi SA (2012) Water quality indices. Elsevier

Ahmed F, Fakhruddin ANM, Imam MT, Khan N, Abdullah ATM, Khan TA, Rahman MM, Uddin MN (2017) Assessment of roadside surface water quality of Savar, Dhaka, Bangladesh using GIS and multivariate statistical techniques. Appl Water Sci 7:3511–3525

Allison M, Kepple E (2001) Modern sediment supply to the lower delta plain of the Ganges Brahmaputra River in Bangladesh. Geo-Mar Lett 21(2):66–74

Alves RI, Machado CS, Beda CF, Fregonesi BM, Nadal M, Sierra J, Domingo JL, Segura-Muñoz SI (2018) Water quality assessment of the Pardo River Basin, Brazil: a multivariate approach using limnological parameters, metal concentrations and indicator bacteria. Arch Environ Contam Toxicol 75(2):199–212

Bakshi A, Panigrahi A, Mondal A (2016) Seasonal variation of aquatic diversity of a lotic ecosystem–a case study of River Churni, West Bengal. Int J Appl Pune Sci Agric 2(7):9–16

Bhutiani R, Kulkarni DB, Khanna DR, Gautam, A. (2016) Water quality, pollution source apportionment and health risk assessment of heavy metals in groundwater of an industrial area in North India. Expo Health 8:3–18

Bora M, Goswami DC (2017a) Water quality assessment in terms of water quality index (WQI): case study of the Kolong River, Assam, India. Appl Water Sci 7:3125–3135

Bora M, Goswami DC (2017b) Water quality assessment in terms of water quality index (WQI): case study of the Kolong River, Assam, India. Appl Water Sci 7 (6):3125–3135

Boyacioglu H, Boyacioglu H (2008) Water pollution sources assessment by multivariate statistical methods in the Tahtali Basin, Turkey. Environ Geol 54:275–282

Brown RM, McClelland NI, Deininger RA, O'Connor MF (1972) A water quality index—crashing the psychological barrier. In: Indicators of environmental quality: proceedings of a symposium held during the AAAS meeting in Philadelphia, Pennsylvania, December 26–31, 1971. Springer US, pp 173–182

Chakrabarty D, Das SK (2006) Fish community structure and ecological degradation in tropical rivers of India. Web Ecol 6(1):27–36

Chapman GP, Rudra K (2015) Time streams/history and rivers in Bengal. Centre for Archaeological Studies & Training, Eastern India

CPCB (2013) Pollution assessment: River Ganga. Central pollution control board (Ministry of Environment and Forests, Govt. of India) Parivesh Bhawan, East Arjun Nagar, Delhi, 110032. www.cpcb.nic.in

Cui L, Wang X, Li J, Gao X, Zhang J, Liu Z (2021) Ecological and health risk assessments and water quality criteria of heavy metals in the Haihe River. Environ Pollut 290:117971

Das BC (2015) Modeling of most efficient channel form: a quantitative approach. Model Earth Syst Environ 1 (3):15

Das SK, Chakrabarty D (2007) The use of fish community structure as a measure of ecological degradation: a case study in two tropical rivers of India. Biosystems 90(1):188–196

Ding J, Jiang Y, Liu Q, Hou Z, Liao J, Fu L, Peng Q (2016) Influences of the land use pattern on water quality in low-order streams of the Dongjiang River basin, China: a multi-scale analysis. Sci Total Environ 551:205–216

Duan W, He B, Nover D, Yang G, Chen W, Meng H, ... & Liu C (2016) Water quality assessment and pollution source identification of the eastern Poyang Lake Basin using multivariate statistical methods. Sustain 8(2):133

Dutta S, Choudhury AK (2021) An assessment of the temporal alterations in the trophic status and habitat heterogeneity of the anthropogenically influenced Bhagirathi-Hooghly estuary in reference to phytoplankton community and environmental variables. Environ Sci Pollut Res 28(35):48681–48705

Ebqa'ai M, Ibrahim B (2017) Application of multivariate statistical analysis in the pollution and health risk of traffic-related heavy metals. Environ Geochem Health 39:1441–1456

Goodberd S (2003) Response of the ganges dispersal system to climate change: a source-to sink-view since the last interstade. Sediment Geol 162:83–104

Guchhait SK, Islam A, Ghosh S, Das BC, Maji NK (2016) Role of hydrological regime and floodplain sediments in channel instability of the Bhagirathi River, Ganga-Brahmaputra Delta, India. Phys Geogr 37(6):476–510

Haque MR, Ali MM, Ahmed W, Siddique MAB, Akbor MA, Islam MS, Rahman MM (2023) Assessment of microplastics pollution in aquatic species (fish, crab, and snail), water, and sediment from the Buriganga River, Bangladesh: an ecological risk appraisals. Sci Total Environ 857:159344

Islam A, Guchhait SK (2017) Analysing the influence of Farakka Barrage Project on channel dynamics and meander geometry of Bhagirathi river of West Bengal, India. Arabian J Geosci 10(11):245

Janoska O, Gruszecka-Kosowska A (2020) Water quality and human health risk assessment: a case study of the Czarna Przemsza river source in Zawiercie, Poland. Hum Ecol Risk Assess Int J 26(3):757–781

Jolly YN, Rakib MRJ, Kumar R, Sultana S, Rahman SMM, Kabir J, Akter S, Mamun KM, Fatema KJ, Mehnaz M, Pal P, Bhattacharya P (2023) Evaluation of surface water quality near pollution sources in Buriganga River and deciphering their probable emergence, ecological, and health risk aspects. Regional Stud Marine Sci 63:102988

Kanuri VV, Saha R, Raghuvanshi SP, Singh AK, Chakraborty BD, Kumar VK, Mohapatra SC, Vidyarthi AK, Sudhakar A, Saxena RC (2020) Sewage fluxes and seasonal dynamics of physicochemical characteristics of the Bhagirathi-Hooghly River from the lower stretch of River Ganges, India. Chem Ecol 36(1):30–47

Khatri N, Tyagi S, Rawtani D, Tharmavaram M, Kamboj RD (2020) Analysis and assessment of ground water quality in Satlasana Taluka, Mehsana district, Gujarat, India through application of water quality indices. Groundw Sustain Dev 10:100321

Kibena J, Nhapi I, Gumindoga W (2014) Assessing the relationship between water quality parameters and changes in landuse patterns in the Upper Manyame River, Zimbabwe. Phys Chem Earth, Parts A/B/C 67:153–163

Koukal B, Dominik J, Vignati D, Arpagaus P, Santiago S, Ouddane B, Benaabidate L (2004) Assessment of water quality and toxicity of polluted Rivers Fez and Sebou in the region of Fez (Morocco). Environ Pollut 131(1):163–172

Kuehl SA, Levy BM, Moore WS, Allison MA (1997) Subaqueous delta of the Ganges-Brahmaputra river system. Mar Geol 144(1–3):81–96

Kuehl SA, Allison MA, Goodbred SL, Kudrass HERMANN, Giosan L, Bhattacharya JP (2005) The Ganges-Brahmaputra Delta. Special Publ-SEPM 83:413

Medeiros AC, Faial KRF, Faial KDCF, da Silva Lopes ID, de Oliveira Lima M, Guimarães RM, Mendonça NM (2017) Quality index of the surface water of Amazonian rivers in industrial areas in Pará, Brazil. Marine Pollut Bull 123(1–2):156–164

Meher PK, Sharma P, Gautam YP, Kumar A, Mishra KP (2015) Evaluation of water quality of Ganges River using water quality index tool. Environ Asia 8(1)

Mirzaei Aminiyan M, Baalousha M, Mousavi R, Mirzaei Aminiyan F, Hosseini H, Heydariyan A (2018) The ecological risk, source identification, and pollution assessment of heavy metals in road dust: a case study in Rafsanjan, SE Iran. Environ Sci Pollut Res 25:13382–13395

Mitra S, Sarkar SK, Raja P, Biswas JK, Murugan K (2018) Dissolved trace elements in Hooghly (Ganges) River Estuary, India: risk assessment and implications for management. Mar Pollut Bull 133:402–414

Na M, Zhao Y, Rina S, Wang R, Liu X, Tong Z, Zhang J (2023) Residues, potential source and ecological risk assessment of polycyclic aromatic hydrocarbons (PAHs) in surface water of the East Liao River, Jilin Province, China. Sci Total Environ 886:163977

Ogwueleka TC (2015) Use of multivariate statistical techniques for the evaluation of temporal and spatial variations in water quality of the Kaduna River, Nigeria. Environ Monit Assess 187:1–17

Panigrahi AK, Bakshi A (2014) A study on profile of fishing community of the river side villages of river Churni, Nadia, West Bengal with special reference to socio-economic and technological appraisal of fishermen. Int J Res Appl Nat Soc Sci 2(3):97–102

Panigrahi AK, Pattnaik S (2021) Studies to estimate the pollution load of some South Bengal Rivers and its impacts: a survey. In: Proceedings of the zoological society, vol 74. Springer India, pp 83–90

Parua PK (2010) The ganga: water use in the Indian subcontinent (Vol. 64). Springer Science & Business Media

Rahman MM, Rahaman MM (2018) Impacts of Farakka barrage on hydrological flow of Ganges river and environment in Bangladesh. Sustain Water Res Manag 4:767–780

Samanta S, Amrutha K, Dalai TK, Kumar S (2017) Heavy metals in the Ganga (Hooghly) River estuary sediment column: evaluation of association, geochemical cycling and anthropogenic enrichment. Environ Earth Sci 76:1–14

Sanyal T, Kaviraj A, Saha S (2015) Deposition of chromium in aquatic ecosystem from effluents of handloom textile industries in Ranaghat-Fulia region of West Bengal, India. J Adv Res 6(6):995–1002

Sarkar B, Islam A (2019) Assessing the suitability of water for irrigation using major physical parameters and ion chemistry: a study of the Churni River, India. Arab J Geosci 12:1–16

Sarkar B, Islam A (2020) Drivers of water pollution and evaluating its ecological stress with special reference to macrovertebrates (fish community structure): a case of Churni River, India. Environ Monitor Assess 192 (1):45

Sarkar B, Islam A (2021) Assessing poverty and livelihood vulnerability of the fishing communities in the context of pollution of the Churni River, India. Environ Sci Pollut Res 1–24

Sarkar B, Islam A, Das BC (2021) Role of declining discharge and water pollution on habitat suitability of fish community in the Mathabhanga-Churni River, India. J Clean Prod 326:129426

Shammi M, Rahman MM, Islam MA, Bodrud-Doza M, Zahid A, Akter Y, Quaiyum S, Kurasaki M (2017)

Spatio-temporal assessment and trend analysis of surface water salinity in the coastal region of Bangladesh. Environ Sci Pollut Res 24:14273–14290
Sharma M, Kansal A, Jain S, Sharma P (2015) Application of multivariate statistical techniques in determining the spatial temporal water quality variation of Ganga and Yamuna Rivers present in Uttarakhand State, India. Water Qual Expo Health 7:567–581
Rudra K (2014) Changing river courses in the western part of the Ganga-Brahmaputra delta. Geomorphology 227:87–100
Rudra K (2018) Rivers of the Ganga–Brahmaputra–Meghna Delta: an overview. In: Rivers of the Ganga-Brahmaputra-Meghna Delta. Springer, Cham, pp 1–14

9 Water Quality Assessment by Using Landsat Images in Urban Wetland: A Case Study of Deepor Beel, Assam

Rajib Tarani Das, Mrinalendra Narayan Dutta, and Shukla Acharjee

Abstract

The health of Deepor Beel in Assam has declined since the unprecedented rapid growth of the Guwahati Urban area in the past few decades. This urban water body has immense importance as it acts as a growth corridor of the different governmental projects and supports thousands of livelihoods around. To keep enduring the economic benefit and maintain the ecosystem of the lake, the lake's health has to be kept nontoxic. Hence frequent monitoring and evaluation of the water quality of the lake are essential. The present work attempts to analyze the water quality of the wetland with remote sensing techniques by monitoring the effect of urbanization on Deepor Beel. To determine the current state of the water quality in Deepor Beel, statistical studies were carried out on biological (Chl-a) and physical (turbidity, TDS, conductivity, pH) indicators. Linear and non-linear regression models between water quality parameters and reflectance of Landsat imagery were fitted based on band combinations. For 2004 and 2020 two different periods the result shows, in 2020 the turbidity and TDS in Deepor Beel have increased significantly (>50%) due to the direct influx of urban sewages to the lake's water. Adequate measures to reduce the non-point source of pollutant inflow by targeting the source areas (Waste dumping areas) near the city need to be priorities for stakeholders. Moreover, reducing the recession urbanization around the Deepor Beel and wetland effective management could be a key strategy for improving water quality in the lake.

Keywords

TDS · Chl-a · Conductivity · Turbidity · pH · Band

R. T. Das (✉) · S. Acharjee
Centre for Studies in Geography, Dibruragh University, Dibrugarh, Assam, India
e-mail: rajibtaranidas@gmail.com

M. N. Dutta
Department of Applied Geology, Dibrugarh University, Dibrugarh, Assam, India

9.1 Introduction

Wetland ecosystems are important natural landscapes for the survival of various endangered species of the biosphere are under serious pressure from a variety of stressors, including human activities like changes in land use, agricultural activities, pollution, and climate change. At present wetlands cover only 4–5% of the Earth's surface (Mitsch 2010), still, wetlands provide 30–40% of global ecosystem services (Zedler and Kercher 2005). Besides its vast ecological importance, in the last few decades wetlands

S. Talukdar et al. (eds.), *Environmental Risk and Resilience in the Changing World*, GIScience and Geo-environmental Modelling, https://doi.org/10.1007/978-3-031-62442-1_9

have recorded the fastest rate of loss than any other ecosystems on earth and it is estimated that 50% of world wetlands have been lost or degraded since 1980 (Cutter and Finch 2008) due to the anthropogenic pressures (Marti-Cardona et al. 2013; Prasad et al. 2002). Reclamation of wetland area by agricultural land expansion (Pal and Osoundu 2009), the extension of the urban area, mismanaged solid waste disposal and draining of untreated household and Industrial sewage to wetlands have brought a notable change in wetland's water quality and related ecological connectivity in wetland ecosystem (Pal 2015). Most of the wetlands including all surface waters across the world are impaired by mainly non-point source pollutants through run-off. This affects the water quality of wetlands along with the aquatic life in the surface waters such as a river, lakes. The situation is more devastating in developing countries like India, where the water quality management is not satisfactory and existing facilities become worse. Therefore one of the challenges in assessing the efficiency of management policy is to monitor the water quality and evaluate the trends in water quality data (Mozumder and Tripathi 2012). It has been observed that Deepor Beel is a freshwater wetland and the only Ramsar site in Assam, which has experienced rapid water quality degradation, and there is evidence of a continuing decline of areal extent over the last two decades (Mozumder and Tripathi 2014). Relentless wetland degradation and steady water quality dilapidation in Deepor Beel have posed challenges in environmental sustaining in the surrounding area of the lake (Bhattacharyya and Kapil 2009). Several laws have already been executed for the restoration of the Deepor Beel ecosystem; still, this valuable ecological heritage of Assam is in danger as there is a huge gap between policymaking and its ground-level implementation (MoEF 2008). Satellite imagery-based analysis shows in the last three decades trend of wetland conversion in the Deepor Beel area has been highly explicit. Previous studies by Bhattacharyya and Kapil (2009) have indicated that the lake water quality shows signs of degradation. Incessant adverse human activities like the reclamation of lands for agriculture, encroachment of Guwahati urban area for habitation purposes, cutting the sides of wetlands, pollution from various sources of waste dumping sites in the northeastern side, pressure from illegal overfishing by local communities, illegal poaching of migratory birds and animal, excessive fertilizer use in agricultural land creates eutrophication like situation, which strongly influenced the Deepor Beel area's ecosystem condition especially water quality (Bhattacharyya and Kapil 2009; Mozumder et al. 2014). The degradation of water quality in Deepor Beel has reduced its capacity to support aquatic life, and thus, influences sustainable utilization of the lake water resources (Dash and Kalamdhad 2021). The previous study further supports the facts that incidental occurrence and spread of water hyacinths in the lake since 1999 (After the extension of the North-Easternstern Frontier Railway network along the southern boundary of the lake) is attributed to the increased nutrient inflow and sediment concentration into the lake (Gohain and Bordoloi 2021). Studies specify that eutrophication from various anthropogenic drivers is the key cause for the fast spreading of water hyacinth over the lake (Bhattacharjya et al. 2022). To deal with such a problem it is required to carry out water quality assessment and management, in which the monitoring of water quality plays a significant role in restoring the water quality. The present paper attempted to investigate the water quality of Deepor Beel with selected criteria, make a comparison with the previous study, and investigate associated reasons behind such degradation in Deepor Beel. The Deepor Beel is selected because it is one of the largest freshwater lakes in the northeastern part of India and is increasingly impaired by pollution caused by anthropogenic causes. Therefore the main objective of this study is to analyze the water quality of the lake and to investigate the impacts of rapid urbanization on the water quality of the Deepor Beel. There are several ways to assess water quality; one method is conventional techniques that depend on in situ data or sequential laboratory analysis of the samples, which may give precise measurements but they are usually expensive, and time-consuming tasks (Mushtaq and Nee Lala 2017). Another drawback of this method is, it's the incapability to access a real-time spatial and temporal overview of water quality (Wang et al. 2006). But the indirect way of satellite imagery-based water quality assessment provides an effective approach for analysis of water quality and may offer an appropriate method to assimilate water quality data collected from in situ measurements (Chipman et al. 2009; Khattab and Merkel 2014). Remote Sensing techniques reduce labor-intensive and expensive in situ data process by providing better spatial and temporal coverage of environmental processes (Khattab and Merkel 2014). The spectral brightness of satellite imagery is correlated with water quality variables (i.e., turbidity, pH, chl-a, etc.) that affect a wetland's optical properties (Nas et al. 2009). Numbers of previous studies have shown good relationships between brightness data from satellite imagery and wetland's water quality parameters (Duan et al. 2007; Tyler et al. 2006). Regression analysis is widely used to validate relationships between the spectral reflectance measured by sensors and water quality variables collected in the wetlands. The main essence of regression analysis is to select a suitable regression technique and independent variables. An R^2 value approaching 0 indicates a poor correlation between the predicted and measured data whereas R^2 value 1 indicates vice-versa (Mushtaq and Nee Lala 2017).

9.2 Study Area

Deepor Beel, a freshwater wetland located (Fig. 9.2) about 10 km southwestern part of Guwahati city, is one of the largest and vital wetlands in Assam with the latitudinal extension of 26° 05′ N to 26° 11′ N and Longitudinal extension of 91° 35′ E to 91° 43′ E (Fig. 9.1). This type of wetland is prominent in the bio-geographic region of Burma Monsoon Forest (Das and Saikia 2011). It was recognized as one of the most important wetland systems in India under the Ramsar International Convention on wetlands (In 2002). It is assumed that the wetland is a discarded channel of the Brahmaputra River situated in a U-shaped valley between two cliffs on the north and south (MoEF 2008; Mozumder and Tripathi 2014). The wetland stands 4–5 m deep in the rainy season and up to about 1 m deep in the dry summer season, where the main sources of water to the wetland are monsoonal rainfall and some inflows from Kalman and Bharalu rivers to the wetland (Mozumder et al. 2014). Ecologically Deepor Beel acts as a lung of Guwahati city and economically approximate ten thousand population directly or indirectly reliant on this valuable wetland (Fig. 9.2).

9.3 Database and Methodology

Various data used here are in the tabulated form (Table 9.1).

9.3.1 Delineation of Wetland Boundary

Landsat satellite (LANDSAT TM and LANDSAT 8 OLI, path/row: 137/42,43; spatial resolution:30 m.) imagery of 2004 and 2020 have been collected from USGS and used for wetland areas. Landsat image bands 3–5 are usually used for categorizing waterbody extent, as these bands have the ability to differentiate land–water boundaries. Google Earth imagery (2020) and a Topographical map from the Survey of India (SOI) are used to prepare the base map and identify the wetland's areal extent.

- Normalized Differences Water Index (NDWI) was selected for wetland identification (McFeeters 1996) among various potential image base indices like WRI, NDVI, MNDWI etc., because of its maximum reflectance of a water body in the green band and it reduces the same in the near-infrared band, and therefore water is recognized from others object. NDWI is calculated using Eq. 9.1.

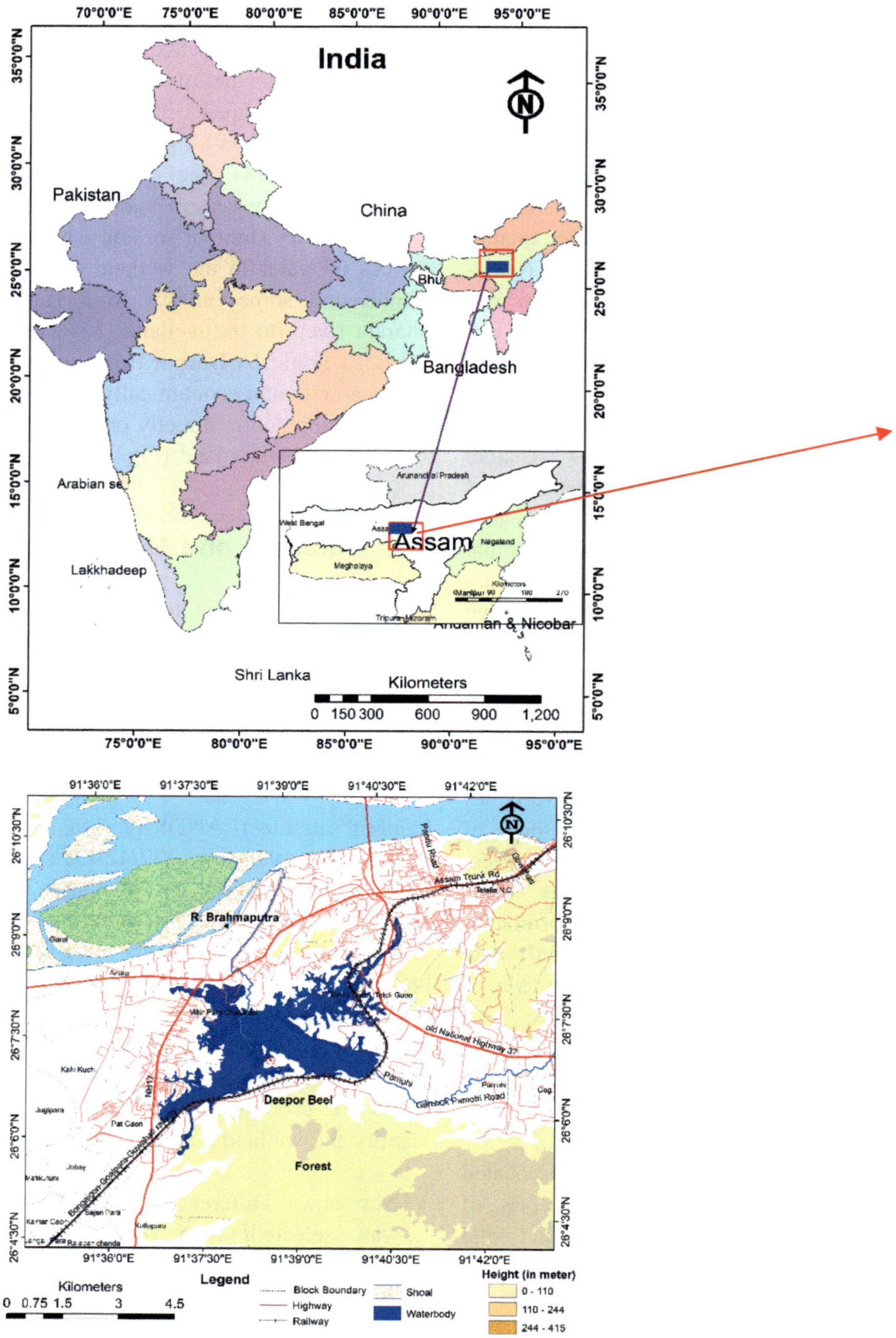

Fig. 9.1 Study area Deepor Beel

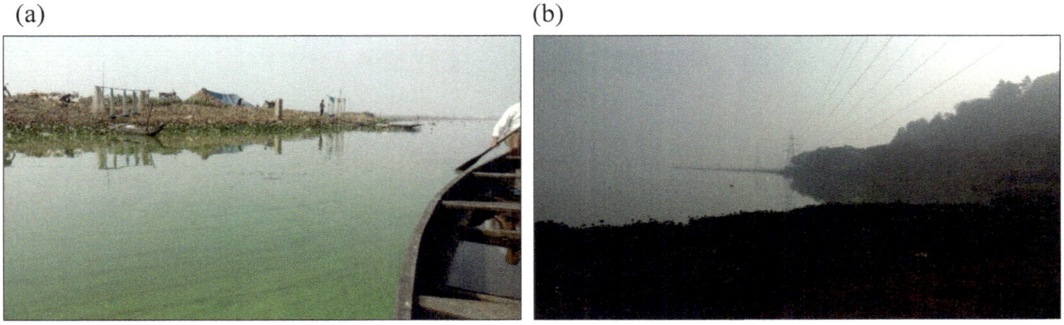

Fig. 9.2 General view of Deepor Beel: fishing in Deepor Beel (**a**) and water hyacinth in water (**b**)

Table 9.1 Data and its sources

S. No.	Data type	Sources	Date/time	Resolution	Category of data
1	Topographical map	Survey of India (SOI), University of Texas	1989–90, 1955	1:50000, 1:250000	Secondary
2	District resource map	Survey of India (SOI)		1:250000	Secondary
4	Google Earth imagery	Google Earth	2018	0.6 m	Secondary
5	Satellite imageries (landast MT & OLI TIRS)	glovis.usgs.gov	1988–2020	30 m	Secondary
6	SRTM DEM	urs.eosdis.nasa.gov.in	2000	30 m	Secondary
7	Water quality Data	Field survey and secondary sources	2020, 2004	–	Primary
8	Population data	Census of India	2001, 2011	–	Secondary
9	Litholog data ground water data		1996, 2017	–	Secondary
11	Wetland depth	Field survey	2018		Primary

$$\text{NDWI} = \frac{\text{Green} - \text{NIR band}}{\text{Green} + \text{NIR band}}. \quad (9.1)$$

where a water body is spotted when the value is positive (0–1) (Fig. 9.3).

9.3.2 Water Sampling and Laboratory Measurements

In situ water sample collected on 10th of January, 2020 from different 13 locations over clearwater area only and due to hiding true reflectance of water body, aquatic vegetation area was not taken into consideration. Sample locations were determined by a GPS receiver (Table 9.2). 13 sampling locations (Fig. 9.4) covered the different parts of the Deepor Beel with various ecological aspects like roads site, waste dumping ground, agricultural field area. pH (Digital pH meter) was analyzed in the field immediately after the collection of water. At least a 1000 ml water sample was collected at each sample site. All samples were analyzed within 24 h at Geography Lab, Guwahati University using standards procedures. The water quality parameter contains TDS, TSS, Chl-a, turbidity, and electric conductivity. The results of the analysis

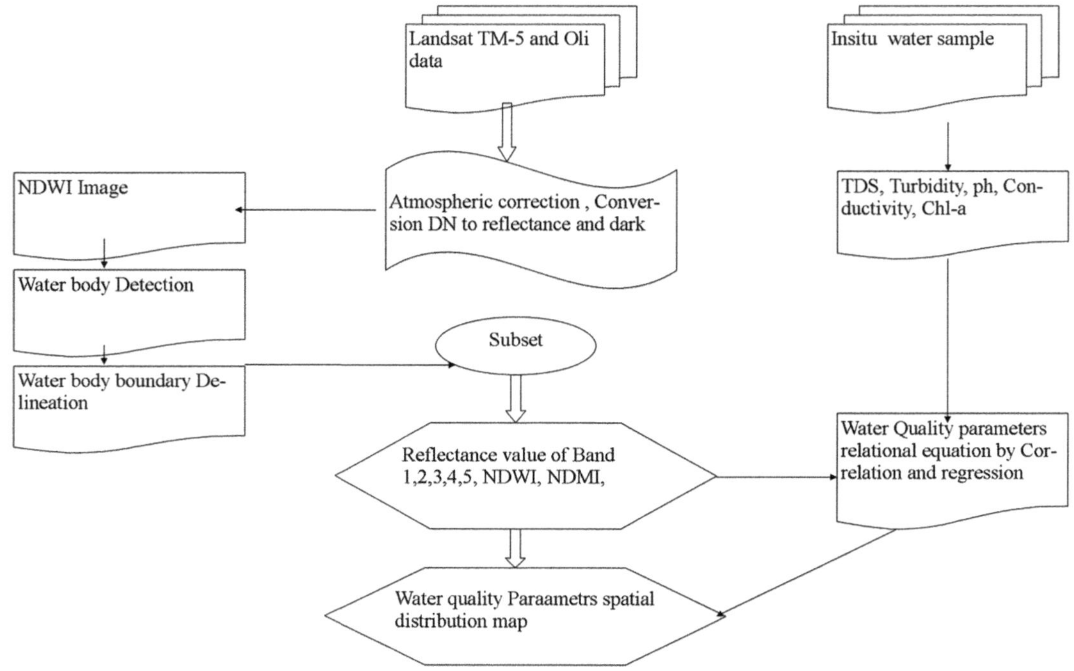

Fig. 9.3 Methodological proceeding for study

are shown in Table 9.1. For the understanding of changes in water quality, a comparison of present water quality (2020) data is compared with previous data (Table 9.3), collected by Dr. Nibedita Kapil for her Ph.D. thesis (2004). Both the locations site shown in Fig. 9.4.

9.3.3 Remotely Sensed Data Processing

Landsat- TM and 8 OLI imagery (path: 137 and row: 42) acquired on specific dates for 2004 and 2020 are used in this study for water quality assessment. The necessary geometric, radiometric correction is done to the images and true surface reflectance is calculated by converting digital numbers (DNs) to spectral radiance and spectral radiance to reflectance (Lu et al. 2002; Yang and Lo 2000). USGS (2013) proposed equation is used to convert DNs to spectral radiance (Chander and Markham 2003). The general form of the equation is shown in Eq. 9.2.

$$L_\lambda = \left(\frac{\mathrm{LMAX}_\lambda - \mathrm{LMIN}_\lambda}{Q_{\mathrm{CALMAX}}}\right) Q_{\mathrm{CALMAX}} + \mathrm{LMIN}_\lambda \tag{9.2}$$

where

L_λ spectral radiance at the sensor's aperture in W/(m^2 * sr * mm).

Q_{CAL} quantized calibrated pixel value in DN.

Q_{CALMAX} maximum quantized calibrated pixel value (DN = 255), corresponding to.

LMIN_λ spectral radiance scaled to Q_{CALMIN} in W/(m^2 * sr * mm).

LMIN_λ spectral radiance scaled to Q_{CALMAX} in W/(m^2 * sr * mm).

Exo-atmospheric radiance changes throughout the year, and depends on the earth-sun distance. For this reason, any study that occurs over time usually relies on reflectance. A pseudo-invariant object technique is often employed to adjust the radiance for multiple scenes over time. This study uses the varying earth-sun distance and sun zenith angle to adjust values for time stability. The equation for converting radiance to reflectance is shown in Eq. 9.3.

Table 9.2 Geographic location with corresponding water sampling data and reflectance for Deepor Beel (2020 January)

	Geographic location		Water quality parameters					Reflectance values			
Sample ID	Longitude	Latitude	pH	Conductivity (mS/cm)	TDS (mg/l)	Chlorophyll	Turbidity	*b2*	*b3*	*b4*	*b5*
1	91.656	26.116	7.81	0.3	215	0.0867	220.9	0.154686406	0.132492	0.099132873	0.146726087
2	91.661	26.117	7.49	0.4	331	0.0552	46.7	0.160319343	0.130468	0.104664616	0.07660114
3	91.664	26.116	7.45	0.4	360	0.0686	14.56	0.158936396	0.129659	0.105676517	0.075420588
4	91.651	26.123	7.59	0.3	250	0.0793	126.4	0.158565372	0.133774	0.105238028	0.095085256
5	91.654	26.12	7.56	0.4	221	0.0779	143.8	0.157519743	0.136067	0.103720173	0.113569371
6	91.655	26.117	7.88	0.3	260	0.1477	162.7	0.156170532	0.133504	0.100380883	0.11684119
7	91.644	26.123	7.78	0.3	285	0.147	86.47	0.154146731	0.13428	0.100077316	0.126825318
8	91.64	26.125	7.59	0.3	150	0.1892	36.69	0.155259818	0.135899	0.101730093	0.151380852
9	91.639	26.123	7.6	0.3	211	0.1084	25.57	0.151448309	0.12244	0.095017791	0.102269776
10	91.642	26.121	7.89	0.3	261	0.0806	112.5	0.153168544	0.135629	0.099402711	0.282253772
11	91.646	26.119	7.87	0.3	155	0.1022	99.2	0.153573304	0.134988	0.09943644	0.316827178
12	91.648	26.119	7.8	0.3	231	0.1131	155.4	0.15384315	0.136675	0.100144774	0.351333112
13	91.633	26.129	7.89	0.4	330	0.665	56	0.152494	0.120518	0.099335253	0.09798605

Source Primary field survey, 2020, January

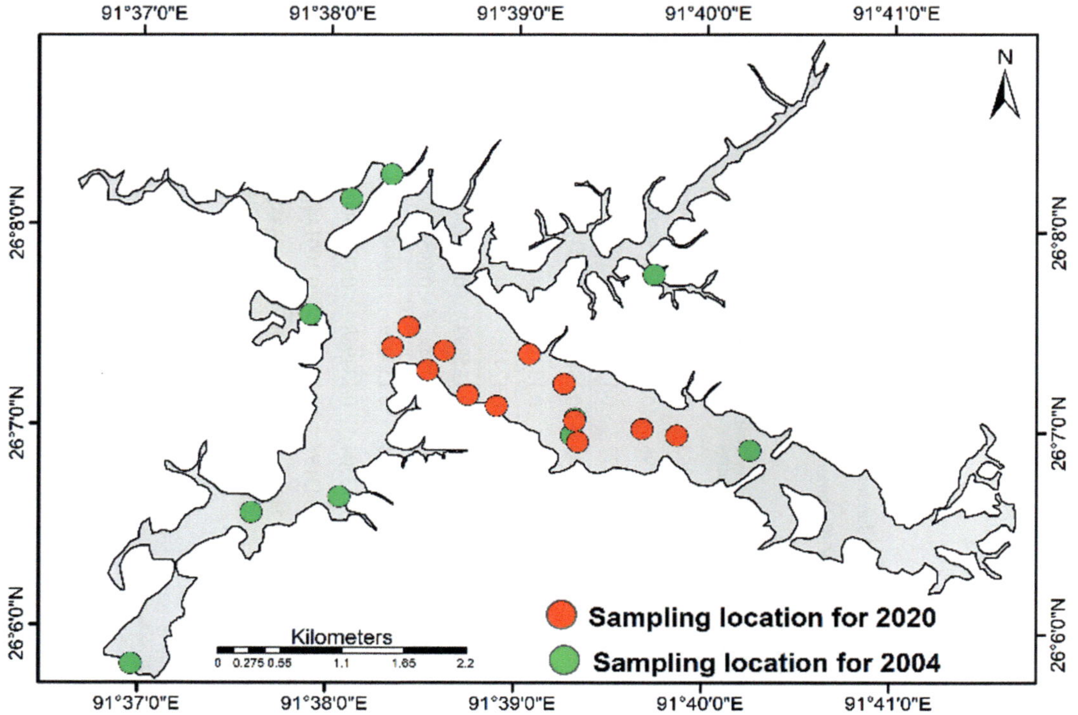

Fig. 9.4 Location of water sampling

Equation 9.3:

$$\text{Unitless Reflectance} = \left(\frac{\pi * L_\lambda * d^2}{\text{ESUN}_\lambda * \text{coscos}Z}\right) \quad (9.3)$$

where

$L_\lambda$spectral radiance at the sensor's aperature.

d Earth-sun distance in astronomical units.

ESUN_λ mean solar exoatmospheric irradiances.

Z solar zenith angle in degrees.

9.3.4 Regression Modeling

Regression analysis is used to assess the relationship between the reflectance values of different bands and in situ data water quality parameters. In order to avoid (probable) GPS measurement-related errors during sample collections, the mean reflectance values of imagery are calculated using a 3 × 3 array average reflectance values of pixels surrounding the sample pixel. For the final result total of 13 samples for 2020 and 10 samples for 2004 are used for regression modeling. MNDWI, NDVI, Band 2, Band 3, Band 4, Band 5, etc. area used as independent variables in the regression. SPSS 24 statistical software is used for calculating the level of significance. The R^2 value (coefficients of determination) is calculated for each parameter and the band yielding the highest coefficient of determination value is selected for final modeling. To decide the best fitting curve—logarithmic, polynomial, and linear fit of the second degree are tried.

9.4 Result and Discussion

9.4.1 Explanation of the Univariate Regression Results

The relationships between spectral reflectance and water quality parameters are explored by

Table 9.3 Geographic location with corresponding water sampling data and reflectance for Deepor beel (2004 February)

Object_ID	Geographic location		Water quality parameters					Reflectance Values					
	Longitude	Latitude	Turbidity (NTU)	pH	Conductivity (mS/cm)	TDS (mg/L)	Chl-a	MNDWI	NDWI	*B*5	*B*4	*B*3	*B*2
1	91.67083	26.115	15	6.8	0.13	145	0.22	−0.10922	−0.44941	0.135151	0.205712	0.091084	0.108535
2	91.655	26.11611	12	6.8	0.13	126	0.12	0.386634	0.125994	0.04801	0.084246	0.091084	0.108535
3	91.66139	26.12806	13	6.8	0.13	126	0.76	0.362925	−0.37864	0.156936	0.232262	0.087811	0.104681
4	91.63194	26.12583	9	6.7	0.12	120	0.65	−0.26126	−0.23885	0.148767	0.182923	0.094357	0.115389
5	91.65528	26.1175	11	6.54	0.1	107	0.44	−0.19974	−0.279	0.137874	0.199369	0.091084	0.115389
6	91.63444	26.11083	12	6.5	0.09	102	0.23	−0.1393	−0.3698	0.175999	0.261042	0.100903	0.120097
7	91.62667	26.10944	17	6.4	0.09	97	0.57	−0.10183	−0.28838	0.140597	0.203481	0.094357	0.112389
8	91.615	26.09583	22	6.5	0.08	128	0.63	0.181506	0.079929	0.050733	0.092469	0.087811	0.109348
9	91.63639	26.13639	46	7	0.16	145	0.23	−0.1115	−0.00059	0.072518	0.104803	0.087811	0.104681
10	91.64028	26.13861	52	7.1	0.14	131	0.33	−0.18879	−0.29228	0.178722	0.191146	0.091084	0.100681

Source Ph.D research work by Dr. Nibedita Kapil, 2004

Table 9.4 Results of regression algorithms for landsat TM-5 bands and water quality parameters (Feb. 2004) for Deepor beel

Water quality parameter	Regression equation	R^2	Significance level
Turbidity(NTU)	$y = -0.0003x + 0.1157$	0.0.503	0.013
pH	$y = -0.0155x + 0.2136$	0.5417	0.00002
Conductivity	$y = -1.6848x^2 + 0.3112x + 0.0964$	0.563	0.0137
TDS (mg/L)	$y = -0.0002x + 0.1367$	0.566	0.0016
Chl-a	$y = -0.0597x + 0.1394$	0.711	0.0004

Univariate and Bivariate analysis. It's worthy to note that algorithms are quite different in the selected bands when paralleled with those used in other studies. It is found that linear and Polynomial fit with univariate regression model shows a good correlation for selected parameters. The models have the highest $R2$ value from various regression models, which is selected for estimation of water quality and to predict the spatial distribution in Deepor Beel (Table 9.4 and Figs. 9.5 and 9.6). In most of the cases, parameters exhibit well correlation with single Landsat TM & OLI bands. In the present study for both the years (2004 and 2020), the result of the analysis shows that all the parameters, i.e., pH, turbidity, TDS, and conductivity are strongly correlated with Band 2 for Landsat TM & OLI. The R^2 value in all analyses ranges above 0.5 with a good significance level in the linear fit model. Only chl-a shows a very significant correlation with band 3 with $R2 = 0.71$ and significance equal to 0.002 for both the years 2004 and 2020 (Tables 9.4 and 9.5). The value $R2$ for turbidity, pH, conductivity, and TDS are 0.503, 0.541, 0.563, and 0.566 significance equals to 0.013, 0.00002, 0.0137, and 0.0016, respectively, for the year of 2004. Whereas the value $R2$ for turbidity, pH, conductivity, and TDS are 0.645, 0.55, 0.570, and 0.559 significance equals to 0.0001, 0.00002, 0.009, and 0.0000005, respectively, for the year of 2020. This empirical relationship is further applied to the entire study area for producing a series map to represent the spatial pattern of water quality parameters.

9.4.2 Assessment of Spatial Pattern of Water Quality

The raster layers are generated to show the spatial distribution pattern of various water quality parameters of deepor beel, (Figs. 9.7 and 9.8). The inference showed that the values of pH in Deepor Beel in 2020 are alkaline in nature as pH value in every location is above 7 (2020). The pH value in the past (2004) was quite strange as except for two locations, the water samples showed acidic condition in the lake and a very negligible area of waterbody showed neutral pH value. The value of alkalinity in most portions of the waterbody is ranging from 7.4–7.9 except for the southern part of the wetland where the value is less than 7 (in 2020). The spatial distribution of values of conductivity (for 2020) in Deepor Beel is in the range of 0.266–0.382 which is very high than the previous value (0.08–0.159) in 2004 (Fig. 9.8). Such massive changes of these two physical characteristics of water infer that the aquatic life of the lake is under stress. The spatial distribution of turbidity clearly indicates that the Deepor Beel is turbid and the increased value of TDS in the lake infers about the impact of pollutant influx in the lake. The TDS showed a

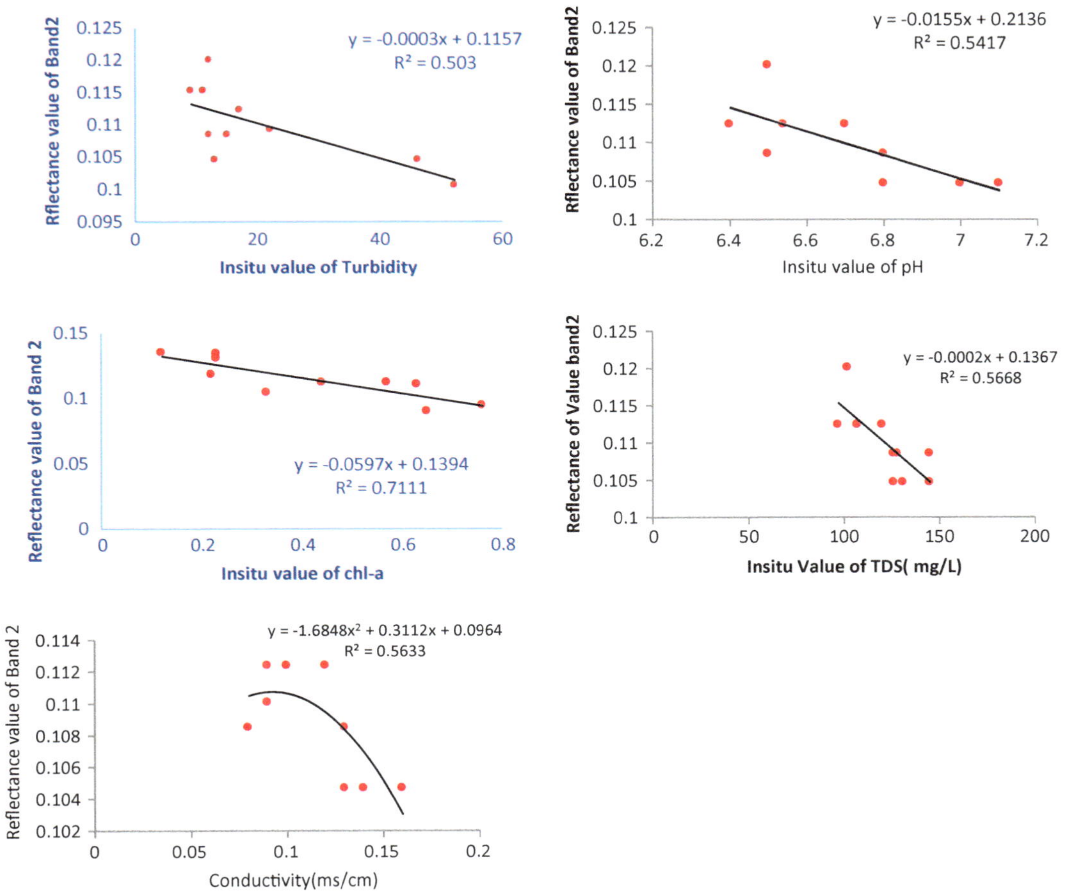

Fig. 9.5 The predicted values of water quality parameters for 2004

high concentration in the southeastern parts, whereas a higher concentration chl-a value was found in the southwestern part of the lake (for 2020).

9.5 Conclusion

The application of remote sensing techniques for water quality assessment resulted as a very useful tool and the estimated water quality parameters can be used as a supplement to the traditional measurements. The univariate regression analysis using various bands can be the best appropriate method to envisage the water quality for Deepor Beel. It is clear that most of the parameters are well correlated with single Landsat TM & OLI bands 2 with R^2 greater than 0.5. The parameter like chl-a and TDS showed a good relation with OLI Band 4 and 5, respectively. Comparison of all the parameters, the amount of pH, turbidity, chl-a, conductivity, and TDS present in water infers the intense human-induced effect due to various organic and inorganic pollutants. At last, it can be concluded that the technique is able to establish an effective method for the tedious monitoring of polluted wetlands. However, some drawbacks still exist in this technique and it can possibly be improved in the future. Problems like seasonal variability and spatial pattern of water quality are not analyzed with 100% reliability because the spatial resolution of TM images is

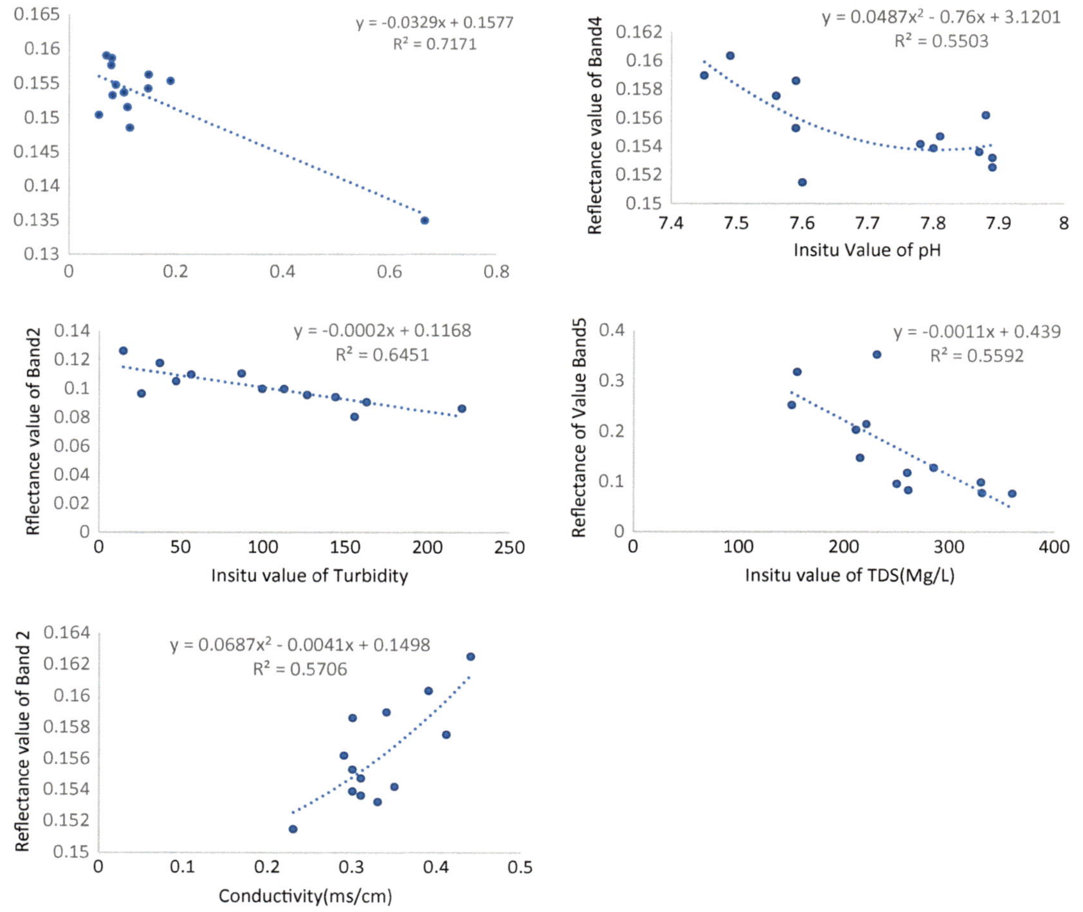

Fig. 9.6 The predicted values of water quality parameters for 2020

Table 9.5 Results of regression algorithms for Landsat TM-5 bands and water quality parameters (Jan. 2020) for Deepor Beel

Water quality Parameter	Regression equation	R^2	Significance level
Turbidity(NTU)	$y = -0.0002x + 0.1168$	0.6451	0.0001
pH	$y = 0.0487x^2 - 0.76x + 3.1201$	0.5503	0.00002
Conductivity	$y = 0.0687x^2 - 0.0041x + 0.1498$	0.570	0.009
TDS (mg/L)	$y = -0.0011x + 0.439$	0.559	0.0000005
Chl-a	$y = -0.0329x + 0.1577$	0.717	0.0001

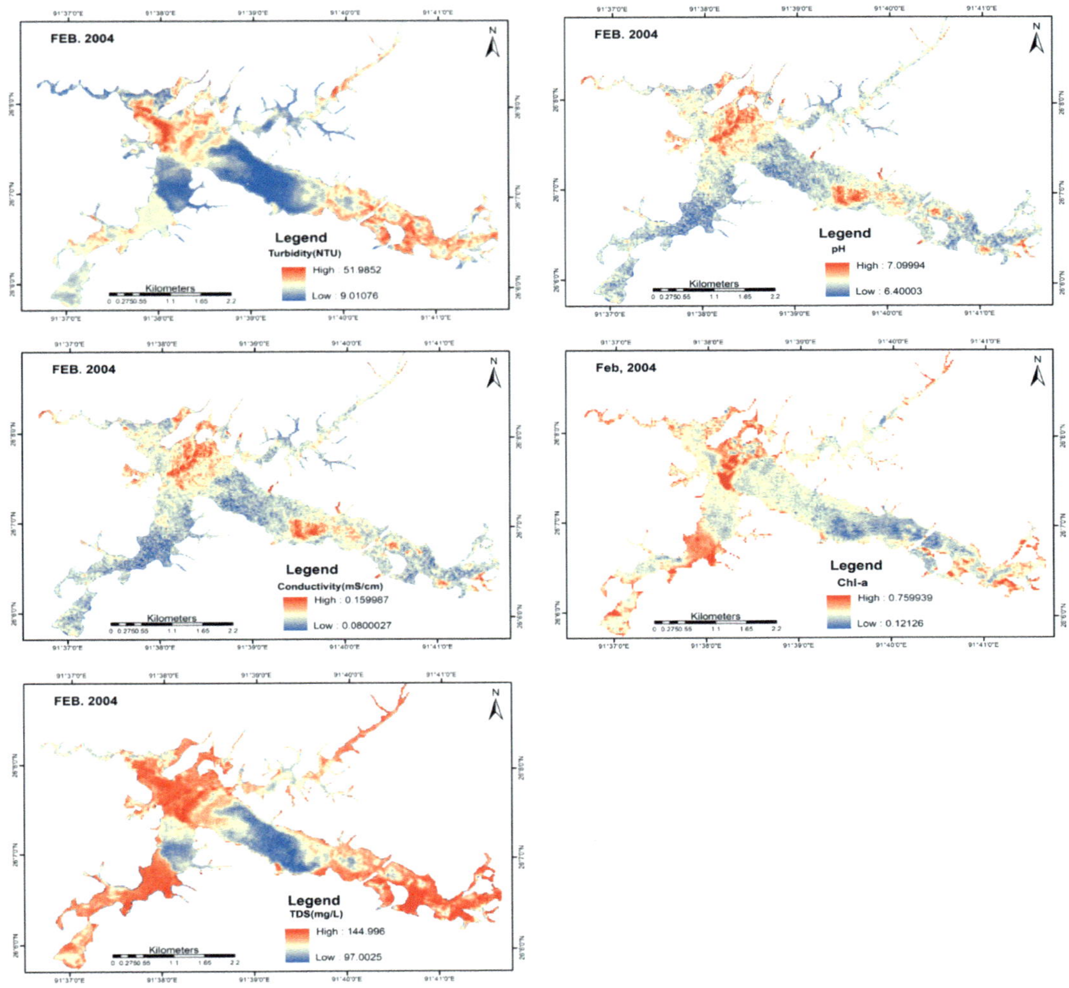

Fig. 9.7 Spatial distribution of turbidity, pH, conductivity, Chl-a, and TDS (in 2004)

not high enough for some sampling stations, the spatial resolution of TM & OLI images is not high enough for some small sampling stations. The improved modified equation for water quality estimation needs to be applied to study areas to eliminate the loopholes.

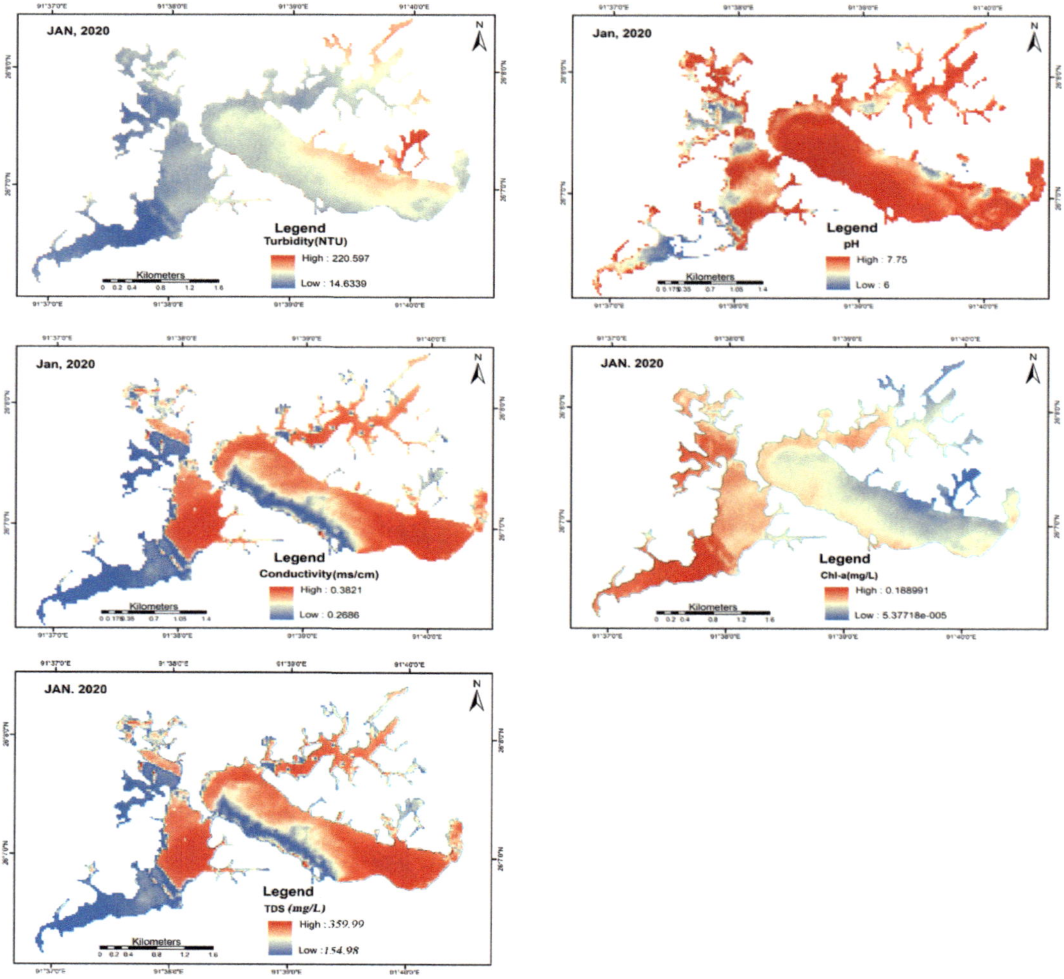

Fig. 9.8 Spatial distribution of turbidity, pH, conductivity, Chl-a, and TDS (in 2020)

References

Bhattacharyya KG, Kapil N (2009) Impact of urbanizationon the quality of water in a natural reservoir: a case study with the Deepor Beel in Guwahati City, India. Water Environ J 24:83–96

Bhattacharjya BK, Saud BJ, Borah S, Saikia PK, Das BK (2022) Status of biodiversity and limno-chemistry of Deepor Beel, a Ramsar site of international importance: conservation needs and the way forward. Aquat Ecosyst Health Manage 24(4):64–74

Chander G, Markham B (2003) Revised Landsat-5 TM radiometric calibration procedures and post calibration dynamic ranges. IEEE Trans Geosci Remote Sens 41 (11):2674–2677

Chipman JW, Olmanson LG, Gitelson AA (2009) Remote sensing methods for Lake management: a guide for resource managers and decision-makers. In: Developed by the North American Lake management society in collaboration with Dartmouth College, University of Minnesota, University of Nebraska and University of Wisconsin for the United States Environmental Protection Agency

Cutter SL, Finch C (2008) Temporal and spatial changes in social vulnerability to natural hazards. PNAS 105:2301–2306. https://doi.org/10.1073/pnas.0710375105

Das J, Saikia PK (2011) Species diversity of water birds in DeeporBeel, Assam. J Res Biol 5:363–369

Dash S, Kalamdhad AS (2021) Understanding the dynamics of heavy metals in a freshwater ecosystem through their toxicity and bioavailability assay. Environ Dev Sustain 23(11):16381–16409

Duan H, Zhang Y, Zhang B, Song K, Wang Z (2007) Assessment of chlorophyll-a concentration and trophic state for Lake Chagan using Landsat TM and field spectral data. Environ Monit Assess 129:295–308

Gohain SB, Bordoloi S (2021) Impact of municipal solid waste disposal on the surface water and sediment of adjoining wetland Deepor Beel in Guwahati, Assam, India. Environ Monitor Assess 193(5):1–19

Khattab MFO, Merkel BJ (2014) Application of Landsat 5 and Landsat 7 images data for water quality mapping in Mosul Dam Lake, Northern Iraq. Arabian J Geosci 7:3557–3573. https://doi.org/10.1007/s12517-013-1026-y

Lu D, Mausel P, Brondizio E, Moran E (2002) Assessment of atmospheric correction methods for Landsat TM data applicable to Amazon basin LBA research. Int J Remote Sens 23:1671–2651

Marti-Cardona B, Dolz-Ripolles J, Lopez-Martinez C (2013) Wetland inundation monitoring by the synergistic use of ENVISAT/ASAR imagery and ancilliary spatial data. Remote Sens Environ 139:171–84. https://doi.org/10.1016/j.rse.2013.07.028

McFeeters SK (1996) The use of the normalized difference water index (NDWI) in the delineation of open water features. Int J Remote Sens 17(7):1425–1432. https://doi.org/10.1080/01431169608948714

Mitsch WJ (2010) Conservation, restoration and creation of wetlands: a global perspective. In: Comin F (ed) Ecological restoration: a global challenge. Cambridge University Press, Cambridge, pp 175–188

MoEF (2008) Report on visit to Deepor Beel in Assam: a wetland included under national wetland conservation management programme of the ministry of environment and forests. Government of India

Mozumder C, Tripathi NK (2012) Use of multiple satellite images for feature extraction and image classification: a case study of Ramsar wetland in North East India. In: Asian conference on remote sensing. Asian association of remote sensing, 26–30 Nov, Pattaya, Thailand, p 7

Mozumder C, Tripathi NK (2014) Geospatial scenario based modelling of urban and agricultural intrusions in Ramsar wetland Deepor Beel in Northeast India using a multi-layer perceptron neural network. Int J Appl Earth Obs Geoinf 32:92–104

Mozumder C, Tripathi NK, Tipdecho T (2014) (2014) Ecosystem evaluation (1989–2012) of Ramsar wetland DeeporBeel using satellite-derived indices. Environ Monit Assess 186:7909–7927

Mushtaq F, Nee Lala MG (2017) Remote estimation of water quality parameters of Himalayan lake (Kashmir) using Landsat 8 OLI imagery. Geocarto Int 32 (3):274–285

Nas B, Karabork H, Ekercin S, Berktay A (2009) Mapping chlorophyll-a through in-situ measurements and Terra ASTER satellite data. Environ Monit Assess 157:375–382

Pal S (2015) Impact of Massanjore Dam on hydrogeomorphological modification of Mayurakshi River, Eastern India. Environ Dev. Sus-Tain 18(3):921–944. https://doi.org/10.1007/s10668-015-9679-1

Pal S, Osoundu CA (2009) Water scarcity in wetland area within Kandi Block of West Bengal: a hydroecological assessment. Ethiopian J Environ Stud Manage 2(3):1–17

Prasad SN, Ramachandra TV, Ahalya N, Sengupta T, kumar A, Tiwari AK, Vijayan VS, Vijayan L (2002) Conservation of wetlands of India—A review. Tropical Ecol 43(1):173–186

Tyler AN, Svab E, Preston T, Presing M, Kovacs WA (2006) Remote sensing of the water quality of shallow Lakes: a mixture modelling approach to quantifying phytoplankton in water characterized by high suspended sediment. Int J Remote Sens 27(8):1521–1537

Wang F, Luoheng H, Hsiang TK, Roy BVA (2006) Applications of Landsat-5 TM imagery in assessing and mapping water quality in Reelfoot Lake, Tennessee. Int J Remote Sens 27(23):5269–5283

Yang X, Lo CP (2000) Relative radiometric normalization performance for change detection from multi-date satellite images. Photogramm Eng Remote Sens 66:967–980

Zedler JB, Kercher S (2005) Wetland resources: status, trends, ecosystem services, and restorability. Annu Rev Environ Resour 30:39–74

Part III

Impact of Environmental Risk on Society

The Impact of Waste Management Practices on the Performance of the Apparel, Textile, and Leather Industry: The Indian Case

10

Soumen Ghosh and Srija Mitra

Abstract

In the era of globalisation, India became one of the major manufacturing hubs for the global north. In terms of the export share, apparel, textiles, and leather are the leading sectors in the Indian economy. Though the export orientation of this sector has increased rapidly, the production still takes place in an informal set-up, which is less efficient than the formal one in several ways. However, in recent times, compliance with environmental standards has become necessary for manufacturing firms to be competitive in the international market, as many national and international organisations have made efforts to maintain a cleaner production framework. It is challenging for these small, informal enterprises to adhere to these environmental standards, because it will raise the cost of production, drive up the price of the product, and make it difficult to compete in the market. Given this complex milieu, this paper aims to investigate the waste management scenario in the apparel, textile, and leather industries in India, especially for the unincorporated enterprises. Also, the study seeks to understand how waste management practices (both solid waste and liquid waste) impact the performance of these unincorporated enterprises. For the empirical analysis, this study used the NSSO 73rd round (2015–16) microdata for unincorporated enterprises. The study tried to portray a comprehensive picture of waste management and its impact on the firm performance of informal enterprises in the above-stated sectors by using figures and bivariate and multivariate analysis. The study reveals a positive relationship between the adoption of waste management practices and an enterprise's performance only in the leather industry.

Keywords

Waste management · Textile and leather · Apparel · Informal firm · Solid waste · Liquid waste

10.1 Introduction

Over the last few decades, the importance of sustainable production processes has massively increased across the globe. In 1987, the United Nations Environment Programme (UNEP) first coined the term 'sustainable development' in their report 'Our Common Future,' which is popularly known as the 'Brundtland Commission

S. Ghosh (✉)
School of Management and Labour Studies, Tata Institute of Social Sciences (TISS), Mumbai, India
e-mail: soumeng592@gmail.com

S. Mitra
Centre for Economic and Social Studies (CESS), Hyderabad, India

S. Talukdar et al. (eds.), *Environmental Risk and Resilience in the Changing World*, GIScience and Geo-environmental Modelling, https://doi.org/10.1007/978-3-031-62442-1_10

Report,' which mentioned the three pillars of sustainability, i.e. environmental, social, and economic (Paul 2021; Serreli et al. 2021). Environmental sustainability is gaining importance day by day due to increasing climatic disorder. Climatic disorder is induced by rising pollution from greenhouse gas (GHG) emissions, untreated wastewater discharge, and hazardous waste disposal (Haines et al. 2006; Patz et al. 2000). For instance, the amount of waste produced worldwide has surpassed 2 billion metric tonnes annually and is estimated to reach 3.4 billion metric tonnes by 2050.[1] Production units or firms are the major contributors to waste generation. So, waste management has become necessary on a global scale for environmental protection.

The existing literature has focused on the sustainable business practices of large firms instead of small-scale enterprises. In the era of globalisation, these tiny enterprises began to play a significant role in the value chain. In the recent decades, the production process has undergone several structural changes and became fragmented across the world. Transnational companies (TNCs) and multinational corporations (MNCs) of the developed nations are propelling the expansion of the cross-border production process by moving or outsourcing the labour-intensive stages of the production process (upstream activities) to the developing regions (Aybar and Ficici 2009; Luo and Tung 2007). The lead firms frequently establish production facilities in developing economies because of cheap and abandoned inputs (both labour and raw materials). Lead firm connects with the first-tier firms or contract manufacturers of developing countries, and these first-tier firms are often outsourced to small firms for production, mostly to reduce their production costs and increase their profit margins. Despite the major contribution of these tiny units to a sizable share of global production, they often operate in an informal manner. Due to existence of large scale of informal firms, these firms always face price competition. Thus, these firms usually strive to cut production costs by allegedly violating the most basic health and safety measures, such as solid waste management, liquid waste management, toilet facilities, etc.

Existing literature has explored that small enterprises contribute significantly to pollution, as these enterprises are responsible for more than 70% of all pollution (Mitra 2022). Furthermore, it is highlighted that these tiny units are responsible for disposing of more than 70% of industrial effluents (Chen et al. 2017). However, only a small proportion of informal enterprises use sustainable business practices in their operations because of the high installation costs associated with engaging in this kind of activity (Bakos et al. 2020; Courrent et al. 2018; Dey et al. 2020; Leonidou et al. 2017; Mitra 2022). According to the stakeholder theory, the adoption of eco-friendly technology (cleaner production) can aid businesses in enhancing their financial performance due to greater societal trust (Donaldson and Preston, 1995). Moreover, if enterprises adhere to the codes of conduct of the higher-tier firms, it will boost their prosperity (Kabeer 2004; Tonoyan et al. 2010). Thus, there is a need to understand how the adoption of waste management practices affects tiny informal firms' performance. And this paper is focusing on this particular issue.

India has played a major role in the international fashion sector in recent decades. India is a major exporter of leather, textile, and garment goods (both intermediate and finished goods) to international markets. India has taken the leading position due to availability of abandonment and cheap inputs (such as labour and raw materials). The small-scale firms usually rely on obsolete technologies and conventional machinery, and the manufacturing procedure produces waste as a result of using these outdated methods. South Asian nations are one of the main producers of waste among emerging nations, and India is no exception to this (Borthakur 2020; Hettige et al. 1996; Ngoc and Schnitzer 2009). Thus, in this chapter, we have concentrated on these three sectors: leather, textiles, and apparel. In other words, this chapter is trying to understand the overall scenario of waste management by the

[1] https://datatopics.worldbank.org/what-A-waste/trends_in_solid_waste_management.html.

unincorporated firms in the three leading exporting sectors (leather, textiles, and apparel) and what the impact of waste management on the performances of these informal enterprises is.

This specific chapter provides a thorough overview of the waste management practices used by India's informal small firms in the garment, textile, and leather sectors. This chapter presented a comparative examination of these three industries in the Indian context. In addition, the study intends to identify how the waste management practices adopted by the informal enterprises influence the performance of these three industries individually. To do the empirical analysis, this study used NSSO 73rd round (2015–16) microdata for unincorporated enterprises.

There are two ways in which this study advances the corpus of knowledge. Firstly, there are very few studies that have concentrated on industry-specific waste management practices. Secondly, the study analyses the impact of the waste management practices of informal firms. Though much of the existing literature has focused on waste management in large-scale enterprises, there have been very few studies that have examined the performance of informal businesses.

This chapter is organised as follows: Sect. 10.2 provides a conceptual framework for the respective study, followed by some discussion on the definition of some concepts related to this study. Section 10.4 talked about the data and methodology of the study. Section 10.5 presented the results and discussion of the respective issues, and the last section presented the concluding remarks.

10.2 Some Definitions

Before we proceed into the main findings, this section has discussed some fundamental concepts from the existing definitions. So, in the last three decades many international and national organisations frequently talk about the sustainable development and cleaner production. Although the production has segregated across the globe, the production become unregulated and often it operates in the informal set-up. Even in India, a large number of tiny firms are informally contributing to the global demand. Moreover, this section defines some basic concepts related to this research for better understanding. Also, recent studies have tried to understand the problem of waste management scenario and its probably remedies. But before that, there is a need to understand the meaning of waste management practices and why it became the central issue of the production economy (Table 10.1).

10.3 Waste Management Practices and the Respective Industries

In the last few decades India became one of the major production hubs for global fashion industry. The share of apparel, textile, and leather export earning is rising over time. But these three industries are fundamentally are different from each other. The existing studies largely concentrated on single industry in some specific location. These studies largely identified the type of waste releases from the production and the remedies. Some scientific studies also put an effort to reuse strategies of the waste from one industry and use it in another industry. Here we will discuss the waste management practices in all three industries in India.

10.3.1 Textile and Apparel Industry

The textile and apparel industry are significantly contributing to the country's export earnings and job creation. There are several stages of production conducted in India and some are adversely impacting the society as well as the environment in terms of waste generation (Nurhayati et al. 2016). From the production of nylon, cotton, polyester, and rayon, dyes release harmful chemical effluents and gaseous wastes into the ecosystem that impacts it in very diverse ways for human life. In the entire production process, there is a huge amount of water used and a large amount of toxic pollutants discharged to

Table 10.1 Definition of some important concepts

Concept	Definition	Author
Global value chain	• This is a production system that links the economic activities of firms with technological and organisational networks and it also involved in the design, production, and marketing of a product (Cattaneo et al. 2010; Gereffi et al. 2005)	Gary Gereffi
Informal Sector	• The unorganised sector consists of all unincorporated private enterprises owned by individuals or households engaged in the sale and production of goods and services operated on a proprietary or partnership basis and with less than ten total workers (NCEUS 2007)	NCEUS
Waste management	• The waste management refers to the collection, transport, treatment, and disposal of waste. It also emphasises the control, monitoring, and regulation of the production, collection, transport, treatment and disposal of waste, and prevention of waste production through in—process modifications, reuse, and recycling	OECD

Source Compiled by the authors

the environment, which is very harmful in nature and eventually degrade the water and soil (Kumar and Pavithra 2019). The wastewater from the textile industry is very alkaline, and it contains biological oxygen demands, chemical oxygen demands, dissolved solids, and alkalinity. Earlier studies also suggested some environmentally friendly effluent treatment methods by which the factory can easily eliminate the detrimental influence of textile effluent as well (Raj et al. 2017; Sivaram et al. 2018). As this sector requires a large amount of electricity, some studies suggest using the combination of biomass and solar photovoltaics to provide the necessary thermal and electrical energy for textile manufacturing that can minimise climate change, ozone depletion, and terrestrial acidification by 90–95% compared to the regular technique (Muthukumarana et al. 2018). However, there is a large noncompliance incident with sustainability identified by recent studies. It has been indicated that Indian-listed firms voluntarily disclose environmental information on their websites and in annual reports. When compared to smaller firms, larger firms usually reveal more environmentally related information (Chatterjee and Zaman Mir 2008; Nurhayati et al. 2016). Apart from this, many studies discuss the recycling and reuse of wastes generated from the production process (Nakhate et al. 2020; Uddin 2021).

10.3.2 Leather Industry

Recently, numerous studies on the leather sector have been carried out and these works are delved into issues including the circular economy, sustainable production methods, environmental impacts, and waste management and treatment. Hazardous waste is produced during several phases of the leather manufacturing process, which has adverse effects on both people and society. The unhairing and fibre opening procedures frequently produce waste during the pretanning phase (Ariram et al. 2022; Durga et al. 2016; Jayanthi et al. 2019). A modern biochemical substitution for this process (the use of proteolytic enzyme and sodium metasilicate) has significantly reduced pollution levels (Bhavan et al. 2008). Again, tanning has considerable negative consequences for the environment in terms of groundwater contamination and pollution from chromium, both of which eventually impact human lives (Bhandari and Maiti 2012; Manivannan and Elango 2017; Musa et al. 2010; Paul et al. 2023). The waste (solid, liquid, and gaseous) is not properly treated before being discharged into the environment (Joseph and Nithya 2009; Kaul et al. 1993; Rossi et al. 2021; Singh and Gupta 2013). In certain research, solutions and cleaner production methods are addressed (Kaul et al. 1993; Musa et al. 2010; Rossi et al. 2021). Although these studies cover a wide range of topics in this domain, they are not very comprehensive. The previous studies are not holistic in the sense that they mostly cover a small geographical area and there is a significant lack of overall pictures.

Despite a sizable number of studies already been done on this specific area, none of them focuses on these three industries in the aggregate level. As the informal tiny units are most vulnerable in terms of waste management, this study provides a comparative picture of these three industries in terms of adoption of waste management practices. Although the reduction of waste from each industry are different, the results would be very interesting. Alongside, the study presents how the waste management practices of these units are impacting the performance.

10.4 Data and Methodology

To conduct this study, NSSO's 73rd round of microdata has been used. NSSO has collected the largest sample of data on the informal sector in India, and the 73rd round has collected data on 2.91 lakh Indian informal enterprises. Here, we have considered the apparel, textile and leather industries as per the NIC 2008 classification. This survey covered the enterprises' solid and liquid waste management practices across all three industries in India. It has focused on the impact of waste management on firm performance. In this regard, multivariate regression has been used. Gross value added (GVA) and net surplus (SURPLUS) of the firm are considered as the dependent variables (Ghosh and Mitra 2023; Paul et al. 2021), which indicate the firm's performance, and the independent variables are the size and age of the firm, capital intensity, ownership, location, and registration status of the firm. Here, the waste management of a firm is captured through dummy variables for both solid and wet waste management.

The comprehensive variable listings are presented in the above table (Table 10.2). Along with this we have structured an econometric model that was derived from the previous table (Table 10.2) to understand the impact of waste management practices on firm's performance across all three industries. This study uses the natural logarithm of all continuous variables (i.e. gross value added, net surplus, total labour, total capital, and firm age) as a proxy measure to reduce the effect of skewness on the dataset (Table 10.3).

Our base ordinary least squares (OLS) regression model is estimated as follows:

$$Y = \alpha + \beta_1 X_{1t} + \beta_2 X_{2t} + \beta_3 X_{3t} + \beta_4 X_{4t} + \beta_5 X_{5t} + \beta_6 X_{6t} + \varepsilon_{0t},$$

where

Y	Performance of the firm (GVA and net surplus)
X_1	Total employed labour
X_2	Total capital
X_3	Age of the firm
X_4	Toilet facility in the workplace
X_5	Solid waste Management practices adopted by the firm
X_6	Liquid waste Management practices adopted by the firm
α, β_i	Parameter
ε	Error term
t	year (2015)

10.5 Results and Discussions

In the globalisation era, production has become fragmented across the globe and it is hard to maintain overall sustainability across the chain. As we mentioned earlier, upstream activities are mostly executed in South Asian countries because of their abundant and cheap workforce. There are multiple intermediaries between supplier firms and the first-tier firms. These intermediaries often try to make their own profit and manufacture in the cheapest possible way. Manufacturing is primarily done by small, informal units, which face fierce competition. As a result, these enterprises frequently violate health and safety standards. Because the cost of such measures frequently increased production

Table 10.2 Framework of NSSO (73rd round) data used in this study

Structure of national sample survey 73rd round (2015–16) data used in this study		
Sl. No	Variable	Descriptions
	Continuous variable	
1	GVA	Total receipt − (total operating expenses + distributive expenses + toll + taxes on products) − in last 365 days
2	NSURPLUS	Net surplus
3	LABOUR	Total fulltime + Total parttime
4	CAPITAL	Hired + owned + net addition net additions to owned assets during last 365 days
5	FIRMAGE	Year of initial operation
	Derived variable (Relative measures)	
1	LAB_PRODUCTIVITY	Gross value added by total employed
2	CAP_LAB_RATIO	Total assets by total employed
3	WAGERATE	Total emoluments by total employed
	Categorical variable (Dichotomous response; Yes = 1, No = 0)	
1	REG_DUM	Is the enterprise registered under any acts?
2	CONTRACT_DUM	Does the enterprise engage in any contract work?
3	TOILET_DUM	Whether the enterprise has toilet facility
4	SOCIAL_DUM	Whether the enterprise owned by socially deprived person
5	LOC_DUM	Whether the enterprise operates within household premises
	Multiple category variable	
1	SOLID_WASTE	Does the enterprise have to maintain solid waste management practices in the workshop? (Ref.—Yes) 1—Yes, having solid waste management practices 2—Does not generate liquid waste
2	LIQUID_WASTE	Does the enterprise have to maintain liquid waste management practices in the workshop? (Ref.—Yes) 1—Yes, having liquid waste management practices 2—Does not generate liquid waste

Source The authors

costs, which in turn caused product prices to rise and ultimately forced them off the market. However, in recent years, the lead firm has put a focus on sustainable parameters such as health, workplace safety, waste management, the minimum wage, and so on.

Informal micro- and small-supply firms often operate in a very vulnerable condition. All environmental and social standards are largely violated. For instance, some of the processes in leather manufacturing produce a lot of harmful solid and liquid waste that eventually impacts the environment as well as the people around it. Therefore, in such case, waste management is essential to carrying out a clean and environmentally friendly production process. In this chapter, we have covered the aggregate picture of the waste management situation in the three different Indian sectors. Afterwards, we tried to find out if the firm's performance was being impacted by waste management practices.

Figure 10.1 describes the solid waste management practices across all three industries. Here, solid waste management has been captured through variables like whether the firm is disposing off the waste for further treatment, whether the enterprise owns any kind of waste treatment system, or whether the firm disposes off the waste

Table 10.3 Structure of econometric models applied to NSSO (73rd round) data

Sl. No	Variable	Scale of measurement
Dependent variables		
1	LNGVA	Continuous
2	LNSURPLUS	Continuous
Independent variables		
1	LNLABOUR	Continuous
2	LNCAPITAL	Continuous
3	LNFIRMAGE	Continuous
Independent variables (Control)		
1	REG_DUM	Dichotomous
2	CONTRACT_DUM	Dichotomous
3	SOCIAL_DUM	Dichotomous
4	LOC_DUM	Dichotomous
5	TOILET	Dichotomous
6	LIQUID_WASTE	Multi Category
7	SOLID_WASTE	Multi Category

Source According to Table 10.2

on the roadside. The following figure shows that the leather industry is much more aware of solid waste management practices compared to the other two industries. According to the data, 43.38% of enterprises in the leather industry dispose of solid waste to the local authority, while 11.44% dispose of waste for further treatment. Also, the intensity of waste generation in the leather industry is higher in its different stages compared to the other two industries. In the apparel sector, 14% of the total enterprises own waste treatment systems, which is higher than the other two sectors. Surprisingly, the data shows that one third of the textile and apparel firms throw their solid waste on the roadside.

All the three sectors generate various amounts of liquid waste. While some parts of the leather-making process discharge a significant amount of hazardous chemicals into the environment, the direct contact of employees with these chemicals during the processes results in a range of skin and other long-term diseases. The other two industries, on the other hand, produce very little liquid waste throughout the course of their whole operations. Figure 10.2 shows that just a very modest volume of liquid was reported by all three industries. This is a rather unexpected outcome considering that, in comparison with the other two industries, the leather industry releases liquid waste during the preponderance of manufacturing stages. One of the reasons behind this striking result might be that the firm is connected with the local common effluent treatment plants (CETP).

Table 10.4 describes the descriptive statistics for all three industries and portrays an overall picture. Here we can see the average GVA and net surplus for the leather industry compared to the other two industries, followed by the textile and apparel industries. Again, the same trend is visible in the case of average labour in each firm across the industries. Whereas the leather industry has the highest average number of workers per firm, the apparel sector is at the bottom of the list. A similar trend has followed in terms of the total capital of the firm for each industry. If we look at the derived variables labour productivity, capital-labour ratio, and wage rate, we can see that the leather industry is performing well in comparison with the textile and apparel sectors. So, based on Table 10.4, it is clear that the leather industry is performing better in comparison with the other two industries.

Now in Table 10.5, we have presented the percentage share of categorical variables

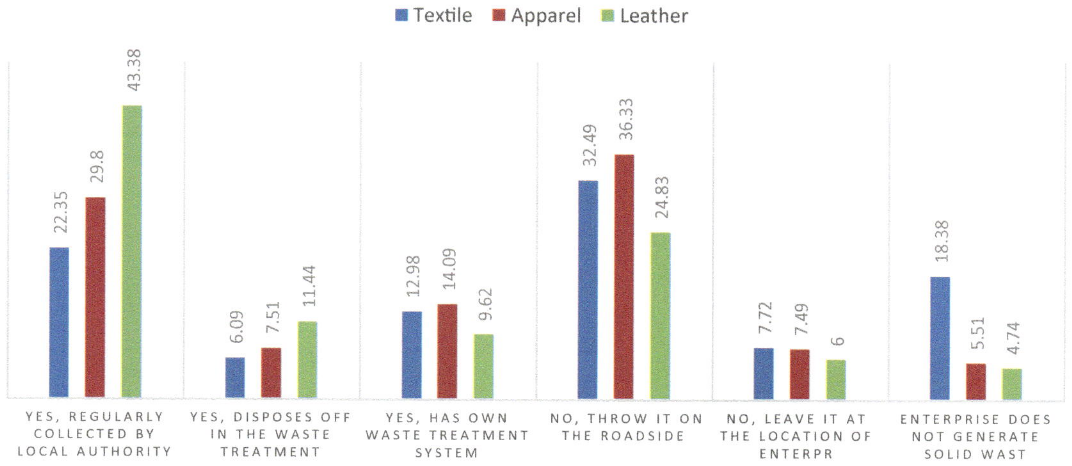

Fig. 10.1 Solid waste management scenario across industries. *Source* NSSO 73rd round unorganised (excluding construction) micro database (2015–16), computed by the authors

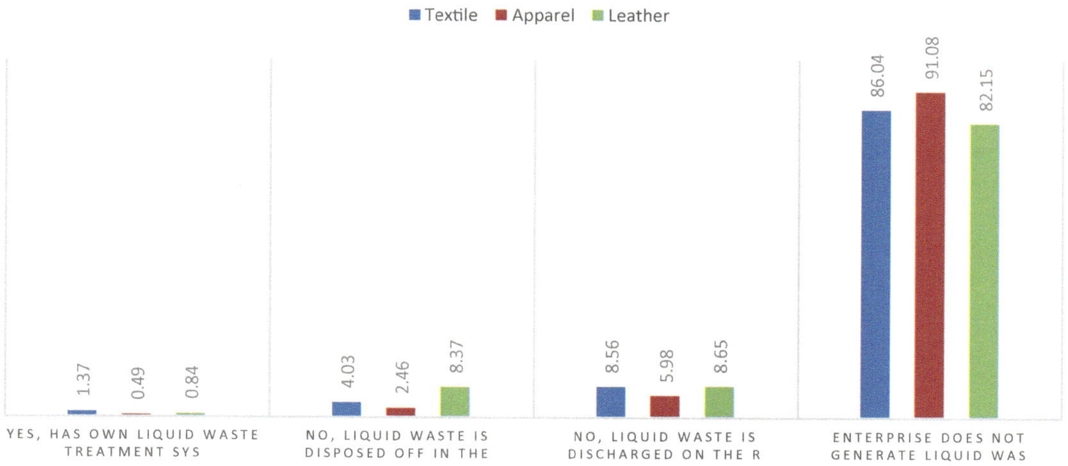

Fig. 10.2 Liquid waste management scenario across industries. *Source* NSSO 73rd round unorganised (excluding construction) micro database (2015–16), computed by the authors

(described in Table 10.2) for all three industries. In terms of adopting solid waste management practices, the leather industry holds the highest share of 64.4%, followed by the apparel and textile industries with 50.7 and 41.4%, respectively. In contrast, the textile sector has the highest share of liquid waste management at 1.4%, followed by leather (0.8%) and the apparel sector (0.5%). The trend is similar for toilet facilities in the workplace across all three industries.

Again, if we look at the socio-economic characteristics of the firm (Table 10.5), we can see that in all three industries, firms are largely owned by the socially backward class. Again, nearly one-third of the leather manufacturing informal firms are located in rural areas, whereas approximately two-thirds of the total textile firms are in the rural sector. So, from this result, we can say that a large number of leather firms are urban informal firms that eventually impact performance.

Table 10.4 Descriptive statistics of the variables in textile, apparel, and leather industry

Industry	Variable	Observations	Mean	Standard deviation	Minimum	Maximum
Textile	LNGVA	5362	9.025	1.494	3.135	16.043
	LNSURPLUS	5349	8.697	1.223	3.219	14.604
	LNLABOUR	5386	0.779	0.821	0	4.19
	LNCAPITAL	218	12.979	1.471	8.795	16.945
	LNFIRMAGE	5387	2.254	0.77	0	4.511
	LNLABPRODUCTIVITY	5362	8.244	0.985	3.091	12.379
	LNCAP LAB RATIO	218	11.665	1.154	8.795	14.38
	LNWAGERATE	2302	7.769	1.143	1.609	12.336
Apparel	LNGVA	27,473	8.718	1.187	2.398	14.559
	LNSURPLUS	27,441	8.508	1.017	2.639	14.201
	LNLABOUR	27,561	0.382	0.579	0	4.635
	LNCAPITAL	2423	12.319	1.016	9.024	17.506
	LNFIRMAGE	27,561	2.092	0.722	0	4.407
	LNLABPRODUCTIVITY	27,473	8.335	0.89	2.398	11.571
	LNCAP LAB RATIO	2423	11.552	0.901	8.756	16.408
	LNWAGERATE	8902	7.545	1.17	2.303	10.94
Leather	LNGVA	716	9.939	1.303	5.193	15.019
	LNSURPLUS	715	9.47	1.038	5.298	13.964
	LNLABOUR	717	1.043	0.944	0	4.382
	LNCAPITAL	60	12.763	1.43	9.551	17.267
	LNFIRMAGE	717	2.141	0.854	0	4.718
	LNLABPRODUCTIVITY	716	8.896	0.705	5.193	11.585
	LNCAP LAB RATIO	60	11.747	1.21	9.551	14.916
	LNWAGERATE	436	8.096	0.9	4.605	10.445

Source NSSO 73rd round unorganised (excluding construction) micro database (2015–16), computed by the authors

Table 10.5 Percentage share of categorical variables in textile, apparel, and leather industry

	Textile industry (%)	Apparel industry (%)	Leather industry (%)
Whether the firm is registered?	25.6	19.4	33.8
The firm undertakes any contractual work	44	6.2	28.5
Whether the owner belongs to backward section	97.7	99	97.9
Location of the firm (Ref.: Rural)	69.6	55.3	37.7
Having solid waste management? (Ref.: Yes)	41.4	50.7	64.4
Having liquid waste management? (Ref.: Yes)	1.4	0.5	0.8%
Having toilet facility? (Ref.: Yes)	80.2	66.2	74.9

Source NSSO 73rd round unorganised (excluding construction) micro database (2015–16), computed by the authors

We are now examining how the firm's waste management practices impact on its performance. We have employed a multivariate regression model in this case (Table 10.6). Furthermore, it is evident from the regression result that waste management practices (both solid and liquid waste management) have no statistically significant impact on the performance of the firm in the textile and clothing industry. Even though these two industries discharge harmful pollutant (both solid and liquid) to the society and environment, the aggregate data shows very shocking results. In the case of leather industry, waste management in terms of liquid waste management has a positive impact on the performance of the firms. Whereas in terms of solid waste management practices adopted by leather industry, firms have no impact on its performance. As we previously mentioned, in the leather manufacturing during the beam house process and tanning generate many hazardous wastes, like solid waste (plastic), liquid (chrome, greases, and syntans), and gaseous wastes (sulphide). These wastes adversely impact the ecology and society (workers and local community).

So far, the adoption of liquid waste manage practices by the firm are concerned, a significant positive relation is present. Moreover, a large number of firms in all three industries recorded non-generation of liquid waste in their operations. First, the cleaner manufacturing has lately received attention from local governments and value chain participants, and most sites have made an attempt to construct a common effluent treatment plant (CETP) for liquid waste. The prime objective of this CETP is to gather effluents from all the member companies. This might be the cause of the fact that many leather manufacturing facilities reported non-generation of liquid waste from their operations. Therefore, over 80% of the leather manufacturing facilities have stated that they do not discharge any liquid waste while operation.

Other reason can be that, small units often hide their environmentally related data, compared to the larger firms. In a recent paper it has been observed that environmental reporting is determined by the effective tax rate, liquidity, and leverage of respective firms (Mukherjee et al. 2010; Nurhayati et al. 2016). In this analysis we are analysing only unincorporated enterprises and by nature these firms are violating basic standard to reduce their production cost as well as bypassing the legal issues related to noncompliance of environmental and social standards. The above two aspects are very significant of such unique result of the enterprises of all three industries.

To comprehend why the aggregate data is producing these undesirable outcomes, this component needs additional in-depth analysis. In our view, the firm is hiding some information in order to avoid legal troubles. Because there is a significant incentive to show clean output, firms may build favourable perceptions of their top rivals.

10.6 Conclusions

In this study, we are trying to understand the waste management practices of three major exporting sectors: textiles, apparel, and leather, and attempting to investigate the impact of waste management on firm performance, particularly for informal enterprises in these three sectors. The informal firms are tiny in nature, though they became one of the most integral parts of the value chain. In recent decades, production processes are no longer geographically limited due to increasing globalisation. As a result of that, developing countries became global manufacturing hubs. Lead firms from developed countries always try to manufacture their products at the lowest possible cost in developing countries. Thus, the supplier firms of the developing countries face price-based competition to capture the international market and try to produce it at the lowest possible cost. Due to this reason, supplier firms engage with informal, small-scale firms through subcontracting (Ghosh and Mitra 2023). To reduce production costs, these

Table 10.6 Determinants of gross value added and net surplus in all three industries

Dependent variable	LNGVA						LNSURPLUS					
VARIABLES	Model 1	Model 2	Model 3	Model 4	Model 5	Model 6	Model 7	Model 8	Model 9	Model 10	Model 11	Model 12
LNLABOUR	0.99***	0.77***	0.94***	1.07***	0.84***	1.14**	0.42***	0.20	0.44***	0.61***	0.37***	0.77
	(0.07)	(0.18)	(0.02)	(0.07)	(0.11)	(0.37)	(0.08)	(0.25)	(0.02)	(0.08)	(0.12)	(0.54)
LNCAPITAL	0.32***	0.39**	0.26***	0.29***	0.11	− 0.03	0.31***	0.54**	0.27***	0.31***	0.20**	0.04
	(0.04)	(0.15)	(0.02)	(0.05)	(0.07)	(0.15)	(0.05)	(0.22)	(0.02)	(0.05)	(0.08)	(0.21)
LNFIRMAGE	0.15***	− 0.01	0.16***	0.17***	0.07	0.16	0.20***	0.13	0.17***	0.16***	0.10	0.22
	(0.05)	(0.14)	(0.02)	(0.05)	(0.09)	(0.26)	(0.06)	(0.20)	(0.02)	(0.06)	(0.10)	(0.37)
TOILET	− 0.17	− 0.55	0.02	− 0.14	0.21	0.42	− 0.04	0.05	− 0.03	− 0.23**	0.07	− 0.01
	(0.11)	(0.38)	(0.03)	(0.09)	(0.17)	(0.60)	(0.13)	(0.51)	(0.03)	(0.10)	(0.19)	(0.86)
SOLID_WASTE (Ref.: Yes)	− 0.04		0.04		− 0.01		− 0.11		0.01		− 0.01	
	(0.10)		(0.03)		(0.17)		(0.12)		(0.03)		(0.18)	
LIQUID_WASTE (REF.: Yes)		− 0.09		0.26		**1.47****		− 0.31		0.31		**2.13****
		(0.42)		(0.35)		**(0.50)**		(0.62)		(0.38)		**(0.73)**
Constant	4.73***	4.80***	5.29***	4.94***	7.57***	8.91**	4.62***	2.00	5.12***	4.68***	6.46***	8.11*
	(0.46)	(1.70)	(0.18)	(0.59)	(0.81)	(2.17)	(0.55)	(2.48)	(0.19)	(0.63)	(0.89)	(3.13)
Observations	203	37	2,343	242	60	10	204	39	2339	241	60	10
R-squared	0.79	0.73	0.63	0.67	0.73	0.90	0.52	0.45	0.38	0.47	0.45	0.78
F Statistics	0.000	0.000	0.000	0.000	0.000	**0.043**	0.000	0.001	0.000	0.000	0.000	**0.164**
Product	Textile	Textile	Apparel	Apparel	Leather	Leather	Textile	Textile	Apparel	Apparel	Leather	Leather

Source NSSO 73rd round unorganised (excluding construction) micro database (2015–16), computed by the authors
Notes see Table 10.4, for model structure standard errors in parentheses
*** $p < 0.01$, ** $p < 0.05$, * $p < 0.1$

small-scale businesses frequently violate several basic health and safety standards, as well as environmental legislation. As a result of this, these firms generate a lot of waste. Thus, there is a need to examine the waste management system of these unregulated, small-scale firms, and this chapter is focusing on this issue. For the empirical analysis, this study has used NSSO 73rd round of data. It is evident from the analysis that the disposal of waste is not the same for these three sectors. When compared to the apparel and textile industries, the leather industry produces more waste. This is one of the reasons that the firms engaged in the leather industry are more engaged with waste management. And proper waste management helps the firms achieve better performances in the leather sector alone. For the apparel and textile sector, the impact of waste management on firm performance is statistically insignificant.

In the era of globalisation where the work becomes flexible in its highest extent, the government and other social bodies should come forward to protect ecology and people. The compliance of basic environment and social standard should be upheld for sustainable and cleaner production among all three industries further. As these three industries have immense potential to further grow and prosper, it is important to make a sustainable business ecosystem. From earlier studies and reports mentioned that the workers and local people often suffer from many long-term skin and heart diseases due to the direct and indirect contact of such hazardous chemical wastes. The government should ensure safe and secure environment in the surroundings of leather clusters in all. Apart from that firms also protect those workers who are exposed by the chemical wastes every day.

Appendix

See Tables 10.7, 10.8 and 10.9.

Table 10.7 Matrix of correlations (textile)

Variables	(1)	(2)	(3)	(4)	(5)	(6)	(7)	(8)	(9)	(10)
(1) LNGVA	1									
(2) LNSURPLUS	0.835	1								
(3) LNLABOUR	0.779	0.417	1							
(4) LNCAPITAL	0.64	0.537	0.487	1						
(5) LNFIRMAGE	− 0.137	− 0.09	− 0.1	− 0.191	1					
(6) LNLABPRODUCTIV ~ Y	0.689	0.845	0.082	0.454	− 0.103	1				
(7) LNCAP_LAB_RATIO	0.243	0.351	− 0.07	0.837	− 0.156	0.467	1			
(8) LNWAGERATE	0.792	0.628	0.543	0.572	− 0.16	0.631	0.314	1		
(9) LNPROFITSHARE	− 0.576	− 0.032	− 0.795	− 0.365	0.117	0.004	0.08	− 0.507	1	
(10) LNWAGESHARE	0.201	− 0.171	0.569	0.193	− 0.08	− 0.338	− 0.135	0.517	− 0.619	1

Source NSSO 73rd round unorganised (excluding construction) micro database (2015–16), computed by the authors

Table 10.8 Matrix of correlations (apparel)

Variables	(1)	(2)	(3)	(4)	(5)	(6)	(7)	(8)	(9)	(10)
(1) LNGVA	1									
(2) LNSURPLUS	0.865	1								
(3) LNLABOUR	0.775	0.524	1							
(4) LNCAPITAL	0.54	0.467	0.491	1						
(5) LNFIRMAGE	0.027	0.076	− 0.067	− 0.069	1					
(6) LNLABPRODUCTIV ~ Y	0.718	0.783	0.118	0.309	0.116	1				
(7) LNCAP_LAB_RATIO	0.039	0.141	− 0.182	0.767	− 0.028	0.262	1			
(8) LNWAGERATE	0.763	0.549	0.544	0.391	0.037	0.601	0.041	1		
(9) LNPROFITSHARE	− 0.559	− 0.067	− 0.675	− 0.303	0.072	− 0.135	0.155	− 0.611	1	
(10) LNWAGESHARE	0.447	0.133	0.597	0.271	− 0.035	0.046	− 0.134	0.826	− 0.668	1

Source NSSO 73rd round unorganised (excluding construction) micro database (2015–16), computed by the authors

Table 10.9 Matrix of correlations (leather)

Variables	(1)	(2)	(3)	(4)	(5)	(6)	(7)	(8)	(9)	(10)
(1) LNGVA	1									
(2) LNSURPLUS	0.843	1								
(3) LNLABOUR	0.807	0.498	1							
(4) LNCAPITAL	0.56	0.546	0.533	1						
(5) LNFIRMAGE	− 0.257	− 0.056	− 0.405	− 0.166	1					
(6) LNLABPRODUCTIV ~ Y	0.366	0.598	− 0.254	0.077	0.217	1				
(7) LNCAP_LAB_RATIO	0.055	0.27	− 0.12	0.776	0.107	0.28	1			
(8) LNWAGERATE	0.673	0.515	0.498	0.379	− 0.033	0.318	0.073	1		
(9) LNPROFITSHARE	− 0.406	0.148	− 0.64	− 0.102	0.378	0.344	0.357	− 0.362	1	
(10) LNWAGESHARE	0.468	0.171	0.656	0.339	− 0.162	− 0.267	− 0.091	0.829	− 0.571	1

Source NSSO 73rd round unorganised (excluding construction) micro database (2015–16), computed by the authors

References

Ariram N, Gopinath A, Madhan B, Rao JR (2022) Supercritical carbon dioxide fiber opening: a new paradigm for cleaner leather manufacture. Clean Technol Environ Policy 0123456789. https://doi.org/10.1007/s10098-022-02286-1

Aybar B, Ficici A (2009) Cross-border acquisitions and firm value: an analysis of emerging-market multinationals. J Int Bus Stud 40(8):1317–1338. https://doi.org/10.1057/jibs.2009.15

Bakos J, Siu M, Orengo A, Kasiri N (2020) An analysis of environmental sustainability in small and medium-sized enterprises: patterns and trends. Bus Strateg Environ 29(3):1285–1296. https://doi.org/10.1002/bse.2433

Bhandari AK, Maiti P (2012) Efficiency of the Indian leather firms: some results obtained using the two conventional methods. J Prod Anal 37(1):73–93. https://doi.org/10.1007/s11123-011-0219-1

Bhavan S, Rao JR, Nair BU (2008) A potential new commercial method for processing leather to reduce environmental impact. Environ Sci Pollut Res 15(4):293–295. https://doi.org/10.1007/s11356-008-0001-8

Borthakur A (2020) Policy approaches on E-waste in the emerging economies: a review of the existing governance with special reference to India and South Africa. J Clean Prod 252:119885. https://doi.org/10.1016/j.jclepro.2019.119885

Cattaneo O, Gereffi G, Staritz C (2010). Global Value Chains in a Postcrisis World. https://doi.org/10.1596/978-0-8213-8499-2

Chatterjee B, Zaman Mir M (2008) The current status of environmental reporting by Indian companies. Manag Audit J 23(6):609–629. https://doi.org/10.1108/02686900810882138

Chen L, Xu J, Zhou Y (2017) Regulating the environmental behavior of manufacturing SMEs: interfirm alliance as a facilitator. J Clean Prod 165:393–404. https://doi.org/10.1016/j.jclepro.2017.07.074

Courrent J-M, Chassé S, Omri W (2018) Do entrepreneurial SMEs perform better because they are more responsible? J Bus Ethics 153(2):317–336. https://doi.org/10.1007/s10551-016-3367-4

Dey PK, Malesios C, De D, Budhwar P, Chowdhury S, Cheffi W (2020) Circular economy to enhance sustainability of small and medium-sized enterprises. Bus Strateg Environ 29(6):2145–2169. https://doi.org/10.1002/bse.2492

Donaldson T, Preston LE (1995) The stakeholder theory of the corporation: concepts, evidence, and implications. Acad Manag Rev 20(1):65–91. https://doi.org/10.5465/amr.1995.9503271992

Durga J, Ranjithkumar A, Ramesh R, Girivasan KTPV, Rose C, Muralidharan C (2016) Replacement of lime with carbohydrases—a successful cleaner process for leather making. J Clean Prod 112:1122–1127. https://doi.org/10.1016/j.jclepro.2015.07.018

Gereffi G, Humphrey J, Sturgeon T (2005) The governance of global value chains. Rev Int Polit Econ 12(1):78–104. https://doi.org/10.1080/09692290500049805

Ghosh S, Mitra S (2023) How does subcontracting affect the performance of the informal enterprise? Source of advancement or exploitation? Vision: J Bus Perspect https://doi.org/10.1177/09722629231164862

Haines A, Kovats RS, Campbell-Lendrum D, Corvalan C (2006) Climate change and human health: impacts, vulnerability and public health. Public Health 120 (7):585–596. https://doi.org/10.1016/j.puhe.2006.01.002

Hettige H, Huq M, Pargal S, Wheeler D (1996) Determinants of pollution abatement in developing countries: evidence from South and Southeast Asia. World Dev 24(12):1891–1904. https://doi.org/10.1016/S0305-750X(96)00076-9

Jayanthi D, Victor JS, Chellan R, Chellappa M (2019) Green processing: minimising harmful substances in leather making. Environ Sci Pollut Res 26(7):6782–6790. https://doi.org/10.1007/s11356-018-04111-z

Joseph K, Nithya N (2009) Material flows in the life cycle of leather. J Clean Prod 17(7):676–682. https://doi.org/10.1016/j.jclepro.2008.11.018

Kabeer N (2004) Globalization, labor standards, and women's rights: dilemmas of collective (in)action in an interdependent world. Fem Econ 10(1):3–35. https://doi.org/10.1080/1354570042000198227

Kaul SN, Mukherjee PK, Sihorwala TA, Kulkarni H, Nandy T (1993) Performance evaluation of full scale wastewater treatment facility for finished leather industry. J Environ Sci Health. Part A: Environ Sci Eng Toxicol 28(6):1277–1286. https://doi.org/10.1080/10934529309375942

Leonidou LC, Christodoulides P, Kyrgidou LP, Palihawadana D (2017) Internal drivers and performance consequences of small firm green business strategy: the moderating role of external forces. J Bus Ethics 140(3):585–606. https://doi.org/10.1007/s10551-015-2670-9

Luo Y, Tung RL (2007) International expansion of emerging market enterprises: a springboard perspective. J Int Bus Stud 38(4):481–498. https://doi.org/10.1057/palgrave.jibs.8400275

Manivannan V, Elango L (2017) Leather processing and its possible impact on groundwater quality in Silk Road sites: a case study from Chennai, India. Environ Earth Sci 76(1):1–8. https://doi.org/10.1007/s12665-016-6352-4

Mitra S (2022) An exploratory study of sustainability and firm performance for Indian manufacturing small and medium enterprises. J Clean Prod 371:133705. https://doi.org/10.1016/j.jclepro.2022.133705

Mukherjee K, Sen M, Pattanaya JK (2010) Firm characteristics and corporate environmental disclosure practices in India. IUP J Account Res Audit Practices 9 (4):24–41

Musa AE, Madhan B, Kanth SV, Raghava Rao J, Chandrasekaran B, Gasmelseed GA (2010) Cleaner tanning process for the manufacture of upper leathers. Clean Technol Environ Policy 12(4):381–388. https://doi.org/10.1007/s10098-009-0216-5

Muthukumarana TT, Karunathilake HP, Punchihewa HKG, Manthilake MMID, Hewage KN (2018) Life cycle environmental impacts of the apparel industry in Sri Lanka: analysis of the energy sources. J Clean Prod 172:1346–1357. https://doi.org/10.1016/j.jclepro.2017.10.261

Nakhate PH, Moradiya KK, Patil HG, Marathe KV, Yadav GD (2020) Case study on sustainability of textile wastewater treatment plant based on lifecycle assessment approach. J Clean Prod 245:118929. https://doi.org/10.1016/j.jclepro.2019.118929

NCEUS (2007) Report on conditions of work and promotion of livelihoods in the unorganised sector. In Dolphin Printo Graphics. https://doi.org/10.1017/CBO9781107415324.004

Ngoc UN, Schnitzer H (2009) Sustainable solutions for solid waste management in Southeast Asian countries. Waste Manage 29(6):1982–1995. https://doi.org/10.1016/j.wasman.2008.08.031

Nurhayati R, Taylor G, Rusmin R, Tower G, Chatterjee B (2016) Factors determining social and environmental reporting by Indian textile and apparel firms: a test of legitimacy theory. Soc Responsib J 12(1):167–189. https://doi.org/10.1108/SRJ-06-2013-0074

Patz JA, Engelberg D, Last J (2000) The effects of changing weather on public health. Annu Rev Public Health 21(1):271–307. https://doi.org/10.1146/annurev.publhealth.21.1.271

Paul B, Patnaik U, Jaganth G, Murari K (2021) Employment, technology and value chain: a case of Indian leather industry. Sci Technol Soc Vc:1–24. https://doi.org/10.1177/09717218211010618

Paul B, Shaji R, Patnaik U, Pulikkamath A, Ahmed I, Ghosh S (2023) Transition to sustainable development in the tanning industry: evidence from leather value chain in Tamil Nadu, India. Sustain Dev March:1–12. https://doi.org/10.1002/sd.2560

Paul B (2021) Why is social sustainability indispensable? Indian J Soc Work 82(3):283–290. https://doi.org/10.32444/IJSW.2021.82.3.283-290

Raj D, Ma YJ, Gam HJ, Banning J (2017) Implementation of lean production and environmental sustainability in the Indian apparel manufacturing industry: a way to reach the triple bottom line. Int J Fashion Des Technol Educ 10(3):254–264. https://doi.org/10.1080/17543266.2017.1280091

Rossi M, Papetti A, Marconi M, Germani M (2021) Life cycle assessment of a leather shoe supply chain. Int J Sustain Eng 14(4):686–703. https://doi.org/10.1080/19397038.2021.1920643

Senthil Kumar P, Grace Pavithra K (2019) Water and textiles. In: Water in textiles and fashion. Elsevier, pp 21–40. https://doi.org/10.1016/B978-0-08-102633-5.00002-6

Serreli M, Petti L, Raggi A, Simboli A, Iuliano G (2021) Social life cycle assessment of an innovative industrial wastewater treatment plant. Int J Life Cycle Assess 26(9):1878–1899. https://doi.org/10.1007/s11367-021-01942-w

Singh SC, Gupta D (2013) Sustainability: a challenge for Indian leather industry. J Supply Chain Manag Syst 2(4):37–43

Sivaram NM, Gopal PM, Barik D (2018) Toxic waste from textile industries. Elsevier Ltd., In: Energy from toxic organic waste for heat and power generation. https://doi.org/10.1016/B978-0-08-102528-4.00004-3

Tonoyan V, Strohmeyer R, Habib M, Perlitz M (2010) Corruption and entrepreneurship: How formal and informal institutions shape small firm behavior in transition and mature market economies. Entrepreneurship: Theory and Practice 34(5):803–831. https://doi.org/10.1111/j.1540-6520.2010.00394.x

Uddin F (2021) Environmental hazard in textile dyeing wastewater from local textile industry. Cellulose 28(17):10715–10739. https://doi.org/10.1007/s10570-021-04228-4

11 Converging Realities: Unveiling the Intersection of COVID-19 and Climate Change Through Scientific Literature Mapping

Pasarul Islam, Enayat Bano,
Shakil Ahmad, Syed Naushad Ahmad,
and Isam Mohammed Abdel-Magid

Abstract

In recent years, the intersection of COVID-19 and climate change has garnered significant interest and concern. While many articles have been published on this topic during the COVID-19 pandemic, there is a lack of specific bibliometric studies that focus on the relationship between COVID-19 and climate change. To address this gap, the study conducted a comprehensive analysis using the Web of Science Core Collection bibliographic database. They retrieved 1970 publications authored by 8194 individuals from 1002 different sources up until June 10, 2023. The objective of this article is to provide a summary of the current scientific literature that maps the connections between COVID-19 and climate change. The study aimed to analyze publishing trends in literature, including top-cited documents, productive countries, institutions, journals, authorship patterns, collaborations, commonly used keywords, funding bodies, and concerns related to climate change such as climate emergency, climate crisis, climate breakdown, global heating, global warming, and COVID-19/SARS-CoV-2/coronavirus. By conducting an extensive review of peer-reviewed articles, reports, and expert opinions, the study seeks to understand the complex interplay between the COVID-19 pandemic and climate change. We utilized various bibliometric software for data analysis and integrated statistical parameters and indexes to provide a comprehensive ranking of entities. Overall, this research endeavor contributes to filling the research gap in understanding the connection between COVID-19 and climate change through bibliometric analysis. It sheds light on the current state of research, identifies influential publications and entities, and provides insights into the areas of focus within this field.

Keywords

COVID-19 · Climate change · SARS-CoV-2 · Web of science · Bibliometric analysis

P. Islam (✉)
Indian Council of Social Science Research, Ministry of Education, Government of India, New Delhi, India
e-mail: islam.pasarul@gmail.com

E. Bano · S. N. Ahmad
Department of Geography, Faculty of Science, Aligarh Muslim University, Aligarh, India

S. Ahmad
Directorate of Library Affairs, Imam Abdulrahman Bin Faisal University, Dammam, Saudi Arabia
e-mail: shahmad@iau.edu.sa

I. M. Abdel-Magid
Dean Graduate College & Scientific Studies, Al-Razi University, Khartoum, Sudan

S. Talukdar et al. (eds.), *Environmental Risk and Resilience in the Changing World*, GIScience and Geo-environmental Modelling, https://doi.org/10.1007/978-3-031-62442-1_11

11.1 Introduction

COVID-19 is a contagious illness caused by the SARS-CoV-2 virus, discovered in Wuhan, China in November 2019. It was first detected in Wuhan, China in November 2019. It has since become a global pandemic (Alberca et al. 2020). Symptoms include fever, cough, fatigue, shortness of breath, and, occasionally, diarrhea. Although children and teenagers usually only experience mild symptoms, it can be fatal for the elderly. The COVID-19 pandemic is characterized by a high transmission rate, requiring a comprehensive understanding of preventative measures (Alberca et al. 2020; Brüssow and Timmis 2021). This pandemic has had a profoundly impacted countries worldwide (Wei et al. 2020). Factors such as diabetes, hypertension, and cardiovascular disease, increase the likelihood of severe illness and mortality in COVID-19 patients (De Almeida-Pititto et al. 2020). Recent research by Yildirim et al. (2021) suggests that epigenetic factors, specifically genetic variations in genes related to the SARS-CoV-2 entry mechanism and genes involved in the host's innate immune response, play a pivotal role in determining disease severity. Moreover, individuals with malignancies or those who have undergone solid organ transplants may be at a heightened risk of experiencing severe symptoms and increased mortality rates when infected with COVID-19 (Omary et al. 2020). Given the serious and potentially life-threatening nature of COVID-19, particularly for those with underlying health conditions, further research is urgently needed to gain a deeper understanding of the factors contributing to disease severity.

Multiple studies have indicated a significantly high global death toll from COVID-19 compared to previous pandemics. Pitlik (2020) highlights that within the first six months of its emergence, COVID-19 had already claimed the lives of over 500,000 people worldwide. According to Chhapare and Bhutada (2020), as of April 13, 2020, global deaths had exceeded 1.1 million. Omary et al. 2020) compare COVID-19 to the Spanish flu of 1918–1919, which caused 675,000 deaths in the US and 50 million worldwide. While high-income countries have been most affected by COVID-19, developing countries may experience a threefold increase in their share of global fatalities. These findings emphasize the severity and urgency of implementing effective measures to control the virus's spread, particularly for vulnerable populations and regions with limited resources. However, numerous studies indicate a significant global death toll from COVID-19, and the pandemic continues (Schellekens and Sourrouille 2020; Simonsen and Viboud 2021; Xie et al. 2020).

COVID-19 has had both positive and negative impacts on the environment and society. The measures taken to combat COVID-19 have resulted in improved air quality and reduced greenhouse gas emissions, but they have also led to increased waste and reduced recycling (Mohialdin 2021). Cheval et al. (2020) demonstrate positive effects on air and water quality but negative impacts on shoreline pollution due to the disposal of sanitary consumables. Soni et al. (2021) suggest that the pandemic has had a positive impact on the environment, leading to environmental and biodiversity recovery. Bashir et al. (2020) highlight the economic consequences of COVID-19, including job losses and economic shutdowns, but they also note that restricted economic activities have contributed to a cleaner environment. In general, COVID-19 has had complex and varied effects on both the environment and society.

The COVID-19 pandemic has significantly impacted the global economy, affecting it through various channels, including measures implemented to slow the epidemic, a decrease in global demand, disruptions in value chains, and deteriorating financial conditions. This pandemic has led to a severe worldwide economic recession and poses unique challenges for China's economic development (Xuesong et al. 2021). It has devastated the global economy, plunging a significant proportion of the world's population into poverty and creating uncertainties regarding economic and social policies (Vitenu-Sackey and Barfi 2021). Begovic (2022) provides a comprehensive overview of the economic consequences of the COVID-19 pandemic, highlighting that both the supply and demand sides contributed to the sharp decline in economic growth, and macroeconomic policies prevented a global depression. Governments are investing in health and education improvements while stimulating their economies to create employment opportunities that can promote growth, thus enhancing poverty alleviation and economic development (Vitenu-Sackey and Barfi 2021). The relationship between COVID-19 and the environment is complex and multifaceted. The connection between COVID-19 cases and environmental factors such as temperature and humidity is mostly negligible to moderate (Bera et al. 2021).

COVID-19 has both positive and negative indirect effects on the environment, such as improved air quality but also increased waste (Zambrano-Monserrate et al. 2020). Shakil et al. (2020) conducted a critical review of 57 studies on the nexus between COVID-19 and the environment and identified four research clusters: environmental degradation, air pollution, climate/meteorological factors, and temperature. Furthermore, Biktasheva (2020) suggested a negative correlation between air humidity in a habitat and COVID-19 morbidity and mortality, but this hypothesis requires additional investigation. Alongside the health implications, the relationship between COVID-19 and the environment is complex and warrants further investigation.

The connection between COVID-19 and climate change has been explored by researchers, although the precise nature of this relationship remains unclear. Amnuaylojaroen and Parasin (2021) and Shaffril et al. (2022) discuss the interplay between COVID-19 and air pollution. Shaffril et al. (2022) found that the pandemic has led to reduced air pollution and lower carbon emissions, which could be considered positive outcomes. However, Amnuaylojaroen and Parasin (2021) note that studies have revealed both positive and negative correlations between air pollution and COVID-19. Further research is needed to fully understand this complex relationship. Muktavat and Kumar (2021) highlight that lockdown measures implemented during the pandemic have resulted in decreased pollution levels, offering a temporary respite for the climate change problem. However, they emphasize that long-term solutions are necessary to address both the ongoing COVID-19 crisis and the larger issue of climate change. In terms virus's spread, Ficetola and Rubolini (2020) discovered that COVID-19 growth rates were highest in temperate regions with specific temperature and humidity conditions, suggesting possible seasonal variations in outbreak patterns. This implies that climate factors might play a role in the transmission of the virus, but further exploration is required to fully understand this aspect. It is important to note that the COVID-19 pandemic has significantly impacted on global efforts to address climate change. Loureiro and Alló (2020) found that climate change-related messages on platforms like Twitter have significantly declined during the pandemic, postponing the climate debate worldwide, particularly in vulnerable countries.

The pandemic has not only impacted government actions but also public perception and discussions surrounding climate change. However, the connection between COVID-19 and the environment is complex and diverse, involving aspects such as air pollution, climate factors in virus transmission, and the pandemic's impact on climate change discussions. Further exploration and research are necessary to obtain a more comprehensive understanding of these dynamics. The pandemic has prompted individuals to reflect upon a more sustainable way of life and reconsider the anthropogenic impact on the environment (Ruiu et al. 2021). However, the failure to act in promptly on climate change may have contributed to the emergence and spread of the COVID-19 pandemic by increasing the potential risk of zoonotic diseases (Gorji and Gorji 2021). Additionally, the pandemic has resulted in both positive and negative environmental impacts, but the effects of COVID-19 are mainly determined by anthropogenic factors, which are becoming increasingly apparent as human activity diminishes across the planet (Cheval et al. 2020).

The climate can influence the spread and severity of COVID-19, and conversely, COVID-19 has a significant impact on the climate. The growth rates of Covid-19 peaked in temperate regions of the Northern Hemisphere with mean temperature of around 5 °C, and specific humidity of 4–6 g/m^3 during the outbreak period, while they were lower in both warmer/wetter and colder/drier regions (Ficetola and Rubolini 2020). An increase in temperatures is negatively associated with reported deaths, while higher relative humidity or wind speed and lower precipitation are negatively associated with confirmed cases (Aboura 2022). COVID-19 has caused short-term climate changes, including a reduction in greenhouse gas emissions and air pollution, but it has also lead to increased precipitation in coastal cities, which can contribute to flooding. It has also had positive effects on global climatic conditions, such as a reduction in global warming and air pollution, but it also underscores the need for special attention and efforts to control the pandemic (Chen et al. 2023). Therefore, it is crucial to recognize the interconnection between COVID-19 and climate change and address them together. The intersection of these two global challenges presents a complex and intertwined set of circumstances that require careful consideration.

This review highlights the importance of interdisciplinary research, policy coherence, and international collaboration to address the impacts of both crises. By acknowledging the interdependencies and shared solutions between COVID-19 and climate change, we can work toward building a more resilient and sustainable future for both humanity and the planet. Tackling these challenges presents opportunities to address the immediate health crisis posed by COVID-19 and align with the long-term sustainability goals of combating climate change. Integrated approaches that consider the health, environmental, and social dimensions of these global challenges are crucial. It is vital to adopt policies and take actions that protect human health, promote sustainable practices, reduce greenhouse gas emissions, enhance resilience to future pandemics, and foster equitable and inclusive societies is important. Ultimately, recognizing the interconnectedness of COVID-19 and climate change and striving for comprehensive solutions can prioritize the well-being of both people and the planet.

11.2 Literature Review

The Spanish flu pandemic, which occurred between 1918 and 1919 remains one of the most devastating in history, claiming the lives of an estimated 40–100 million individuals worldwide (Mijwil et al. 2022; Morens and Fauci 2007; Popkin 2005). It stands as an unparalleled tragedy that had a lasting impact on global health. Despite the severity of COVID-19, it is important to note that cardiovascular disease remains the leading cause of death globally, surpassing the toll of any war or pandemic (Wang and Tian 2021). This highlights the ongoing significance of non-pandemic health issues. In the case of COVID-19, Li et al. (2020) provide a comprehensive review of the diverse clinical manifestations associated with the disease, including respiratory, neurological, gastrointestinal, and cardiac symptoms. The research by Nueangnong et al. (2020) emphasizes the high number of confirmed cases and fatalities due to COVID-19. However, it is important to recognize that international preventive measures have contributed to a decrease in the mortality rate over time. When comparing COVID-19 to the Spanish flu, Omary et al. (2020) note that the 1918–1919 pandemic caused approximately 675,000 deaths in the United States alone. This historical context serves as a reminder of the devastating impact that pandemics can have on society. Furthermore, Khan et al. (2022) provide a comprehensive overview of the COVID-19 pandemic, discussing its global consequences and the absence of a definitive treatment or vaccine at the time of their study. While the COVID-19 pandemic is lethal and has widespread consequences, but it is not the deadliest of all pandemics, as exemplified by the Spanish flu. It is crucial to learn from history, understand ongoing health issues, and persist in striving for effective prevention, treatment, and international cooperation to mitigate the consequences of pandemic.

Numerous studies highlight the diverse consequences of COVID-19 and the importance of ongoing research and preventive measures. Brüssow (2021) discusses the long-term effects of COVID-19 and its broader societal impact. Quijije et al. (2020) emphasize the need for special precautions to prevent complications in newborns. Rassy et al. (2020) explore the neurological effects of COVID-19. Xiao et al. (2021) provide a comprehensive overview of COVID-19 and stress the importance of personal preventive measures. Zheng et al. (2021) discover a complex relationship between climate variables and COVID-19 transmission, where higher temperatures may slow down the epidemic but cannot fully account for variability. Ashok et al. (2021) find links between temperature, humidity, wind speed, and COVID-19 death rate and incidence. Briz-Redón et al. (2021) advise that hot weather alone cannot control the pandemic, and governments must continue restrictions. Stavrou et al. (2020) suggest that more research is needed to validate the link between COVID-19 transmission and higher temperature and humidity levels. A wide range of studies emphasizes the need for continued investigation and preventive measures. Research is needed to understand the long-term effects, address vulnerable populations, and study the neurological implications of the disease. The relationship between climate variables and COVID-19 transmission is complex. Ongoing research is crucial to inform effective strategies to combat the pandemic.

The COVID-19 pandemic has resulted in significant social and economic consequences globally. According to Mohseni Afshar et al. (2022), pandemic control measures have created major issues in the economic, social, political, and psychological spheres, impacting billions of people. Hajek et al. (2023) found that individuals with ASD and their families are particularly vulnerable to the psychosocial effects of the pandemic experiencing increased anxiety, distress, depression, financial problems, and loss of a job. Bashir et al. (2020) noted that the pandemic has caused severe demographic changes, unemployment, and economic shutdowns, while also contributing to a cleaner environment. Birimoğlu Okuyan et al. (2020) highlighted inequality, the revival of social values, and changes in daily life spaces as core research topics related to the pandemic's social impacts. Deb et al. (2022) found that containment measures have had a large impact on economic activity, resulting in a loss of approximately 15% in industrial production over a 30-day period following their implementation. Boissay and Rungcharoenkitkul (2020) estimated that global GDP growth for 2020 is around − 4%, with substantial downside risks if containment policies are prolonged. Islam et al. (2021) analyzed the potential impact of the pandemic on the global economy, including financial deeds, economic growth, supply chain, foreign investment, currency exchange ratio, overseas contribution, and identified several dimensions. Padhan and Bhat (2023) explored the effects of the pandemic and proposed potential policy directions to mitigate its impacts, suggesting coordination among monetary, macroprudential, and fiscal policies to mitigate the adverse economic effects of COVID-19.

COVID-19 has had both positive and negative impacts on the environment and climate. Cheval et al. (2020) found that the pandemic has led to enhanced air and water quality in urban areas, but also negative impacts such as shoreline pollution from the disposal of sanitary consumables. However, Chen et al. (2023) also noted that coastal cities experienced higher precipitation during COVID-19, which could lead to potential flooding. Ficetola and Rubolini (2020) found that COVID-19 growth rates peaked in temperate regions of the Northern Hemisphere with a mean temperature of 5 °C, and specific humidity of 4–6 g/m^3 during the outbreak period, while they were lower in warmer/wetter and colder/dryer regions. Kirana and Bhawiyuga (2021) found a significant relation between climate change and the increase in COVID-19 cases, but a previous study predicted that a tropical climate could help inhibit the virus outbreak. Shaffril et al. (2022) identified four emerging themes related to the inter-relationship between COVID-19 and climate change: reduced air pollution, low carbon emission, the impact of temperature and humidity variation on COVID-19, and the impact of air pollution on COVID-19. Zheng et al. (2021) found that COVID-19 has had a positive impact on global climatic conditions, including a significant reduction in greenhouse gas emissions and air pollution. Shakil et al. (2020) stated that the impact of pandemic crisis on achieving sustainable development goals in the context of energy change and pollution is controversial and complex, and that only a joint social effort and international collaboration can move toward a clean energy system.

The COVID-19 pandemic has had short-term effects on global energy demand and emissions, but the long-term consequences are uncertain. OECD (2021) revealed that environmental pressures related to greenhouse gas emissions and air pollutants dropped by around 7% in 2020, but emissions are projected to increase again from 2021. However, there may be a long-term downward impact on environmental pressures of 1–3%. Ortelli et al. (2021) highlight the challenges that the pandemic has introduced to the energy sector, including the integration of renewable energy sources and the need for new tools to support system operators during global health crises. Wojciech and Mateusz (2021) found that past pandemics led to a short-term fall in CO_2 emissions and an increase in the share of electricity generated from renewable sources, particularly in rich countries and countries experiencing economic recessions. They found that during the COVID-19 pandemic, energy consumption-impeded economic growth while economic expansion was the primary driver of energy resource consumption.

The mitigation measures of the COVID-19 pandemic have both positive and negative environmental consequences. Rume and Islam (2020) found that the pandemic situation significantly improves air quality, reduces GHG emissions, lessens water pollution and noise, and reduces the pressure on the tourist destinations. However, Arefi and Poursadeqiyan (2020) found that the COVID-19 pandemic also has negative effects on the environment, such as increasing chemical pollution in the air, water pollution, and the production of non-biodegradable waste. Cheval et al. (2020) argue that the effects of COVID-19 on the environment are mainly determined by anthropogenic factors which become apparent as human activity diminishes across the planet. They suggest that proactive international actions are required to not only save lives but also protect economic prosperity.

The COVID-19 pandemic and government responses have resulted in short-term reductions in environmental pressures, including greenhouse gas and air pollutant emissions (OECD 2021). However, emissions are projected to increase again from 2021 as growth rate recover, gradually approaching pre-COVID level. There is also a long-term downward impact on environmental pressures, potentially ranging from 1 to 3%. The pandemic may accelerate the trends of de-globalization and de-dollarization, posing challenging economic and geopolitical consequences (OECD 2021). The long-term effects of the pandemic on the economy are highly uncertain and may persist for several decades (Pachauri et al. 2021). Maqbool et al. (2023) examine the social, economic, and environmental ramifications of the pandemic, including short-term effects on oil, economic and agricultural policies, long-term implications of monetary and fiscal interventions, investments in green agreements, prospects for further de-globalization and its impact on climate change and nature, as well as intergenerational environmental consequences.

Casado-Aranda et al. (2021) conducted a bibliometric analysis to explore the growth of COVID-19 scientific publications in the field of environmental studies. Their study, which analyzed 440 articles, identified six main themes. These themes included the reduction in air pollution, improvement in water pollution, the relationship between wind speed, ultraviolet radiation, humidity, and infections rate. They also examined the impact of the pandemic on the food supply chain and waste habits, the use of wastewater monitoring as a warning for COVID-19 transmission, the application of artificial intelligence and intelligent devices in monitoring citizen mobilization, and the lessons from the pandemic that can inform actions to mitigate climate change. A similar finding was reported by Zhu et al. (2020).

The main objective of this research is to evaluate the quantity, significance, and leading areas of research in publications that explore the combined effect of COVID-19 on climate change since the beginning of the pandemic. Additionally, it aims to set research agenda for scholars and researchers who focus on the intersection of climate change and COVID-19.

11.3 Methodology and Methods

Searching query signified the main methodology followed in this research work. Main features of the query identified TS = ("Climate Change*" OR "climate emergenc*" OR "Greenhouse effect*" OR "climate breakdown*" OR "global heatin*" OR "global warmin*") AND TS = ("2019-nCoV" OR "Covid 19" OR "COVID-19" OR "SARS-CoV- 2" OR "coronavirus") NOT PY = (2023). The database relied heavily on the Web of Science for the retrieval date was 10th June 2023. Einecker and Kirby (2020) pointed out that using the Web of Science database demonstrates the growth of publications in the field, including papers on mitigation, adaptation, and resilience. Document types included articles and review articles. On the other hand, document types excluded conference papers, editorial material, letters, news items, data papers, book reviews, and corrections for their relatively lower impact as per the research objective and driving criteria. The primary data related to the research topic on COVID-19 and climate change between 2020 and 2022 are represented in Table 11.1. The study considered 1970 publications from 1002 different sources, including journals, books, and other relevant materials. The data's accuracy was ensured by repeating the process by one team member of the research group. Data Analysis was performed using various tools and software, including MS Excel, Biblioshiny (RStudio), ScientoPy, and Graphy.

Table 11.1 Statistics for potential research data and information

Description	Results
Periods	2020:2022
Sources (journals, books, etc.)	1002
Documents	1970
Annual growth rate %	82.21
Document average age	1.68
Average citations per doc	13.79
Document contents	
Keywords plus (ID)	3566
Author's keywords (DE)	5837
Authors	
Authors	8194
Authors of single-authored docs	356
Authors collaboration	
Single-authored docs	364
Co-authors per doc	4.51
International co-authorships %	38.63
Document types	
Article	1713
Review	257

11.4 Data Results and Analysis

The primary data related to the research topic on COVID-19 and climate change between 2020 and 2022 is represented in Table 11.1. The analysis reveals that 1970 publications were published in 1002 different sources (journals, books, etc.), with an average citation of 13.79 per document. These publications were contributed by 8194 authors. It was also revealed that out of the 8194 authors, 356 authored single-authored documents, while the remaining were multi-authored publications. Most of the researchers preferred to publish their research in the form of articles, with 1713 publications, followed by reviews with 257 publications. There are some abbreviation of terms, which is used for data analysis in various columns of the tables as NP stands for a number of publication, TC for total citation, ACPP for average citations per publication, AGR for average growth rate, ADY for average documents per year, PDLY for percentage of documents published in last year and APSY for author publication start year on this specific topic. This study's citation impact was calculated by dividing the total number of citations by the total number of publications. This illustrates the average number of citations a specific publication has received (Clarivate 2018). To determine the AGR, first, calculate the growth rate for each year. Then, calculate the average of all the growth rates obtained. Finally, add up the growth rates for each year and divide the sum by the number of years.

Yearly publication and citation structures on COVID-19 and climate change research are presented in Table 11.2 and Fig. 11.1. The data illustrate that the number of publications increased from 275 in 2020 to 913 in 2022, with an annual growth rate of 82.21%. However, the total number of citations and the average citations per publication have declined. In terms of citations, 2020 reported the highest number (n = 12,835), followed by 2021 (n = 10,456), and 2022 (n = 3868) citations. Similarly, the average citations per publication (ACPP) also decreased from 46.67 to 4.24 per document between 2020 and 2022.

Table 11.3 illustrates that authors studying COVID-19 and Climate Change contributed at least five articles during this study. Lui Z and Sharifi published the highest number of research articles, with ten and eight each, and received 121 and 143 citations respectively. They are followed by Rahman MM and Leal W, who each contributed seven publications and received 240

Table 11.2 Yearly publication and citation structures on Covid-19 and climate change

Year	Publications	Citations	ACPP
2020	275	12,835	46.67
2021	782	10,456	13.37
2022	913	3868	4.24

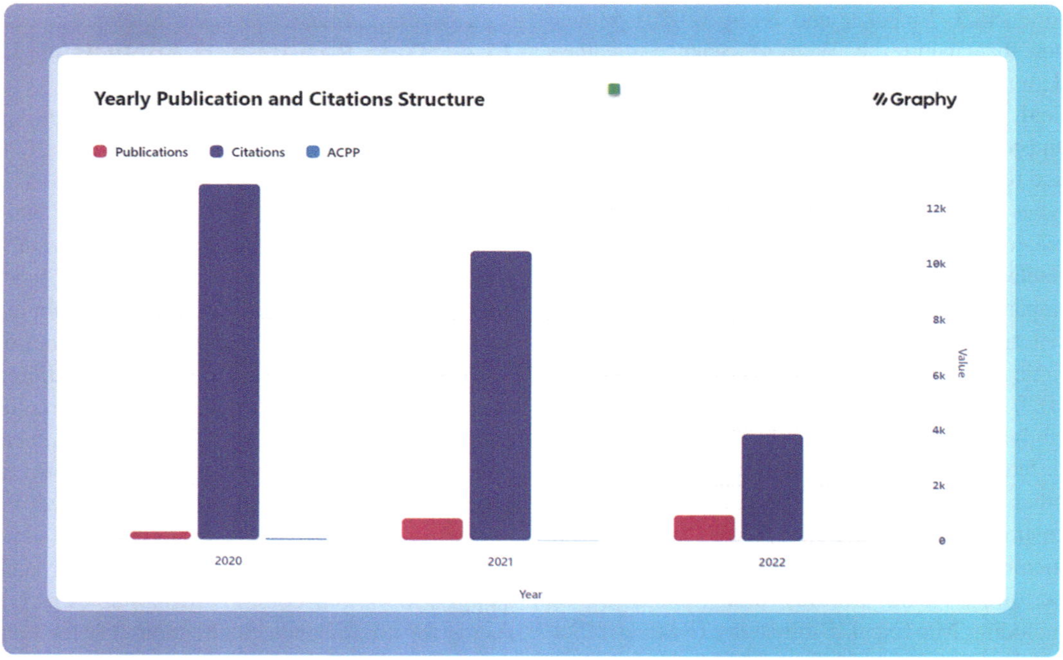

Fig. 11.1 Trends of publications and citations on COVID-19 and climate change

Table 11.3 Top influential author on COVID-19 and climate change

Rank	Authors	NP	TC	ACPP	h-index	g-index	APSY
1	Liu Z	10	121	12.10	8	10	2021
2	Sharifi A	8	143	17.88	4	8	2020
3	Rahman MM	7	240	34.29	7	7	2020
3	Leal W	7	46	6.57	4	6	2020
5	Gossling S	6	256	42.67	6	6	2020
5	Klemes JJ	6	97	16.17	5	6	2021
5	Kumar A	6	116	19.33	4	6	2021
5	Kumar R	6	104	17.33	3	6	2020
5	Liu Y	6	40	6.67	3	6	2021
10	Lee J	5	48	9.60	3	5	2020
10	Zhang Y	5	58	11.60	3	5	2021

and 46 citations respectively. Gossling S, Klemes JJ, Kumar A, Kumar R, and Liu Y published six articles each, with citation counts of 256, 97, 116, 104, and 40 respectively. Lee J and Zhang Y had the lowest contribution, with five articles each, and received 48 and 58 citations, respectively. Gossling S had the highest average citation per publication (42.67), followed by Rahman MM (34.29) and Kumar A (19.33), while Leal W had the lowest (6.57), followed by Liu Y (6.67), Lee J (9.60), Zhang Y (11.60), and Liu Z (12.10). Liu Z achieved the highest h-index and g-index of 8 and 10, respectively.

Table 11.4 displays top 10 countries in COVID-19 and climate change research publications, indicating a growing interest in this field. The United States leads with 547 published documents, followed by the United Kingdom with 337 publications. China ranks third with 243 published documents. Australia and Germany also demonstrate notable engagement with 185 and 166 publications, respectively. The figures in Table 11.4 emphasize global interest and collaborative efforts to understand the relationship between COVID-19 and climate change. This research is crucial for developing effective strategies and policies to address these interconnected global issues. Conversely, France (79), the Netherlands (84), Spain (107), India (116), and Canada (142) exhibited the least interest in COVID-19 and climate change-related publications. The data shows that the AGR and ADY in

Table 11.4 Top ten productive countries on COVID-19 and climate change

Rank	Country	Total	AGR	ADY	PDLY	h-Index	2020	2021	2022
1	USA	547	78.5	231.5	84.6	47	84	222	241
2	UK	337	50	146.5	86.9	37	44	149	144
3	China	243	57	113	93	32	17	95	131
4	Australia	185	32.5	81.5	88.1	22	22	76	87
5	Germany	166	34.5	76.5	92.2	24	13	71	82
6	Canada	142	19	58.5	82.4	24	25	54	63
7	India	116	19.5	52	89.7	21	12	53	51
8	Spain	107	20	47.5	88.8	21	12	43	52
9	Netherlands	84	17.5	38.5	91.7	18	7	35	42
10	France	79	15.5	34	86.1	18	11	26	42

the United States are the highest at 78.5 and 231.5, respectively, whereas France has the lowest at 15.5 and 34, respectively. It is evident that the number of published documents has significantly increased from 84 in 2020 to 241 in 2022 in the United States. Furthermore, the percentage of documents published in the last year (PDLY) is highest in China (93%), followed by Germany (92.2%) and the Netherlands (91.7%). On the other hand, it is lowest in Canada (82.4), followed by the United States (84.6%) and France (86.1%).

The contributions and trends of the top ten institutions in the world regarding COVID-19 and climate change research are represented in Table 11.5 and Fig. 11.2. The Chinese Academy of Sciences has produced the highest number of documents, with 34, and also has the highest average growth rate (AGR) and average number of documents per year, with 7.5 and 15, respectively, between 2020 and 2022. The University of Oxford is the second-highest contributor, with 25 publications and an average of 11 documents per year. The third-largest producer is Tsinghua University, with 20 publications, followed by Stanford University and University College London, each with 17 publications. Columbia University, Aarhus University, and the University of Queensland have contributed 16 documents each. Griffith University has produced only 13 documents and is the least contributor to COVID-19 and climate change-related research, followed by the University of Colorado with 14 publications.

The Chinese Academy of Sciences has acquired the maximum h-index of 12, followed by the University of Oxford, Tsinghua University, and Stanford University, each with a score of 10. The minimum h-index is observed for Columbia University and Griffith University, both with a score of 5, followed by the University of Queensland (6), Aarhus University (7), University College London, and the University of Colorado, both with a score of 8. Tsinghua University, the University of Colorado, and Griffith University are the top institutions in terms of the percentage of documents published in the last year (2021–22), with 100% each, followed by University College London (94.1%), Columbia University (93.8%), the Chinese Academy of Sciences (88.2%), and the University of Oxford (88%). On the other hand, Aarhus University is positioned at the bottom with 75%, followed by Stanford University (82.4%) and Stanford University (87.5%).

Table 11.6 represents the top ten highly influential Web of Science journals on COVID-19 and climate change-related publications. It shows that Sustainability has emerged as a leading journal, publishing 132 documents with the highest average growth rate of 17. The second most influential journal is the International

Table 11.5 Top ten institutions on COVID-19 and climate change

Rank	Institution	Country	Total	AGR	ADY	PDLY	h-Index	2020	2021	2022
1	Chinese Academy of Sciences	China	34	7.5	15	88.2	12	4	11	19
2	University of Oxford	UK	25	4	11	88	10	3	11	11
3	Tsinghua University	China	20	4.5	10	100	10	0	11	9
4	Stanford University	USA	17	3.5	7	82.4	10	3	4	10
5	University College London	UK	17	3.5	8	94.1	8	1	8	8
6	Columbia University	USA	16	4.5	7.5	93.8	5	1	5	10
7	Aarhus University	Denmark	16	3	6	75	7	4	2	10
8	University of Queensland	Australia	16	3	7	87.5	6	2	6	8
9	University of Colorado	USA	14	3	7	100	8	0	8	6
10	Griffith University	Australia	13	4.5	6.5	100	5	0	4	9

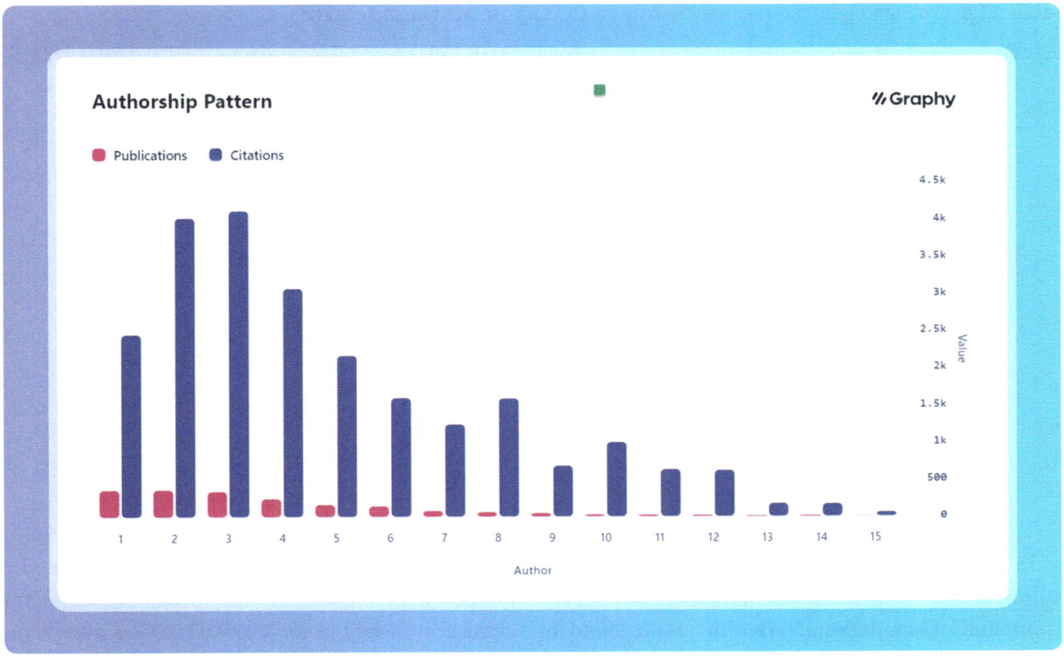

Fig. 11.2 Top 15 Authorship pattern of COVID-19 and climate change

Journal of Environmental Research and Public Health with 55 publications and an average growth rate of 8, followed by Science of the Total Environment with 43 documents and an average growth rate of 2.5. Energies accounted for 29 documents and an average growth rate of 6, and Environmental Science and Pollution Research with 28 publications and an average growth rate of 7.5. On the other hand, Marine Policy is the least productive journal, accounting for 8 publications and an average growth rate of 3, followed by Frontiers in Environmental Science, Frontiers in Sustainable Food Systems, Frontiers in Public Health, and Environmental Research Letters, with their publications ranging between 9 and 19. Sustainability also holds the top position in average documents per year with 55, followed by the International Journal of Environmental Research and Public Health (23.5), and Science of the Total Environment (16.5). The lowest average is observed for Marine Policy and Frontiers in Environmental Science, with 4 and 4.5, respectively. The average documents published per year for the remaining journals vary from 6 to 14. Sustainability again has the highest h-index of 20, while Marine Policy and Frontiers in Environmental Science have an h-index of three each.

Environmental Science and Pollution Research, Frontiers in Sustainable Food Systems, Frontiers in Environmental Science, and Marine Policy have the maximum influence in terms of the percentage of documents published in the last year (2021–22), with 100% each, followed by Energies (96.6%) and Environmental Research Letters (94.7%). On the other hand, Science of the Total Environment is positioned at the bottom, accounting for 76.7%, followed by Frontiers in Public Health (80.0%), Sustainability (83.3%), and the International Journal of Environmental Research and Public Health (85.5%).

Table 11.7 shows the contribution of the top ten subjects mostly associated with COVID-19 and climate change research between 2020 and 2022. It indicates that out of 1970 published documents, Environmental Sciences and Ecology is at the top with 688 documents, accounting for 34.92%. Science and Technology—Other Topics

Table 11.6 Top ten highly productive journals on COVID-19 and climate change

Rank	Name of journals	Total	AGR	ADY	PDLY	h-Index	2020	2021	2022
1	Sustainability	132	17	55	83.3	20	22	54	56
2	International Journal of Environmental Research and Public Health	55	8	23.5	85.5	13	8	23	24
3	Science of the total environment	43	2.5	16.5	76.7	18	10	18	15
4	Energies	29	6	14	96.6	10	1	15	13
5	Environmental Science and Pollution Research	28	7.5	14	100	12	0	13	15
6	Environmental Research Letters	19	2.5	9	94.7	7	1	12	6
7	Frontiers in Public Health	15	2	6	80	6	3	5	7
8	Frontiers in Sustainable Food Systems	12	3.5	6	100	5	0	5	7
9	Frontiers in Environmental Science	9	3.5	4.5	100	3	0	2	7
10	Marine Policy	8	3	4	100	3	0	2	6

is in second position with 14.77% ($n = 291$), followed by the research areas of Public, Environmental Occupational Health with 9.44% ($n = 186$), Business & Economics with 9.13% ($n = 180$), and Engineering with 5.83% ($n = 115$). The impact of research areas varies between 3 and 5% among the subjects of Social Sciences, Other Topics (4.97%), Energy & Fuels (4.72%), and Meteorology and Atmospheric Sciences (4.66%). The lowest impact is observed in the areas of International Relations (2.69%), followed by Psychology (3.81%), and Government and Law (3.81%).

Table 11.8 illustrates the author's keywords on COVID-19 and Climate Change related research contributed to at least forty-one articles during the period of this study. It was found that academicians commonly used "COVID-19" as a keyword when publishing their work on COVID-19 and climate change. This keyword had the maximum number of appearances (716), while "Climate Change" was the second most commonly used keyword with 489 occurrences. Similarly, the h-index of both keywords was much higher compared to others. The keyword "COVID-19" was used the most in 2020 with 309 appearances, while "Climate Change" appeared in 2022 with 224 appearances. Other important keywords included "Pandemic," "Sustainability," "Resilience," "COVID-19 Pandemic," "SARS-CoV-

Table 11.7 Top ten ranked research areas on COVID-19 and climate change

Rank	Research areas	Record count	% of 1970
1	Environmental sciences and ecology	688	34.92
2	Science and technology—other topics	291	14.77
3	Public, environmental occupational, health	186	9.44
4	Business and economics	180	9.13
5	Engineering	115	5.83
6	Social sciences, other topics	98	4.97
7	Energy and fuels	93	4.72
8	Meteorology and atmospheric sciences	88	4.46
9	Government and law	75	3.81
9	Psychology	75	3.81
10	International relations	53	2.69

Table 11.8 Top ten author's keywords on COVID-19 and climate change

Rank	Keywords	Total	AGR	ADY	PDLY	h-Index	2020	2021	2022
1	COVID-19	716	76.5	294.5	82.3	55	127	309	280
2	Climate change	489	71	203.5	83.2	39	82	183	224
3	Pandemic	119	9.5	46	77.3	22	27	46	46
4	Sustainability	88	10	38	86.4	20	12	44	32
5	Resilience	81	10	33.5	82.7	16	14	33	34
6	COVID-19 pandemic	75	18	34.5	92	12	6	27	42
7	SARS-CoV-2	47	3	18	76.6	15	11	19	17
8	Public health	46	5.5	20.5	89.1	11	5	25	16
9	Environment	41	6	18	87.8	11	5	19	17
10	Food security	41	6	18	87.8	11	5	19	17

2," "Public Health," "Environment," and "Food Security" with 119, 88, 81, 75, 47, 46, 41, and 41 appearances, respectively.

Figure 11.1 provides insights into the number of publications and citations on COVID-19 and climate change. It shows that the number of publications has significantly increased, while the number of citations has decreased over time. The burst of publications in the early stages of the Covid-19 pandemic has impacted field-normalized citation scores and will also affect the journal impact factor (Loureiro and Alló 2020). According to Brandt et al. (2022), Covid-19 papers published in the past two years have a median citation rate of 120.79, which is still high but has decreased compared to earlier in the pandemic. Kambhampati et al. (2021) also found that the top 50 most cited articles on Covid-19 were published in the early stages of the pandemic. Overall, while citation rates for Covid-19 papers remain high, they seem to have decreased over time.

Figure 11.2 demonstrates the top fifteen authorship patterns studied in relation to COVID-19 and Climate Change during this study period. The papers suggest that the high number of authors involved in COVID-19 research during the early stages of the pandemic may be attributed to the large volume of publications and the prevalence of secondary articles. Älgå et al. (2020) found a remarkably high number of research articles on COVID-19 published within six months since the novel coronavirus was first detected. Di Girolamo and Meursinge Reynders (2020) discovered that over half of the articles on COVID-19 were secondary articles, with a higher prevalence of reviews and guidance articles compared to in vitro and animal research studies. The COVID-19 pandemic required expertise and collaboration from different scientific disciplines, as it was a new and constantly changing situation. Researchers from different fields, such as virology, epidemiology, immunology, public health, geography and clinical medicine, came together to contribute their knowledge and skills in order to effectively understand and combat the virus.

Table 11.4 and Fig. 11.3 present the top ten highly productive nations in the domains of COVID-19 and climate change. These illustrations provide information on the total publications, average growth rate (AGR), and percentage of documents in recent years (ADY). Wang and Tian (2021) found that the United States has contributed the most published literature on COVID-19, followed by China. Teixeira da Silva et al. (2021) also found that the USA had the largest volume of published papers related to COVID-19. However, Surulinathi et al. (2021) focused on India's contribution to COVID-19 and climate change research and found that the Indian Institute of Technology, Kharagpur, and the Academy of Scientific and Innovative Research, New Delhi, contributed the

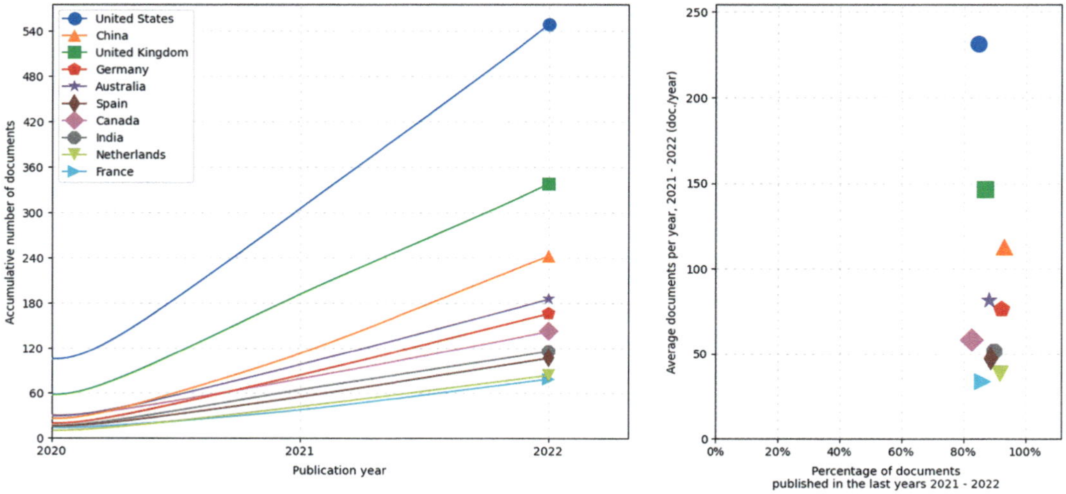

Fig. 11.3 Top ten productive countries on COVID-19 and climate change

largest number of papers. Overall, the United States has the largest publication output on COVID-19 and climate change due to its strong research infrastructure and funding, as well as its early and widespread response to the pandemic.

Figure 11.4 showcases the top ten highly productive institutions, which are located in five countries around the world. Among them, the USA has the maximum number of institutions, accounting for three, followed by China, UK, and Australia, each with two. Denmark is the only country with one institution contributing to COVID-19 and climate change research. Wang and Tian (2021) revealed that the United States has contributed the most published literature, with China following closely behind. Diéguez-Campa et al. (2021) also noted that China had the highest number of publications, with the United States in second place. Teixeira da Silva et al. (2021) highlighted that the top three countries, in terms of the volume of published papers, are the USA, China, and Italy. In contrast, Khan et al. (2022) found that while China has gained prominence in publishing articles related to green energy practices, the University of Cambridge emerged as the most contributing research institute. However, it should be noted that the USA and China are significant players in COVID-19 and climate change research, but the extent to which they are the most productive institutions varies depending on the specific research area.

Figure 11.5a and b depict the contribution of the top ten prominent research areas on COVID-19 and climate change between 2020 and 2022. The findings reveal that COVID-19 has significantly influenced the agenda of climate change research for approximately half of the surveyed researchers and is expected to continue doing so in future. This intersection of COVID-19 and climate change highlights the crucial role of public health nursing in addressing both issues (Zang et al. 2021). The most highly regarded research areas on COVID-19 and climate change include the interrelationship between the two (Shaffril et al. 2022), the environmental impacts of COVID-19 (Shakil et al. 2020), the potential of COVID-19 fiscal recovery packages to either accelerate or impede progress on climate change (Hepburn et al. 2020), and the effects of COVID-19 on climate change (Chen et al. 2023). The study identified several specific research areas, namely 'environmental science & ecology,' 'science & technology-others,' 'public environment & occupational health,' 'business & economics,' 'engineering,' 'social sciences & other,' 'energy & fuels,' 'meteorology & atmospheric sciences,' 'government & law,' and 'psychology'.

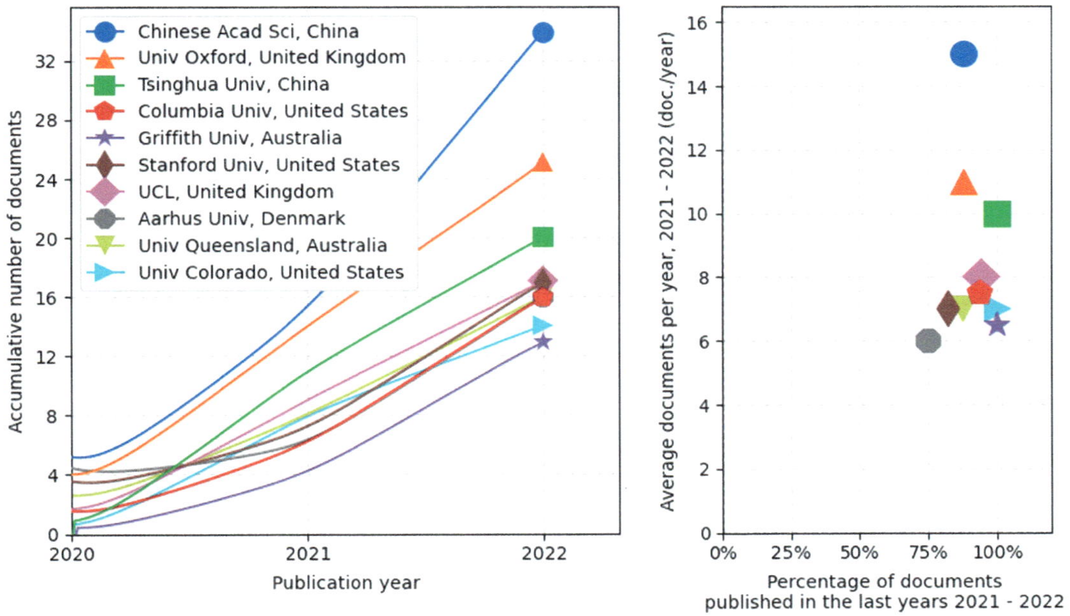

Fig. 11.4 Top ten highly influential institutions on COVID-19 and climate change

The co-occurrence network of author keywords is depicted in Fig. 11.6. The size of each ball represents the strength of the network and the total links with other keywords. Each color represents a separate cluster. The top keywords related to COVID-19 show frequent occurrences and strong linkages. The study of the citation map has contributed to the development of informative and visually appealing thematic networks and sub-themes, which are of great interest to scholars investigating the relationship between COVID-19 and climate change. These results specifically highlight five principal lines of research: COVID-19 and sustainable ecology, COVID-19 and greenhouse emissions with global warming, COVID-19 and transportation, COVID-19 and infectious diseases (Yang et al. 2020) in relation to climate policy, and COVID-19 and food sovereignty. These findings align with those of Sweileh (2020), who identified four research themes in health-related literature: climate change and infectious diseases, climate change, public health, and food security, heatwaves, mortality, and non-communicable diseases, and climate change, air pollution, allergy, and respiratory health.

The three-factor analysis of keywords, source, and country is illustrated in Fig. 11.7. The figure presents the three-factor analysis of the relationship among author keywords (left), affiliated organizations (middle), and country (right). It shows that ten countries (USA, China, United Kingdom, Australia, Canada, Germany, Spain, France, Italy, and India) have mostly published COVID-19 and climate change literature using nine main keywords (COVID-19, climate change, SARS-CoV-2, coronavirus, lockdown, pandemic, sustainability, air quality, and resilience) (Sharifi 2020). These countries and keywords have a strong relationship with well-known organizations and establishments. This finding aligns with the research conducted by Gul et al. (2020) on mapping the scientific literature on COVID-19 and mental health. They identified similar relationships with organizations such as Wuhan University, Huazhong Science & Technology, the University of Paris, King's College, the University of Pisa, Columbia University, and King Saud University.

Table 11.9 and Figure 11.8 provides data about different countries and their characteristics

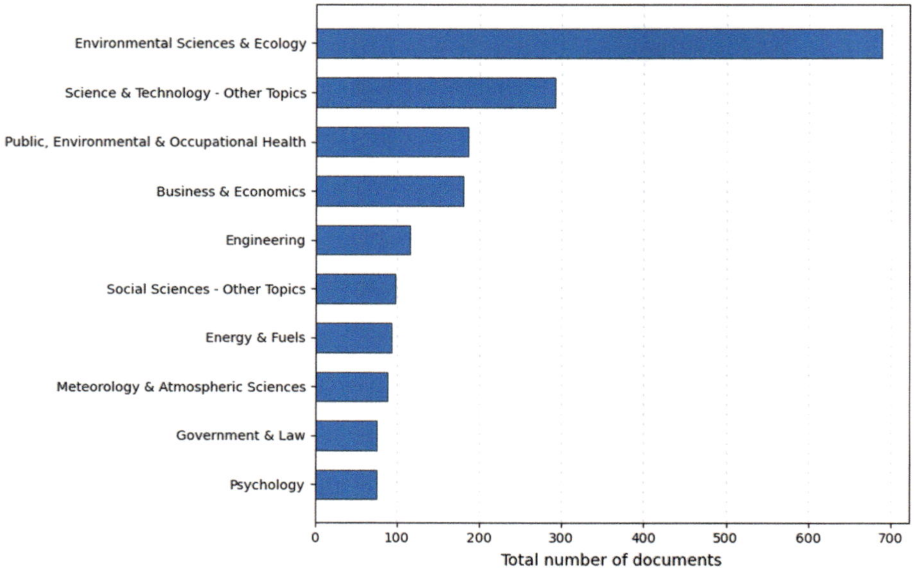

(a)

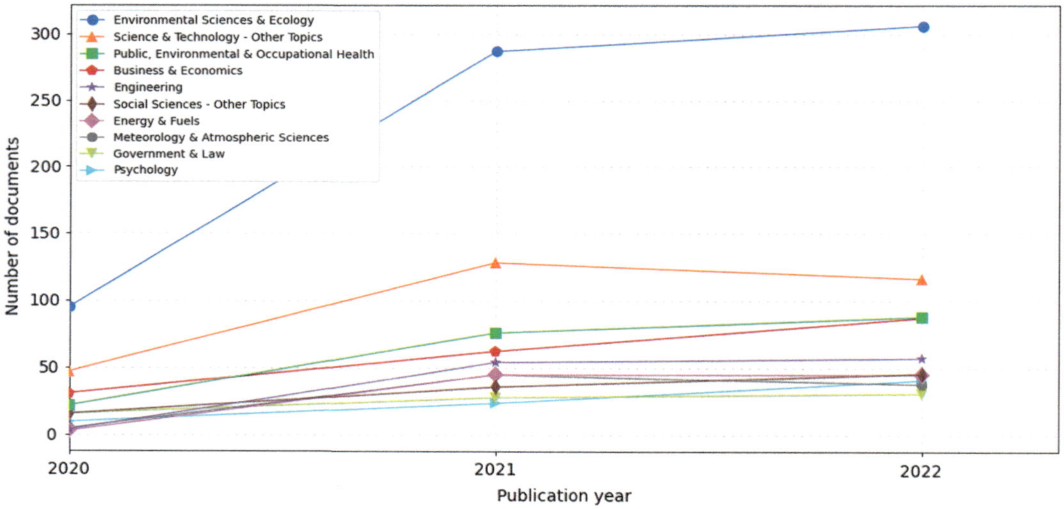

(b)

Fig. 11.5 **a** Top ten research areas on COVID-19 and climate change. **b** Trends of top ten research areas on COVID-19 and climate change

in a network. The countries are grouped into clusters numbered from 1 to 10. The table includes information on betweenness centrality, which measures how much a country acts as a bridge for information exchange between other countries. Closeness centrality indicates how easily information can spread from one country to others in the network. The PageRank metric reflects the importance or influence of a country based on the network's structure. The values in the table represent these metrics for countries like the USA, France, China, India, the United

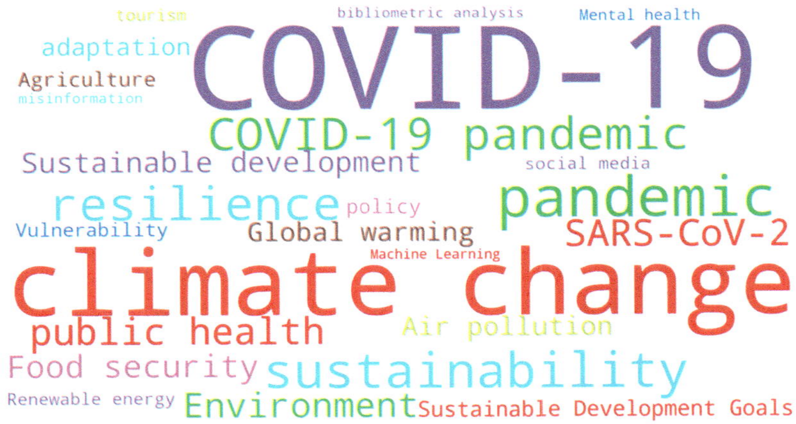

Fig. 11.6 Author's keywords on COVID-19 and climate change

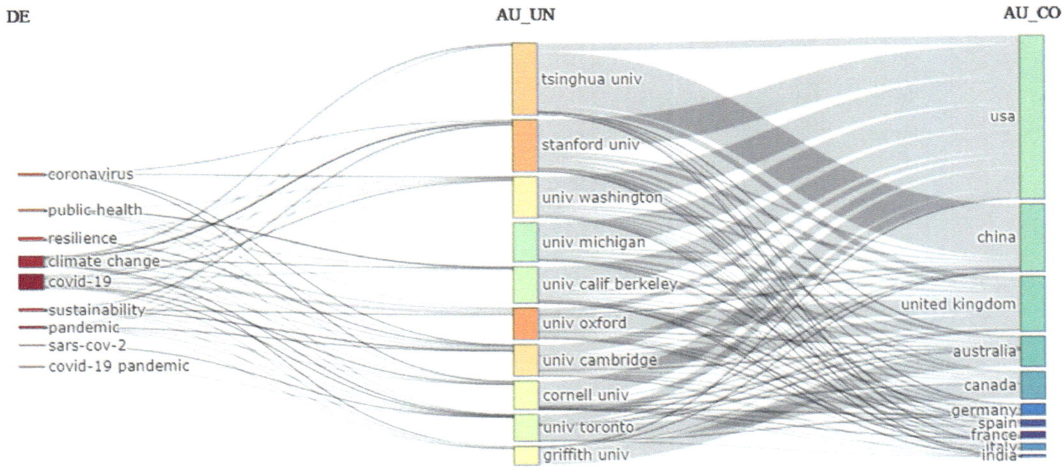

Fig. 11.7 Three-factor analysis of the relationship among keywords (left), source (middle), and country (right)

Kingdom, Canada, Australia, Finland, Italy, Spain, Germany, Netherlands, Brazil, Sweden, South Africa, Switzerland, Japan, Iran, Poland, and Austria.

Table 11.10 shows the top ten Funding Organizations for COVID-19 and Climate Change research. The table indicates that authors working on these topics contributed at least fourteen articles during the study period. Out of a total of 1970 published articles, the National Natural Science Foundation of China ranked first among the funding organizations, having funded 76 articles, which accounts for 3.85% of the total. The European Commission and UK Research Innovation jointly occupied the second position, each financing 2.64% of the published documents, followed by the National Science Foundation at 2.28%. The United States Department of Health and Human Services, CGIAR, National Institutes of Health, and Natural Environment Research Council each provided funding for approximately 1% of the articles, while the remaining three organizations (Social Sciences and Humanities Research Council of Canada, Spanish Government, and Australian Research Council) contributed less than 1%.

Table 11.9 Collaboration network (top 20 countries) on COVID-19 and climate change

Rank	Node	Cluster	Betweenness	Closeness	PageRank
1	USA	1	1.735	0.053	0.118
2	France	1	0.000	0.050	0.052
3	China	2	0.799	0.053	0.065
4	India	2	0.000	0.050	0.040
5	UK	3	0.819	0.053	0.110
6	Canada	3	0.000	0.050	0.059
7	Australia	4	0.381	0.053	0.060
8	Finland	4	0.000	0.050	0.032
9	Italy	5	1.138	0.053	0.047
10	Spain	5	0.000	0.050	0.047
11	Germany	6	1.719	0.053	0.076
12	Netherlands	6	0.626	0.053	0.047
13	Brazil	7	0.258	0.053	0.041
14	Sweden	7	0.000	0.050	0.040
15	South Africa	8	0.000	0.050	0.038
16	Switzerland	8	0.000	0.050	0.036
17	Japan	9	0.244	0.053	0.029
18	Iran	9	0.000	0.037	0.011
19	Poland	10	0.124	0.053	0.024
20	Austria	10	0.159	0.053	0.029

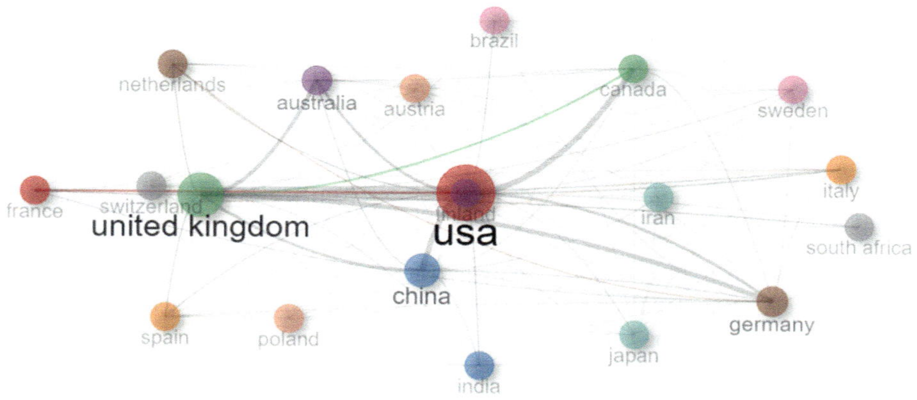

Fig. 11.8 Collaboration network map (top 20 countries) on COVID-19 and climate change

11.5 Conclusion

The current study aims to provide baseline data for researchers and policymakers. The results of the bibliometric analysis, conducted using the Web of Science and Scopus databases, revealed a total of 1970 articles and documents published in peer-reviewed scientific journals and reputable periodicals since the start of the pandemic in January 2020, focusing on the themes of COVID-19 and climate change.

Table 11.10 Funding agencies on COVID-19 and climate change

Rank	Funding agencies	Record count	% of 1970
1	National Natural Science Foundation of China (NSFC)	76	3.85
2	European Commission	52	2.64
2	UK Research Innovation (UKRI)	52	2.64
4	National Science Foundation (NSF)	45	2.28
5	United States Department of Health Human Services	23	1.16
6	CGIAR	20	1.01
6	National Institutes of Health (NIH)	20	1.01
6	Natural Environment Research Council (NERC)	20	1.01
9	Social Sciences and Humanities Research Council of Canada (SSHRC)	16	0.81
9	Spanish Government	16	0.81
10	Australian Research Council	14	0.71

The citation map study identified five main lines of research: COVID-19 and sustainable ecology, COVID-19 and greenhouse emissions and global warming, COVID-19 and transportation, COVID-19 and water quality in relation to infectious diseases and climate policy, and COVID-19 and food sovereignty.

Further research should aim to include all published materials from other platforms and means of disseminating information and findings. This comprehensive research should encompass preprints, early versions of scientific articles, early access materials, commentaries, critical reviews, news stories and narratives, open-access products, review articles, book chapters, proceedings papers, and other independent domains.

The three-factor study, which aimed at examine the relationship between keywords, countries, and organizations, highlighting the importance of resilience and stigma. It indicates that COVID-19 has influenced the agenda of climate change research and is likely to continue shaping it in future.

The intersection of COVID-19 and climate change presents a complex set of challenges and opportunities. This literature mapping underscores the need for interdisciplinary research, policy coherence, and international collaboration to develop effective strategies for mitigating the impacts of both crises. By recognizing the interdependencies and shared solutions, we can work toward building a more resilient and sustainable future for the well-being of both human and the planet.

National and international climate change agencies should actively promote and fund ongoing assessments and research by researchers on the effects of climate change on various aspects of the coronavirus and emerging strains that affect public health.

References

Aboura S (2022) The influence of climate factors and government interventions on the Covid-19 pandemic: evidence from 134 countries. Environ Res 208. https://doi.org/10.1016/J.ENVRES.2021.112484

Alberca GGF, Fernandes IG, Sato MN, Alberca RW (2020) What is COVID-19? Front Young Minds 8. https://doi.org/10.3389/FRYM.2020.00074

Älgå A, Eriksson O, Nordberg M (2020) Analysis of scientific publications during the early phase of the COVID-19 pandemic: topic modeling study. J Med Internet Res 22(11). https://doi.org/10.2196/21559

Amnuaylojaroen T, Parasin N (2021) The association between COVID-19, air pollution, and climate change. Front Public Health 9. https://doi.org/10.3389/FPUBH.2021.662499/PDF

Arefi MF, Poursadeqiyan M (2020) A review of studies on the COVID-19 epidemic crisis disease with a preventive approach. Work 66(4). https://doi.org/10.3233/WOR-203218

Ashok N, Krishnamurthy K, Singh K, Rahman S, Majumder MAA (2021) High COVID-19 vaccine hesitancy among healthcare workers: should such a

trend require closer attention by policymakers? Cureus. https://doi.org/10.7759/cureus.17990
Bashir MF, Ma B, Shahzad L (2020) A brief review of socio-economic and environmental impact of Covid-19. Air Qual Atmos Health 13(12):1403–1409. https://doi.org/10.1007/S11869-020-00894-8
Begovic B (2022) The global economic effects of the Covid-19 Pandemic: a bird's-eye view. Anali Pravnog Fakulteta u Beogradu 70(4):977–1015. https://doi.org/10.51204/ANALI_PFBU_22402A
Bera B, Bhattacharjee S, Shit PK, Sengupta N, Saha S (2021) Significant impacts of COVID-19 lockdown on urban air pollution in Kolkata (India) and amelioration of environmental health. Environ Dev Sustain 23 (5):6913–6940. https://doi.org/10.1007/S10668-020-00898-5
Biktasheva IV (2020) Role of a habitat's air humidity in Covid-19 mortality. Sci Total Environ 736. https://doi.org/10.1016/J.SCITOTENV.2020.138763
Birimoğlu Okuyan C, Karasu F, Polat F (2020) The effect of COVID-19 on health anxiety levels of nursing students. Van Med J 13(Special Issue)
Boissay F, Rungcharoenkitkul P (2020) Macroeconomic effects of Covid-19: an early review. BIS Bull 7(20)
Brandt MD, Ghozy SA, Kallmes DF, McDonald RJ, Kadirvel RD (2022) Comparison of citation rates between Covid-19 and non-Covid-19 articles across 24 major scientific journals. PLoS ONE 17(7 July). https://doi.org/10.1371/journal.pone.0271071
Briz-Redón Á, Belenguer-Sapiña,C, Serrano-Aroca Á (2021) Changes in air pollution during COVID-19 lockdown in Spain: a multi-city study. J Environ Sci (China) 101. https://doi.org/10.1016/j.jes.2020.07.029
Brüssow H, Timmis K (2021) COVID-19: long Covid and its societal consequences. Environ Microbiol 23(8):4077–4091. https://doi.org/10.1111/1462-2920.15634
Brüssow H (2021) What we can learn from the dynamics of the 1889 'Russian flu' pandemic for the future trajectory of COVID-19. Microbial Biotechnol 14(6). https://doi.org/10.1111/1751-7915.13916
Casado-Aranda LA, Sánchez-Fernández J, Viedma-del-Jesús MI (2021) Analysis of the scientific production of the effect of COVID-19 on the environment: a bibliometric study. Environ Res 193(110416)
Chen Y-H, Cheuh Y-N, Chen C-M, Kuo H-W (2023) Epidemiological characteristics of the three waves of COVID-19 epidemic in Taiwan during April 2022 to March 2023. J Form Med Assoc. https://doi.org/10.1016/J.JFMA.2023.05.027
Cheval S, Adamescu CM, Georgiadis T, Herrnegger M, Piticar A, Legates DR (2020) Observed and potential impacts of the COVID-19 Pandemic on the environment. Int J Environ Res Public Health 17(11):1–25. https://doi.org/10.3390/IJERPH17114140
Chhapare S, Bhutada RS (2020) Covid-19: a Pandemic situation—review article. Int J Res Pharm Sci 11 (Special Issue 1):1110–1115. https://doi.org/10.26452/IJRPS.V11ISPL1.3545
De Almeida-Pititto B, Dualib PM, Zajdenverg L, Dantas JR, De Souza FD, Rodacki M, Bertoluci MC (2020) Severity and mortality of COVID 19 in patients with diabetes, hypertension and cardiovascular disease: a meta-analysis. Diabetol Metab Syndr 12(1). https://doi.org/10.1186/S13098-020-00586-4
Deb P, Furceri D, Jimenez D, Kothari S, Ostry JD, Tawk N (2022) The effects of COVID-19 vaccines on economic activity. Swiss J Econ Stat 158(1). https://doi.org/10.1186/s41937-021-00082-0
Di Girolamo N, Meursinge Reynders R (2020) Characteristics of scientific articles on COVID-19 published during the initial 3 months of the pandemic. Scientometrics 125(1). https://doi.org/10.1007/s11192-020-03632-0
Diéguez-Campa CE, Pérez-Neri I, Reyes-Terán G, Flores-Apodaca IA, Castillo-Ledón-Pretelini J, Mercado-Bautista O, Álvarez-Santana R, Zenteno MA, Bowles B, Lee Á (2021) The 2020 research pandemic: a bibliometric analysis of publications on COVID-19 and their scientific impact during the first months. Archivos de Cardiologia de Mexico 91(COVID). https://doi.org/10.24875/ACM.20000370
Einecker R, Kirby A (2020) Climate change: a bibliometric study of adaptation, mitigation and resilience. Sustainability 12(17):6935
Ficetola G, Rubolini D (2020) Climate affects global patterns of COVID-19 early outbreak dynamics. MedRxiv. https://doi.org/10.1101/2020.03.23.20040501
Gorji S, Gorji A (2021) COVID-19 pandemic: the possible influence of the long-term ignorance about climate change. Environ Sci Pollut Res 28(13):15575–15579. https://doi.org/10.1007/S11356-020-12167-Z
Gul S, Rehman SU, Ashiq M, Danubina AK-P, 2020 undefined (2020) Mapping the scientific literature on COVID-19 and mental health. Hrcak.Srce.Hr 32 (4):463. https://doi.org/10.24869/psyd.2020.463
Hajek A, Kretzler B, Walther C, Aarabi G, Zwar L, König HH (2023) Neighbourhood cohesion, loneliness and perceived social isolation prior and during the COVID-19 pandemic. Longitudinal evidence from the German ageing survey. Soc Psychiatry Psychiatr Epidemiol. https://doi.org/10.1007/s00127-023-02447-7
Hepburn C, O'Callaghan B, Stern N, Stiglitz J, Zenghelis D (2020). Will COVID-19 fiscal recovery packages accelerate or retard progress on climate change? Oxf Rev Econ Policy 36. https://doi.org/10.1093/oxrep/graa015
Islam MS, Rahman ME, Banik R, Emran MGI, Saiara N, Hossain S, Hasan MT, Sikder MT, Smith L, Potenza MN (2021) Financial and mental health concerns of impoverished urban-dwelling Bangladeshi people during COVID-19. Front Psychol 12. https://doi.org/10.3389/fpsyg.2021.663687
Kambhampati SBS, Vasudeva N, Vaishya R, Patralekh MK (2021) Top 50 cited articles on Covid-19 after the first year of the pandemic: a bibliometric analysis. Diab Metabol Syndr Clin Res Rev 15(4). https://doi.org/10.1016/j.dsx.2021.05.013
Khan KI, Nasir A, Rashid T (2022) Green practices: a solution for environmental deregulation and the future of energy efficiency in the post-COVID-19 era. Front

Energy Res 10. https://doi.org/10.3389/fenrg.2022.878670

Kirana AP, Bhawiyuga A (2021) Coronavirus (COVID-19) pandemic in Indonesia: cases overview and daily data time series using Naïve forecast method. Indonesian J Electron Electromed Eng Med Inf 3(1). https://doi.org/10.35882/ijeeemi.v3i1.1

Li LQ, Huang T, Wang YQ, Wang ZP, Liang Y, Huang TB, ... Wang Y (2020) COVID-19 patients' clinical characteristics, discharge rate, and fatality rate of meta-analysis. J Med Virol 92(6):577–583

Loureiro ML, Alló M (2020) Climate affects global patterns of COVID-19 early outbreak dynamics. MedRxiv 124: 451–460. https://doi.org/10.1016/J.ENVSCI.2021.07.011

Maqbool M, Akhter T, Faheem M, Nadeem S, Park CH (2023). Erratum: CO_2 free production of ethylene oxide via liquid phase epoxidation of ethylene using niobium oxide incorporated mesoporous silica material as the catalyst. RSC Adv 13(8):1779–1786 https://doi.org/10.1039/D2RA07240H. https://doi.org/10.1039/d3ra90009f

Mijwil MM, Al-Mistarehi A-H, Zahran DJ, Alomari S, Doshi R (2022) Spanish flu (Great Influenza) 1918: the tale of the most deadly pandemic in history. Asian J Appl Sci 10(2). https://doi.org/10.24203/AJAS.V10I2.6949

Mohialdin SN (2021) Review on the positive and negative impact of Covid-19 pandemic on environment and society. Eur J Nat Sci Med 4(2):1. https://doi.org/10.26417/573VGE34V

Mohseni Afshar Z, Hosseinzadeh R, Barary M, Ebrahimpour S, Alijanpour A, Sayad B, Hosseinzadeh D, Miri SR, Sio TT, Sullman MJM, Carson-Chahhoud K, Babazadeh A (2022) Challenges posed by COVID-19 in cancer patients: a narrative review. Cancer Med 11(4). https://doi.org/10.1002/cam4.4519

Morens DM, Fauci AS (2007) The 1918 influenza pandemic: insights for the 21st century. J Infect Dis 195(7):1018–1028. https://doi.org/10.1086/511989

Muktavat K, Kumar V (2021) Testing link of climatic factors and air pollution with COVID-19 amid the second wave in India. J Environ Prot 12(12):1069–1085. https://doi.org/10.4236/JEP.2021.1212063

Nueangnong V, Hasan Subih AAS, Al-Hattami HM (2020) The 2020's world deadliest pandemic: Corona Virus (COVID-19) and International Medical Law (IML). Cogent Social Sci 6(1). https://doi.org/10.1080/23311886.2020.1818936

OECD (2021) The long-term environmental implications of COVID-19. OECD, May

Omary MB, Eswaraka J, Kimball SD, Moghe PV, Panettieri RA, Scotto KW (2020) The COVID-19 pandemic and research shutdown: staying safe and productive. J Clin Investig 130(6):2745–2748. https://doi.org/10.1172/JCI138646

Ortelli P, Ferrazzoli D, Sebastianelli L, Maestri R, Dezi S, Spampinato D, Saltuari L, Alibardi A, Kofler M, Quartarone A, Koch G, Oliviero A, Versace V (2021) Fatigue and "brain fog" in the aftermath of mild COVID-19: a neuropsychological and TMS study. J Neurol Sci 429. https://doi.org/10.1016/j.jns.2021.119854

Pachauri S, Poblete-Cazenave M, Aktas A, Gidden MJ (2021) Access to clean cooking services in energy and emission scenarios after COVID-19. Nat Energy 6(11). https://doi.org/10.1038/s41560-021-00911-9

Padhan L, Bhat S (2023) Interrelationship between trade and environment: a bibliometric analysis of published articles from the last two decades. Environ Sci Pollut Res 30(7). https://doi.org/10.1007/s11356-023-25168-5

Pitlik SD (2020) COVID-19 compared to other pandemic diseases. Rambam Maimonides Med J 11(3). https://doi.org/10.5041/RMMJ.10418

Popkin BM (2005) Using research on the obesity pandemic as a guide to a unified vision of nutrition. Public Health Nutr 8(6a):724–729. https://doi.org/10.1079/PHN2005776

Quijije R, Vargas G, Olaya Pincay ME, Cando N (2020) EVALUACIÓN DE ESTRÉS EN EL PERSONAL MÉDICO HOSPITALARIO ANTE EL COVID-19. Universidad Ciencia y Tecnología 24(106). https://doi.org/10.47460/uct.v24i106.401

Rassy E, Khoury-Abboud RM, Ibrahim N, Kattan C, Assi T, Kattan J (2020) What the oncologist needs to know about COVID-19 infection in cancer patients. Future Oncol 16(17). https://doi.org/10.2217/fon-2020-0312

Ruiu G, Ruiu ML, Ragnedda M (2021) How the COVID-19 pandemic impacted the perception of climate change in the UK. Am Behav Sci. https://doi.org/10.1177/00027642221085885

Rume T, Islam SMDU (2020) Environmental effects of COVID-19 pandemic and potential strategies of sustainability. Heliyon 6(9). https://doi.org/10.1016/j.heliyon.2020.e04965

Schellekens P, Sourrouille D (2020) Covid-19 mortality in rich and poor countries: a tale of two pandemics? World Bank: Infrastruct (topic). https://doi.org/10.1596/1813-9450-9260

Shaffril HAM, Samah AA, Samsuddin S.F, Sahharon H (2022) The emerging evidences related to the Interrelationship between COVID-19 and climate change: a scoping review perspective. J Trop Resour Sustain Sci (JTRSS) 10(2):41–49. https://doi.org/10.47253/JTRSS.V10I2.1006

Shakil MH, Munim ZH, Tasnia M, Sarowar S (2020) COVID-19 and the environment: a critical review and research agenda. Sci Total Environ 745. https://doi.org/10.1016/J.SCITOTENV.2020.141022

Sharifi A (2020) Urban resilience assessment: mapping knowledge structure and trends. Sustainability 12(15). https://doi.org/10.3390/su12155918

Simonsen L, Viboud C (2021) A comprehensive look at the COVID-19 pandemic death toll. ELife 10. https://doi.org/10.7554/ELIFE.71974

Soni M, Verma S, Jethava H, Payra S, Lamsal L, Gupta P, Singh J (2021) Impact of covid-19 on the air quality over china and India using long-term (2009–2020) multi-satellite data. Aerosol Air Qual Res 21(3). https://doi.org/10.4209/aaqr.2020.06.0295

Stavrou V, Chatziprodromidou I, Vantarakis A (2020) The battle against Legionella. Disinfection in Man-made water systems: a systematic review. J Environ Sci Public Heal 04(03). https://doi.org/10.26502/jesph.96120098
Surulinathi SM, Gupta GBM, Bhatkal SN, Bansal M (2021) Covid-19 and environmental sciences: a scientometric assessment of India's publications. J Young Pharm 13(3s). https://doi.org/10.5530/jyp.2021.13s.81
Sweileh WM (2020) Bibliometric analysis of peer-reviewed literature on climate change and human health with an emphasis on infectious diseases. Global Heal 16(1). https://doi.org/10.1186/S12992-020-00576-1
Teixeira da Silva JA, Tsigaris P, Erfanmanesh M (2021) Publishing volumes in major databases related to Covid-19. Scientometrics 126(1). https://doi.org/10.1007/s11192-020-03675-3
Vitenu-Sackey PA, Barfi R (2021) The impact of Covid-19 pandemic on the global economy: emphasis on poverty alleviation and economic growth. https://doi.org/10.17632/B2WVNBNPJ9.1
Wang P, Tian D (2021) Bibliometric analysis of global scientific research on COVID-19. J Biosaf Biosecur 3 (1). https://doi.org/10.1016/j.jobb.2020.12.002
Wei J, Liang G, Alex J, Zhang T, Ma C (2020) Research progress of energy utilization of agricultural waste in China: bibliometric analysis by citespace. Sustainability 12(3):812. https://doi.org/10.3390/SU12030812
Wojciech, K., & Mateusz, B. (2021). Effect of covid-19 pandemic on state funding of Polish football clubs. J Phys Educ Sport 21. https://doi.org/10.7752/jpes.2021.s2154
Xie Y, Wang Z, Liao H, Marley G, Wu D, Tang W (2020) Epidemiologic, clinical, and laboratory findings of the COVID-19 in the current pandemic: systematic review and meta-analysis. BMC Infect Dis 20(1). https://doi.org/10.1186/S12879-020-05371-2
Xuesong L, Hongju W, Ming F, Shuangshuang L, Binbin Z (2021) Economic impact of the COVID-19 pandemic. Econ Pandemic Weathering Storm Restoring Growth 26–45. https://doi.org/10.4324/9781003184447-2
Yang W, Zhang J, Ma R (2020) The prediction of infectious diseases: a bibliometric analysis. Int J Environ Res Public Heal 17(17):6218. https://www.mdpi.com/809586
Yildirim Z, Oyku | Sahin S, Seyhan Yazar | Vildan | Cetintas B (2021) Genetic and epigenetic factors associated with increased severity of Covid-19. Wiley Online Library 45(6):1158–1174. https://doi.org/10.1002/cbin.11572
Zambrano-Monserrate MA, Ruano MA, Sanchez-Alcalde L (2020) Indirect effects of COVID-19 on the environment. Sci Total Environ 728. https://doi.org/10.1016/J.SCITOTENV.2020.138813
Zang SM, Benjenk I, Breakey S, Pusey-Reid E, Nicholas PK (2021) The intersection of climate change with the era of COVID-19. Public Heal Nurs 38(2). https://doi.org/10.1111/phn.12866
Zheng B, Zhang Q, Geng G, Chen C, Shi Q, Cui M, Lei Y, He K (2021) Changes in China's anthropogenic emissions and air quality during the COVID-19 pandemic in 2020. Earth Syst Sci Data 13(6). https://doi.org/10.5194/essd-13-2895-2021
Zhu J, Kang S, Zhao W, Li Q, Xie X, Land XH (2020) A bibliometric analysis of food–energy–water nexus: Progress and prospects. Mdpi. Com 9(12):514

12 Evaluating Ecosystem Health: A Study on Asansol–Durgapur Development Region, West Bengal, India

Rumki Khatun, Sekh Sayeed, and Somen Das

Abstract

For effective ecosystem governance on a regional and global level, dynamic ecosystem health (EH) monitoring and assessment are essential. The vigor–organization–resilience (V–O–R) model is used in the current study to evaluate the Asansol–Durgapur Development Region's ecosystem health. The landscape's vigor, organization, and elasticity were measured using landscape metrics. The findings show an increase in settlement areas from 3.29% in 1990 to 7.20% in 2020 and an increase in agricultural croplands from 57.63% in 1990 to 73.44% in 2020. On the contrary, vegetation decreased from 28.84% in 1990 to 17.35% in 2020. EH decreased from 0.86 to 0.38 between 1990 and 2020, and good EH decreased from 33.66 to 3.51%. Alternately, weak EH increased during this period, reaching 66.70%. Increasing agricultural lands and urban areas is responsible for the deterioration of EH in this region. The findings of this study also provide policymakers with a better understanding of EH in the Asansol–Durgapur Development Area, which will help them develop eco-environmental management strategies.

Keywords

Asansol Durgapur development region · Ecosystem health · Vigor organization resilience (VOR) · Health degradation

12.1 Introduction

Ecosystems offer a variety of crucial advantages for human well-being. Natural ecosystems worldwide have undergone unprecedented change, degrading ecosystem services (ES), and ecosystem health (EH) due to urbanization and industrialization (Zhang et al. 2017). Ecosystem health (EH) is the capacity of an ecosystem to satisfy the reasonable demands of society and preserve quality (Lackey 2001). A healthy ecosystem depends on three major aspects, i.e., ecosystem vigor; ecosystem organization; and resilience (Costanza et al. 1997). Ecosystem vigor denotes activity and primary productivity. Ecosystem organization is defined as diversification and interactions among various components, and ecosystem resilience is the capacity of an ecosystem to preserve or rebuild its structure (Wang et al. 2020; Ekumah et al. 2020). One of the significant essential bearings of environmental assessment and a critical factor in ecosystem monitoring is considered to

R. Khatun · S. Sayeed · S. Das (✉)
Department of Geography, Kazi Nazrul University, Asansol, West Bengal 713340, India
e-mail: das.somen93@gmail.com

S. Talukdar et al. (eds.), *Environmental Risk and Resilience in the Changing World*, GIScience and Geo-environmental Modelling, https://doi.org/10.1007/978-3-031-62442-1_12

be ecosystem health (Wilkins 1999). In recent years, pertinent research on assessing ecosystem health has been conducted. The ecosystem health assessment also considers global, regional, watershed, administrative, and specific ecoregional levels. EH is vital for determining regional ecological traits from two central angles: ecosystem functions and structure.

In the literature, two primary techniques have been broadly determining an ecosystem's health: the index method and the indicator species method (Kang et al. 2018). Many index-based framework models have been developed, and these models were applied to monitor ecological health. Examples of the models include Pressure State Response (PSR) (Li et al. 2021), Driving Force State Response (DSR) (Wu et al. 2022), Vitality Organization Resilience (V–O–R) (Peng et al. 2017), and Driving Force Pressure State Impact Response (DPSIR) (Cheng et al. 2018; Niekerk et al. 2013). In India, studies on EH assessment are limited. Therefore, the present research focuses on evaluating ecosystem health in Indian context.

12.2 Study Area

The research area covers an area of around 1651 km^2 and consists of 2 municipal corporations, 3 municipalities, and 8 CD blocks of the region. This region's northern and southern sides are surrounded by the Ajay River and the Damodar River, respectively. Latitude and longitude of this region are 23° 22′ 20″ to 23° 53′ 59″ and 86° 46′ to 87° 29′, respectively (Fig. 12.1). There were 28, 82,031 residents in 2011 as opposed to 25, 52,781 in 2001. The population expanded by 12.90% per decade between 2001 and 2011. Hot summers, moderate to heavy monsoon rains, and dry winters define this area's climate. In the summer, the highest temperature is 44 °C, while in the winter, it is 5 °C. This area receives 1408 mm of rainfall on average per year.

12.3 Methodology

The study uses Landsat 4–5 TM images (1990, 2000, and 2010), and Landsat 8 OLI images (2020) with a 30 m resolution were collected for NDVI, elasticity, fragmentation, and LULC mapping.

12.3.1 Land Use Land Cover

This study analyzed land use and land cover evolution over time using the maximum likelihood method, a widely employed supervised classification technique for processing remote sensing image data. The basis of this method is the probability that a pixel is a member of a particular class. The underlying theory assumes that the input bands have normal distributions and that their probabilities are the same for all classes. Compared to other supervised algorithms, it requires the least processing time (Ahmadi and Hames, 2009). Several signatures have been gathered for each of the seven land use classes.

The accuracy of LULC has been assessed using Kappa coefficient (k) which measures the degree of agreement between user and producer ratings.

12.3.2 Ecosystem Vigor

Ecosystem vigor is a term used to describe an environment's primary production, metabolic capability, and activity (EV). Utilizing the NDVI, various studies have looked at ecosystem productivity and spatial distribution. NDVI has a strong positive correlation with primary productivity, typically used to gauge the health of an ecosystem.

12.3.3 Ecosystem Organization

Landscape shape, landscape connectedness, and spatial heterogeneity are measuring indicators that can be applied to analyze an ecological organization (EO). It covers landscape aggregation/fragmentation, heterogeneity, and fractal

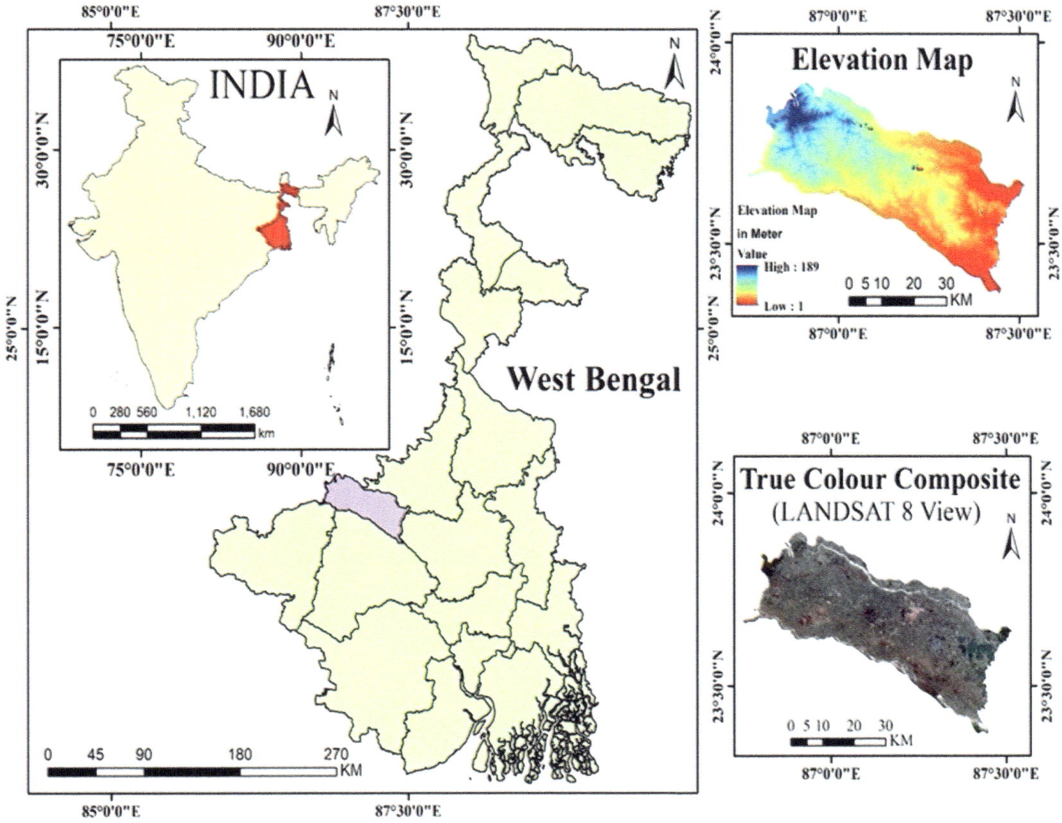

Fig. 12.1 Spatial location, elevation, and images of study area

dimension (Turner 1989). Figure 12.2 lists the criteria used to determine ecosystem organization.

$$EO = (0.35 \times LH) + (0.3 \times LC) + (0.35 \times CC), \quad (12.1)$$

$$\begin{aligned} EO = {} & (0.25 \times SHDI) + (0.18 \times PARA) \\ & + (0.2 \times FRAC) + (0.1 \times AI) \\ & + (0.25 \times FRAC - mean) \\ & + (0.1 \times COHESION). \end{aligned} \quad (12.2)$$

12.3.4 Ecosystem Elasticity

Ecosystem elasticity gauges a system's resiliency and resistance to outside forces (disturbances). Based on their research, Peng et al. (2015) and Xiao et al. (2019) calculated ecosystem elasticity (Eq. 12.3)

$$EE = (0.3 * Resil) + (0.7 * Resist). \quad (12.3)$$

Where, EE denotes ecosystem elasticity, while Resil and Resist denote ecosystem resilience and ecosystem resistance. Ecosystem elasticity describes an ecosystem's capacity to recover from disturbances caused by people or the environment to reach its pre-disturbance state. A robust ecosystem is resilient enough to withstand different types of interference. Resistance and resilience to outside disturbances can be used to describe ecosystem elasticity. Given the rapid expansion of the Asansol and Durgapur Municipality area and the resulting rise in interferences and disruptions, this study focused on the relevance of resilience rather than resistance.

Shannon's Diversity Index (SHDI)	measure of diversity in community ecology
Aggregation Index (AI)	frequency of neighbouring patches of various sorts
Fractal Dimension Index (FRAC)	shape complexity across multiple spatial scales
Patch Cohesion Index (COHESION)	measures the physical interconnectedness of the relevant patch type
Perimeter-Area ratio (PARA)	assessing shape complexity (e.g., square) without standardizing to a basic Euclidean shape can be done simply by using the perimeter-area ratio

Fig. 12.2 Parameters for ecosystem organization

According to Peng et al. (2017), the coefficients of resistance and resilience utilized for various land use categories, such as agricultural land (resilience coefficient: 0.4; resistance coefficient: 0.6); settlement (resilience coefficient: 0.2; resistance coefficient: 0.3); fallow land (resilience coefficient: 1; resistance coefficient: 0.2); vegetation (resilience coefficient: 0.8; resistance coefficient: 1); water bodies (resilience coefficient: 0.7; resistance coefficient: 0.8); and mining area (resilience coefficient: 1; resistance coefficient: 0.3).

12.3.5 Applying AHP for Ecosystem Organization Parameters

One of the multi-criteria decision-making processes for handling a complicated problem is the Analytical Hierarchy Process (AHP) (Saaty 1980). Using the AHP technique, this study examined EH and classified it as an entity stratum. The object layer was then used to create the index and sub-index levels. The importance of the layers concerning the dominance of the landscape pattern was ranked. Using the AHP method, the qualities were ranked according to their relative relevance.

12.3.6 Ecosystem Health Calculation

According to Santos-Burgoa et al. (2001), two factors determine the health of an ecosystem: (1) its ability to benefit human beings and (2) its power to assert structure and functionality over time. A key indicator of EH is the supporting capacity of an ecosystem. In this study, the EH of the region is determined by applying the V–O–R model, which considers three leading indices: (1) vigor, (2) organization, and (3) resilience. Each indicator depends on a wide range of factors. The values of the indicators should be standardized because different measurement scales are used for each component. The formula for the V–O–R model is as follows:

$$EH = \sqrt[3]{V \times O \times R}. \quad (12.4)$$

EH stands for ecosystem health, V for vitality, O for an organization, and R for resilience. The integrated EH of a region is based on the condition of an ecosystem's health. High-biodiversity ecosystems can offer various services in qualitative and quantitative terms (Speldewinde et al, 2015).

12.4 Result and Analysis

12.4.1 Dynamics of Land Use Land Cover Change

Result indicates that vegetation zones and water bodies disappear as a result of significant changes in LULC brought about by the expansion of agriculture and human habitation (Fig. 12.3). Built-up area increased progressively by 3.29%, 3.68%, 4.44%, and 7.21% in 1990, 2000, 2010, and 2020, respectively. Besides that, agricultural lands increased to 73.44% in 2020 from 57.63% in 1990 (Table 12.1). Mining areas grew from 1990 (0.048%) by 0.262%, 0.985%, and 1.398% in 2000, 2010, and 2020, respectively.

Vegetation declined from 28.84% in 1990 to 26.42%, 18.29%, and 17.35% in 2000, 2010, and 2020, respectively. Significant declines in vegetation, particularly in terms of ecosystem health, are linked to increases in agricultural land and built-up regions.

12.4.2 Ecosystem Vigor–Organization–Resilience

Figure 12.4 depicts that the vigor index ranges from 0 to 0.68 in last three decades. The areas that displayed high vigor in 1990 decreased in 2020. It suggests that since 1990, the ecosystem vigor has dropped. Conversely, the organization index falls between 0.33 and 1 (Fig. 12.5) and has risen since 1990, indicating that external disturbances are expanding quickly. The range of the elasticity index is from 0.01 to 1 (Fig. 12.6). As time goes on, elasticity also decreases.

12.4.3 Dynamics of Ecosystem Health

The ecosystem's health has been assessed using Eq. 12.4. Ecological health is divided into five categories: very weak, weak, moderate, strong, and very strong. EH has seen a considerable alteration since 1990 (Fig. 12.7). As seen in Fig. 12.7, previously stronger zones have turned into weak or severely weak zones. On the other hand, the growth in weak areas was 9.43% in 1990, 28.71% in 2000, 32.27% in 2010, and 17.94% in 2020. The weak area increased from only 2.25% in 1990 to 12.25% in 2000, 2.5% in 2010, and 66.7% in 2020 (Table 12.2). The entire region experienced a higher decline in terms of ecosystem health. Additionally, vegetation health has been considerably degraded (Fig. 12.7).

Table 12.1 Different types of land cover and land usage for 1990, 2000, 2010 and 2020

LULC	1990		2000		2010		2020	
	Area (km^2)	Area (%)	Area (km^2)	Area (%)	Area (km^2)	Area (%)	Area (km^2)	Area (%)
Water bodies	89.85	4.84	87.38	4.71	46.10	2.48	43.04	2.32
Settlement	61.02	3.29	68.35	3.68	82.35	4.44	133.71	7.2
Agricultural crop land	1069.56	57.63	1138.26	61.33	1311.84	70.68	1362.99	73.44
Vegetation	535.27	28.84	490.27	26.43	339.47	18.29	322.03	17.35
Mining area	0.89	0.048	4.86	0.26	18.28	0.98	25.95	1.4
Sand	53.68	2.89	53.09	2.86	49.62	2.67	37.97	2.05

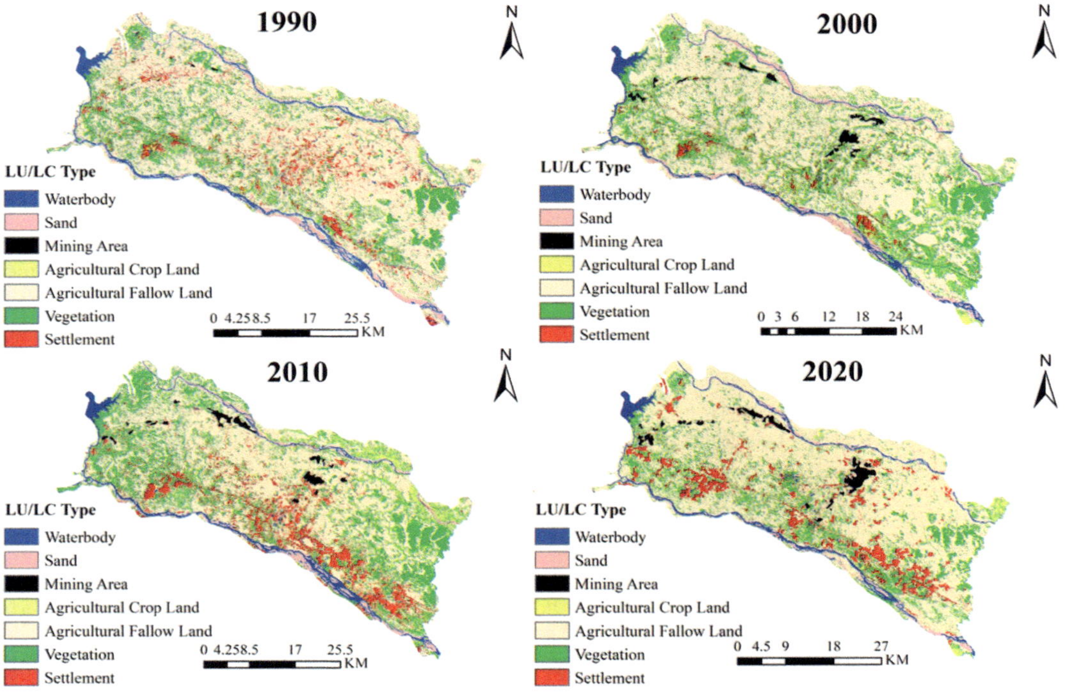

Fig. 12.3 LULC of Asansol–Durgapur Development Region for 1990, 2000, 2010, and 2020

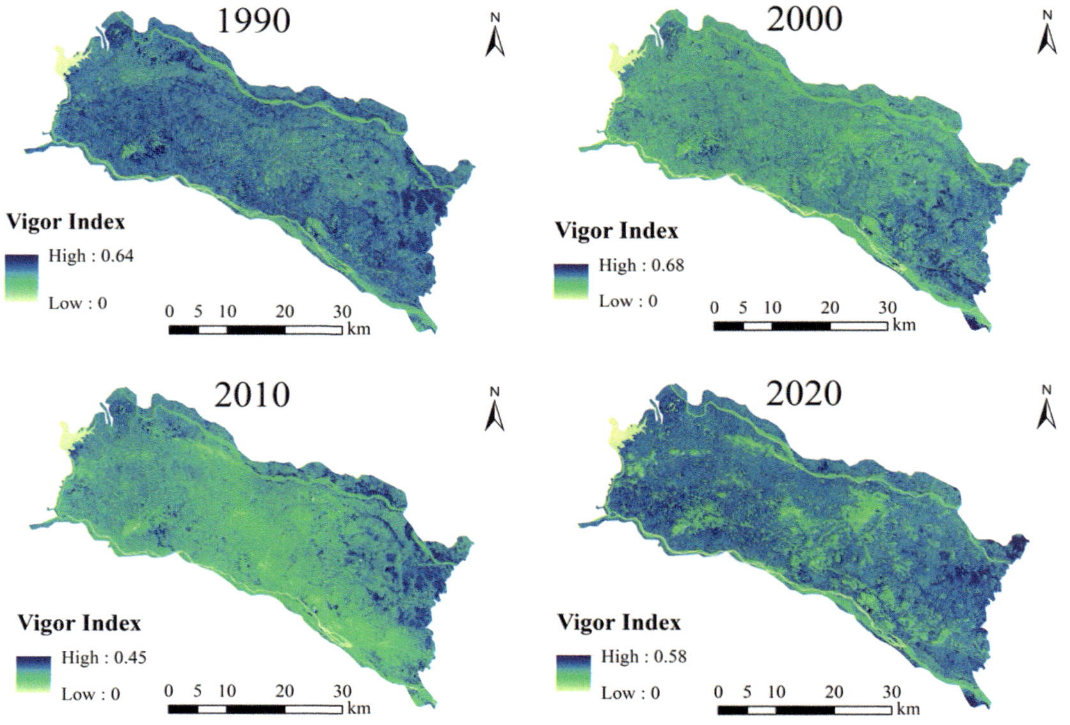

Fig. 12.4 Vigor index of Asansol–Durgapur Development Region for 1990, 2000, 2010, and 2020

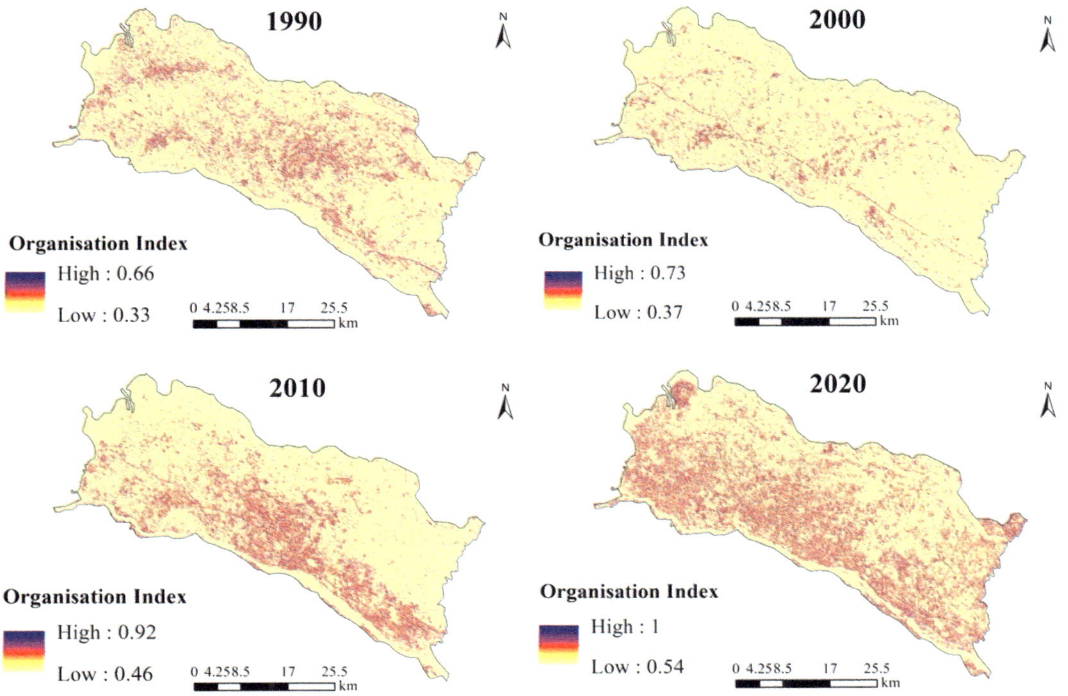

Fig. 12.5 Organization index of Asansol–Durgapur Development Region for 1990, 2000, 2010, and 2020

12.5 Discussion

Built-up areas in Asansol–Durgapur Development Region significantly increased by 7% and forest cover dropped by 17% between 1990 and 2020. Therefore, the primary cause of EH degradation in the Asansol–Durgapur Development Region can be attributed to LULC change. Previous studies showed that the LULC change significantly impacts ES supply (Peng et al. 2016; Wu et al. 2018; Fiquepron et al. 2013). LULC change, mainly urban sprawl, lowers the global ES's value (Costanza et al. 2014). Urbanization in this situation became one of the leading causes of EH deterioration. Urbanization expansion and land consumption were accelerated by population growth, economic growth, and rural exodus, which resulted in severe environmental demotion (Chen et al. 2014).

In earlier works, ES approach was applied in urban management plans (Albert et al. 2016; Salzman et al. 2014; Colding 2007). Planning and decision-making for urban landscapes must consider the urban areas providing ES functional value. Biggs et al. (2017) assert that ES can be functional in creating urban policies. In India, unsustainable urban growth is linked to environmental issues like biodiversity loss and ES deterioration (Kumar et al. 2010). Effective ecological landscape management and sustainable urban planning have received little attention. Therefore, the ES approach must be incorporated into urban plans. EH evaluation is essential to the conservation and sustainable management of urban terrain. Uncontrolled urbanization has a significant negative direct and indirect impact on the quality of life for city dwellers. Therefore, understanding EH is essential for a sustainable ecosystem and enhanced human well-being. EH plays a crucial role in the environment. Urbanization-related environmental degradation accounts for the numerous losses to the aquatic and terrestrial environments in developing nations like India (Sharma et al. 2020; Das et al. 2020). Thus, rapid urban growth causes ES loss,

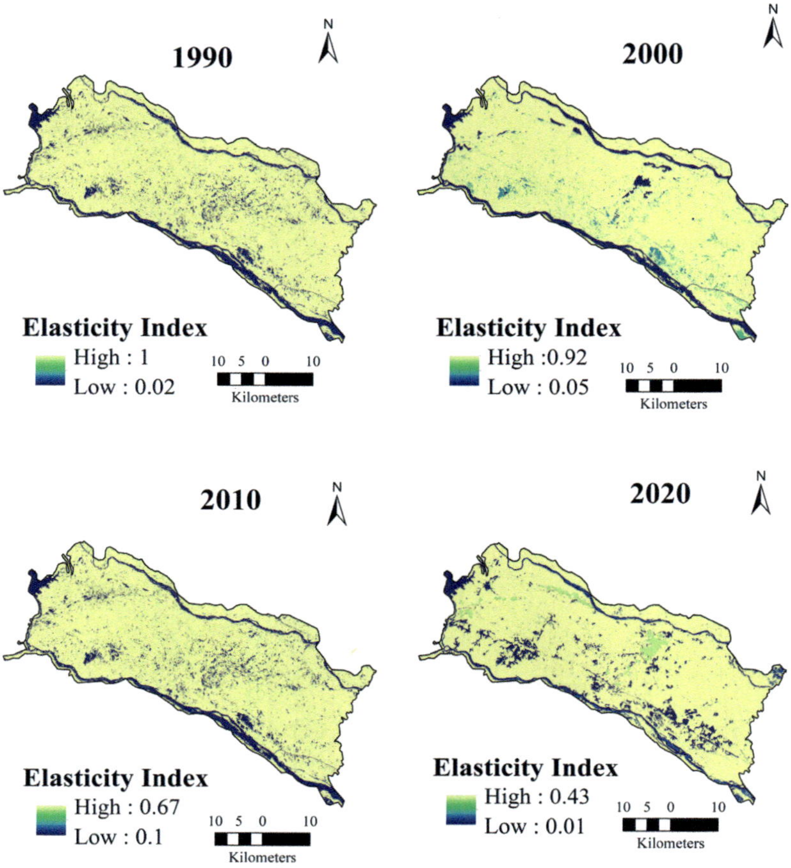

Fig. 12.6 Elasticity index of Asansol–Durgapur Development Region for 1990, 2000, 2010, and 2020

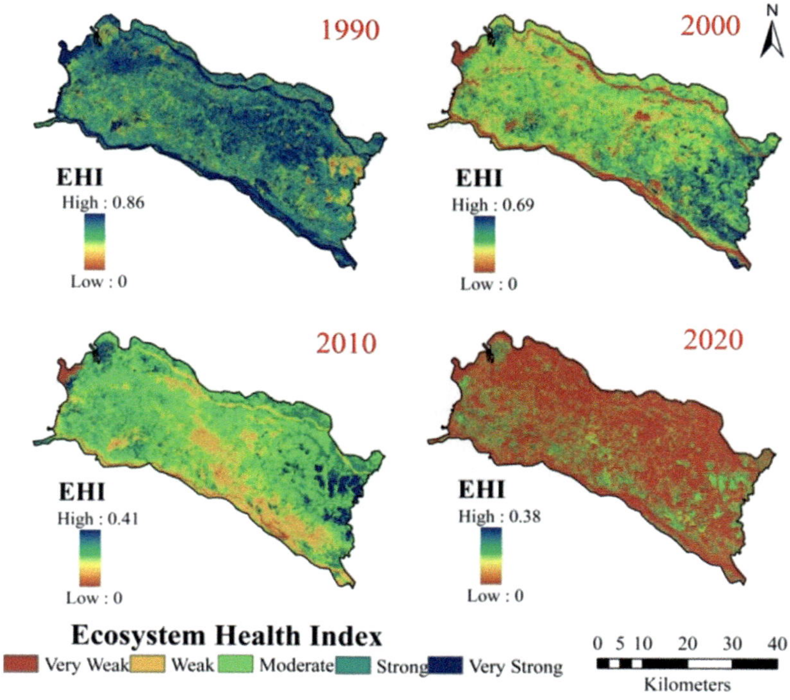

Fig. 12.7 Ecosystem health of Asansol–Durgapur Development Region for 1990, 2000, 2010, and 2020

Table 12.2 Area coverage of different ecosystem health categories for 1990, 2000, 2010, and 2020

Ecosystem health category/year	1990		2000		2010		2020	
	Area (km^2)	Area (%)	Area (km^2)	Area (%)	Area (km^2)	Area (%)	Area (km^2)	Area (%)
Very weak	41.49	2.25	225.95	12.25	46.63	2.53	1230.99	66.70
Weak	173.99	9.43	529.96	28.71	595.57	32.27	331.03	17.94
Moderate	123.01	6.67	561.49	30.42	741.12	40.15	104.69	5.67
Strong	885.86	47.99	375.33	20.34	316.72	17.16	114.23	6.19
Very strong	621.33	33.66	152.98	8.29	145.71	7.89	64.75	3.51

which may impact urban residents' quality of life directly or indirectly. Therefore, it is essential to evaluate EH. Urban areas in India are expanding without a plan, dramatically worsening the state of the country's ecosystems (such as water bodies and croplands). This study will significantly contribute to our understanding of the effects of rapidly urbanizing areas on environmental health because there is a lack of information about EH in India. It will also serve as a foundation for future research in India and other nations. The method used in the current study is applicable to other contexts.

12.6 Conclusion

This study used the V–O–R model to assess EH from 1990 to 2020 in the Asansol–Durgapur Development Region. Ecosystem health deteriorates over time; for example, 1990 good EH was 33% and has since dropped by 3.51% in recent times. Regional EH quantitative study outcomes may show how changing LULC patterns affect the environment. Asansol–Durgapur Development Region's EH conditions have generally been worsen over time. The V–O–R model and the EH results demonstrate that the current urban growth model is unsustainable. It is crucial to undertake various programs to reduce the rate of EH deterioration to preserve and develop EH in this area.

Acknowledgements The authors want to express their gratitude to USGS for delivering satellite images. This study received no funding.

References

Al-Ahmadi FS, Al-Hames AS (2009) Comparison of four classification methods to extract land use and land cover from raw satellite images for some remote arid areas, Kingdom of Saudi Arabia. Earth Sci 20(1)

Albert C, Galler C, Hermes J, Neuendorf F, Von Haaren C, Lovett A (2016) Applying ecosystem services indicators in landscape planning and management: the ES-in-planning framework. Ecol Ind 61:100–113

Biggs J, Von Fumetti S, Kelly-Quinn M (2017) The importance of small waterbodies for biodiversity and ecosystem services: implications for policy makers. Hydrobiologia 793:3–39

Chen J, Sun BM, Chen D, Wu X, Guo LZ, Wang G (2014) Land use changes and their effects on the value of ecosystem services in the small Sanjiang plain in China. Sci World J

Cheng X, Chen L, Sun R, Kong P (2018) Land use changes and socio-economic development strongly deteriorate river ecosystem health in one of the largest basins in China. Sci Total Environ 616:376–385

Colding J (2007) Ecological land-usecomplementation'for building resilience in urban ecosystems. Landscape Urban Plann 81(1–2):46–55

Costanza R, De Groot R, Sutton P, Van der Ploeg S, Anderson SJ, Kubiszewski I et al (2014) Changes in the global value of ecosystem services. Glob Environ Chang 26:152–158

Costanza R, d'Arge R, De Groot R, Farber S, Grasso M, Hannon B et al (1997) The value of the world's ecosystem services and natural capital. Nature 387 (6630):253–260

Das DN, Chakraborti S, Saha G, Banerjee A, Singh D (2020) Analyzing the dynamic relationship of land surface temperature and landuse pattern: a city level analysis of two climatic regions in India. City Environ Interact 8:100046

Ekumah B, Armah FA, Afrifa EK, Aheto DW, Odoi JO, Afitiri AR (2020) Geospatial assessment of ecosystem health of coastal urban wetlands in Ghana. Ocean Coast Manag 193:105226

Fiquepron J, Garcia S, Stenger A (2013) Land use impact on water quality: valuing forest services in terms of the water supply sector. J Environ Manage 126:113–121

Kang P, Chen W, Hou Y, Li Y (2018) Linking ecosystem services and ecosystem health to ecological risk assessment: a case study of the Beijing-Tianjin-Hebei urban agglomeration. Sci Total Environ 636:1442–1454

Kumar M, Mukherjee N, Sharma GP, Raghubanshi AS (2010) Land use patterns and urbanization in the holy city of Varanasi, India: a scenario. Environ Monit Assess 167:417–422

Lackey RT (2001) Values, policy, and ecosystem health: options for resolving the many ecological policy issues we face depend on the concept of ecosystem health, but ecosystem health is based on controversial, value-based assumptions that masquerade as science. Bioscience 51(6):437–443

Li W, Qi J, Huang S, Fu W, Zhong L, He BJ (2021) A pressure-state-response framework for the sustainability analysis of water national parks in China. Ecol Ind 131:108127

Peng J, Liu Y, Wu J, Lv H, Hu X (2015) Linking ecosystem services and landscape patterns to assess urban ecosystem health: a case study in Shenzhen City, China. Landsc Urban Plan 143:56–68

Peng WF, Zhou JM, Fan SY, Yang CJ (2016) Effects of the land use change on ecosystem service value in Chengdu, western China from 1978 to 2010. J Indian Soc Remote Sens 44:197–206

Peng J, Liu Y, Li T, Wu J (2017) Regional ecosystem health response to rural land use change: a case study in Lijiang City, China. Ecolog Indicat 72:399–410

Saaty TL (1980) The analytic hierarchy process McGraw-Hill. New York 324

Salzman J, Arnold CA, Garcia R, Hirokawa K, Jowers K, LeJava J et al (2014) The most important current research questions in urban ecosystem services. Duke Envtl L & Pol'y F 25:1

Santos-Burgoa C, Rios C, Mercado LA, Arechiga-Serrano R, Cano-Valle F, Eden-Wynter RA et al (2001) Exposure to manganese: health effects on the general population, a pilot study in central Mexico. Environ Res 85(2):90–104

Sharma S, Nahid S, Sharma M, Sannigrahi S, Anees MM, Sharma R et al (2020) A long-term and comprehensive assessment of urbanization-induced impacts on ecosystem services in the capital city of India. City Environ Interact 7:100047

Speldewinde PC, Slaney D, Weinstein P (2015) Is restoring an ecosystem good for your health? Sci Total Environ 502:276–279

Turner MG (1989) Landscape ecology: the effect of pattern on process. Annu Rev Ecol Syst 20(1):171–197

Van Niekerk L, Adams JB, Bate GC, Forbes AT, Forbes NT, Huizinga P et al (2013) Country-wide assessment of estuary health: an approach for integrating pressures and ecosystem response in a data limited environment. Estuar Coast Shelf Sci 130:239–251

Wang Z, Yang Z, Shi H, Han F, Liu Q, Qi J, Lu Y (2020) Ecosystem health assessment of world natural heritage sites based on remote sensing and field sampling verification: Bayanbulak as case study. Sustainability 12(7):2610

Wilkins DA (1999) Assessing ecosystem health. Trends Ecol Evol 14(2):69

Wu J, Lu J, Li L, Min X, Luo Y (2018) Pollution, ecological-health risks, and sources of heavy metals in soil of the northeastern Qinghai-Tibet Plateau. Chemosphere 201:234–242

Wu S, Tian C, Li B, Wang J, Wang Z (2022) Ecological environment health assessment of lake water ecosystem system based on simulated annealing-projection pursuit: a case study of plateau lake. Sustain Cities Soc 86:104131

Xiao R, Liu Y, Fei X, Yu W, Zhang Z, Meng Q (2019) Ecosystem health assessment: a comprehensive and detailed analysis of the case study in coastal metropolitan region, eastern China. Ecol Ind 98:363–376

Zhang D, Huang Q, He C, Wu J (2017) Impacts of urban expansion on ecosystem services in the Beijing-Tianjin-Hebei urban agglomeration, China: a scenario analysis based on the shared socioeconomic pathways. Resour Conserv Recycl 125:115–130

13 Groundwater Arsenic Contamination in the Indo-Gangetic Plain: A Major Environmental Issue for Stress, Anxiety, and Depression

Das Ambika Bharti and Shishu Kesh Kumar

Abstract

Background Human beings have always been exposed to environmental hazards like polluted air, water, and soil through many micropollutants and heavy metals. These and many other exposures persist today and even increasing day by day due to industrialization and globalization. Several kinds of micropollutants (like arsenic, lead, mercury, cadmium, etc.) affect human physical as well as psychological health. Groundwater arsenic contamination is a major environmental issue and a major public health concern throughout the Indo-Gangetic plain. The situation is worst in the state of Bihar, where 22 out of 38 districts are affected by arsenic and cancer is the most prevalent disease in this area. Unfortunately, the psychological consequences of groundwater As contamination is scantly studied and found in the literature. *Objective* To explore the arsenic menace in the middle Indo-Gangetic plain from a psychological perspective and assess the level of stress, anxiety, and depression of arsenic-induced cancer patients. *Design* Correlational survey research. *Sample* Through the purposive sampling method, 267 participants (age ranges from 32 to 60 years) were taken from Bihar. Eighty-nine (89) cancer patients were from non-arsenic-affected regions, 89 arsenic-induced cancer patients were from arsenic-affected middle Gangetic region, and rest 89 healthy counterparts were from non-arsenic-affected districts of Bihar. *Method* Hindi-adapted version of a self-report measure of depression, anxiety, and stress scale was used to evaluate the level of three emotional states. Quantitative data analysis was carried out with descriptive statistics (SD and mean) and inferential statistics (Tukey and one-way ANOVA). *Results* The three groups were significantly different on stress ($F = 61.048$, $p < 0.001$), anxiety ($F = 45.701$, $p < 0.001$), and depression ($F = 40.851$, $p < 0.001$). The Tukey test shows that stress and depression were comparatively higher in the arsenic-induced cancer patients as compared to the other two counterparts. *Conclusion* Arsenic-induced cancer patients undergo comparatively greater levels of anxiety and stress and thus warrant constructive attention from the service providers and policymakers toward this end. *Implications* Baseline findings for reporting unfortunate psychological outcomes of arsenic-induced cancer and so for, exposure to high As concentrated groundwater.

D. A. Bharti (✉) · S. K. Kumar
Department of Psychological Sciences, Central University of South Bihar, SH-7 Gaya-Panchanpur Road, P. O.-Fatehpur, P.S.-Tekari, Gaya, Bihar 824236, India
e-mail: ambika@cub.ac.in

S. Talukdar et al. (eds.), *Environmental Risk and Resilience in the Changing World*, GIScience and Geo-environmental Modelling, https://doi.org/10.1007/978-3-031-62442-1_13

In addition, a call to action has been made for academics, researchers, social workers, and policymakers to plan and promote psychological interventions for good health and well-being.

Keywords

Arsenic-induced cancer · Stress · Anxiety · Depression · Groundwater arsenic contamination · Middle Gangetic plain

13.1 Introduction

Human beings have always been exposed to environmental hazards like water contamination, air pollution, and soil contamination through many micropollutants and heavy metals. These and many other exposures persist today, and even increasing day by day due to industrialization, urbanization, and globalization (Ahmed et al. 2022). Several kinds of micropollutants (like arsenic, lead, mercury, cadmium, etc.) affect human physical as well as psychological health. Diseases of the kidney, skin, liver, respiratory, gastrointestinal, cardiology, nervous system disorders, neurodegenerative defects, immune system dysfunction, and cancer are some examples that could be caused by heavy metals toxic effects (Balali-Mood et al. 2021; Kulka 2016; Mitra et al. 2022). These diseases put an extreme social and economic burden on human beings and may also have adverse psychological consequences like frequent experiences of depression, stress, anxiety, etc.

Lead, mercury, cadmium, and arsenic are widespread in the environment as these can be found in organic as well as inorganic forms. Arsenic is the most affecting element for human health as it is found in groundwater which is the most common need of human survival. The most common public health problems emerging from arsenic contamination are skin lesions and many types of diseases like cancer not only limited to skin, cardiovascular diseases, and respiratory diseases. Two standards are proposed to limit the amount of As in drinking water, one is proposed by the World Health Organization (WHO 2011) which is 0.01 mg/l, and another proposed by the Bureau of Indian Standards (BIS 2012) which is 0.05 mg/l.

Arsenic contamination is a concern for the whole world as 180 countries are suffering from an arsenic catastrophe (Shaji et al. 2021). Asian regions are much more affected than any other country. In India only, approximately 7 crore people are affected of which Ganga Basin has the majority of people situated in it (Chatterjee 2014). Bihar is the most adversely affected Indian state; 22 districts are identified as arsenic-affected, out of 38 districts (Mishra et al. 2016).

Arsenic affects physical health adversely. Skin lesions (keratosis, melanosis, and hyperkeratosis) (Guha Mazumdar 2003; Rahman et al. 2009.), different types of cancer (lung, liver, skin), gastrointestinal diseases, kidney and respiratory diseases, etc., are common effects of arsenic. Cancer is the most prevalent and fatal disease condition resulting from arsenic contamination (Kumar and Bharti 2021; Marshall et al. 2007; Morales et al. 2000; Smith and Steinmaus 2009;). Apart from cancer, arsenic is responsible for cognitive impairment, synaptic changes, and neurotransmitter imbalance (Garza-Lombó et al. 2019), non-pitting edema, skin lesions, dorsum, gangrene, and other malignant neoplasms skin cancer (Ahsan et al. 2000; Argos et al. 2011; Chakraborti et al. 2003; Lindberg et al. 2008), high blood pressure (WHO 2011), reproductive disorders, diabetes, and peripheral artery disease, (Chakraborti et al. 2003; Chattopadhyay et al. 2002; Chowdhury et al. 2016; Hopenhayn et al. 2003; Mayer and Goldman 2016; Milton et al. 2004).

A person's psychological health is also affected by arsenic contamination. Chronic diseased conditions frequently make people anxious, fearful, and disorganized (Aspinwall 2004). Arsenicosis patients often report psychosomatic symptoms like weakness, insomnia, loss of appetite, depression, restlessness, and drowsiness (Khan et al. 2006). Not just physical pain arsenicosis patients also suffer from social prejudice and discrimination. Patients as well as their families are forced to be isolated by quitting their job, marriage, social gathering, etc. (George et al. 2013; Hassan et al. 2005; Rahman et al. 2018). Arsenicosis patients have been found to have 3–5 times greater major depressive disorder than non-cancer patients (McDaniel et al. 1995; Spiegel and Giese-Davis 2003). According to ATSDR (Agency for Toxic Substances and Disease Registry 2000), people experience memory loss and emotional liability due to acute inorganic arsenic toxicity while long-term arsenic exposure to the neurotransmitters causes depression.

Literature establishes groundwater As contamination as an environmental health hazard and of serious public health concern worldwide (Chowdhury 2004; Khan et al. 2003; Mukherjee et al. 2006; Thakur et al. 2011) but the situation is catastrophic in Asian countries, particularly in Bihar (Chakraborti et al. 2003; Ghosh et al. 2007, 2009). Most of the research endeavors (Chakraborti et al. 2003; Chattopadhyay et al. 2002; Chowdhury et al. 2016; Milton et al. 2004) have attempted only either to know the concentration/endemic areas of arsenic or its effects on individual's physical health. The psychological aspects have always been grossly neglected in research pursuits and only a few studies could be found reporting the psychosocial aspects of high groundwater As concentration, while the exploration of the arsenic menace from the psychological perspective is significant for its effective remediation. Groundwater As contamination is thus a leading environmental issue in the Indo-Gangetic Plain and needs to be studied holistically and sustainably to meet the visions of the United Nations Sustainable Development Goals of good health and well-being for all (Bundschuh et al. 2022). Henceforth, this study aimed to explore the arsenic menace in the middle Indo-Gangetic plain from a psychological perspective and assess the level of stress, anxiety, and depression of arsenic-induced cancer patients.

13.2 Methods

13.2.1 Hypothesis

There would be no significant difference in the levels of stress, anxiety, and depression experienced and reported by the arsenic-induced cancer patients, cancer patients from non-arsenic-affected areas, and their healthy counterparts.

13.2.2 Study Design and Sample

A correlational survey research design was used to study 267 participants (aged 32–60 years, mean age = 47.86) drawn from Mahavir Cancer Sansthan and Research Centre, Patna (Bihar). The sampling was done with a purposive sampling method; 89 were cancer patients from non-arsenic-affected regions of Bihar, 89 were arsenic-induced cancer patients from arsenic-affected middle Gangetic plain of Bihar and rest 89 were healthy comparative counterparts drawn from different non-arsenic-affected districts of Bihar state.

Sampling was done by purposive sampling method; 89 cancer patients were from non-arsenic contaminated areas of Bihar, 89 arsenic-induced cancer patients were from arsenic-contaminated middle Gangetic plains of Bihar and the remaining 89 were healthy comparison counterparts who came from different non-arsenic-affected districts of Bihar state.

13.2.3 Measures

The following standardized instruments were used to gather the desired data:

a. **Socio-demographic and Clinical Profile Sheet**: A partially structural format was prepared for this study purpose to obtain relevant socio-demographic and clinical data from the participants. Information on the following parameters was collected: age, gender, residential area, education, father education, father's occupation and father's annual income, number of family members, and domicile of the participants.

b. **Depression, Anxiety, and Stress Scales: Hindi Adaptation** (Singh et al. 2013): It is a brief version of its original 42-item scale (Lovibond and Lovibond 1995) intended to measure the three related negative emotional states of stress, anxiety, and depression. It is a 4-point rating scale (0–3), "3" meaning "applies to me a lot/most of the time" and "0" meaning "does not apply to me at all." The reliability of the scale was tested by Cronbach alpha. The Cronbach alpha for stress (0.63), depression (0.70), and anxiety (0.74) were acceptable.
It is a 4-point rating scale (0-3), with "0" meaning "does not apply to me at all" and "3" meaning "applies to me a lot/most of the time." The reliability of the scale was tested by the Cronbach alpha. Cronbach's alphas for stress (0.63), anxiety (0.74), and depression (0.70) were acceptable.

13.2.4 Procedure

Participants were drawn through a purposive sampling method from the state of Bihar, India. The data were collected in three phases. In the *first phase* of data collection, arsenic-induced cancer cases were identified based on their differential diagnosis done by clinicians at Mahavir Cancer Sansthan and Research Centre. The differential diagnosis of arsenic-induced cancer was based on the clinical history of arsenicosis. The following biomarkers were used in the process: arsenic in blood, nail, and hair and arsenic in drinking water used by the participants. The study participants must have been the inhabitants of the middle Gangetic plains for more than 10 years and not suffering from any chronic diseases other than cancer. In the *second phase* of data collection, comparative cancer patients were drawn from the non-arsenic-affected region of Bihar. In the *third phase* of data collection, a comparative healthy counterpart was drawn from the non-arsenic regions of Bihar. The data collection was initiated with rapport building, and the study objectives were briefly explained to each participant. Verbal as well as oral informed consent was taken, and the suggestive ethical guidelines given by American Psychological Association (APA 2010) were followed during the study. Quantitative data analysis was carried out with the datasets to test the hypothesis.

13.2.5 Ethical Approval

This study was given ethical approval from two reputed bodies. Firstly, it was received from the concerned Departmental Research Approval Committee, of the authors' University and subsequently, from the ethical approval committee of Mahavir Cancer Sansthan and Research Center, Patna, to collect data from their esteemed institution.

13.2.6 Statistical Analyses

Descriptive (mean and SD) statistics were used to examine the nature of the sample, while inferential statistics (one-way ANOVA) helped to examine the significant group mean differences. Post hoc tests of Tukey have been used to ascertain the sample groups that significantly differed. All the analysis has been done by using SPSS version 25.

13.3 Results

Data from 267 participants were analyzed through SPSS (version 25). Table 13.1 reveals that arsenic-induced cancer patients scored high on depression (M = 6.0225, SD = 4.791), anxiety (M = 6.3258, SD = 4.337), and stress level (M = 8.9101, SD = 4.764), compared to the cancer patients with scores for depression (M = 5.0899, SD = 3,962), anxiety (M = 5.0337, SD = 3.638), and stress (M = 7.4719, SD = 4.026) and healthy counterparts with scores for

Table 13.1 Descriptive statistics (mean and SD) of the three sample groups on depression, anxiety, and stress ($N = 267$)

Emotional states	Samples	N	Mean	SD
Depression	Cancer patients	89	5.0899	3.962
	Arsenic-induced cancer patients	89	6.0225	4.791
	Healthy participants	89	1.2584	1.736
Anxiety	Cancer patients	89	5.0337	3.638
	Arsenic-induced cancer patients	89	6.3258	4.337
	Healthy participants	89	1.5393	1.943
Stress	Cancer patients	89	7.4719	4.026
	Arsenic-induced cancer patients	89	8.9101	4.764
	Healthy participants	89	2.7640	2.510

depression ($M = 1.2584$, SD = 1.736), anxiety ($M = 1.5393$, SD = 1.943), and stress ($M = 2.7640$, SD = 2.510).

The F test was performed by comparing the F statistic (computed from the data) to the critical value from the F table. The result is significant if the F statistic is larger because this indicates greater differences among the sample averages. The ANOVA Table 13.2 demonstrates that the mean difference between the three sample groups (i.e., cancer patients from non-arsenic-affected areas, arsenic-induced cancer patients, and healthy from non-arsenic-affected regions of Bihar) is statistically significant on all three emotional states of stress ($F = 61.048$, $p < 0.001$), anxiety ($F = 45.701$, $p < 0.001$), and depression ($F = 40.851$, $p < 0.001$). Therefore, the null hypothesis is rejected with 99.9% confidence (note: subtract the p-value threshold from 1 and multiply by 100 to obtain the confidence level) and was concluded that there was very strong evidence that the mean tensile strengths of the different groups were unequal. However, which pair of group mean differences are significantly different was explored by performing the post hoc test (Table 13.3).

Table 13.3 and Fig. 13.1 show that a significant means difference (MD = 3.8314, $p < 0.001$) exists between cancer patients from the non-arsenic-affected area as compared to healthy participants with depression. There is no significant difference (MD = 0.93258, $p < 0.219$) reported between arsenic-induced cancer patients and cancer patients from non-arsenic areas on depression.

Table 13.2 ANOVA to compare the three sample groups (as-induced cancer from arsenic-affected region, cancer from non-arsenic-affected region, and healthy from non-arsenic-affected region) on their depression, anxiety, and stress levels ($N = 267$)

Emotional state		Sum of squares	df	Mean squares	F	Sig.
Depression	Between-group	1134.629	2	567.315	**40.851**	0.001
	Within group	3666.292	264	13.887		
	Total	4800.921	266			
Anxiety	Between-group	1091.468	2	545.734	**45.701**	0.001
	Within group	3152.562	264	11.942		
	Total	4244.030	266			
Stress	Between-group	1839.528	2	919.764	**61.048**	0.001
	Within group	3977.506	264	15.066		
	Total	5817.034	266			

$p \leq 0.001$ means the group mean difference is significant at 0.001 level of significance

Table 13.3 Tukey test for multiple comparisons between cancer patients from non-arsenic-affected regions, arsenic-induced cancer patients, and comparative healthy counterparts from non-arsenic-affected regions of Bihar on depression, anxiety, and stress levels ($N = 267$)

Emotional states	Comparing samples		Mean differences	Sig.
Depression	Cancer patients	Healthy participants	**3.8314**	0.001
	Arsenic-induced cancer patients	Cancer patients	0.93258	0.219
	Healthy participants	Arsenic-induced cancer patients	**4.76404**	0.001
Anxiety	Cancer patients	Healthy participants	**3.49438**	0.001
	Arsenic-induced cancer patients	Cancer patients	**1.29213**	0.035
	Healthy participants	Arsenic-induced cancer patients	**4.78652**	0.001
Stress	Cancer patients	Healthy participants	**4.70787**	0.001
	Arsenic-induced cancer patients	Cancer patients	**1.43820**	0.037
	Healthy participants	Arsenic-induced cancer patients	**6.14607**	0.001

$p \leq 0.001$ means the mean difference is significant at 0.001 level of significance

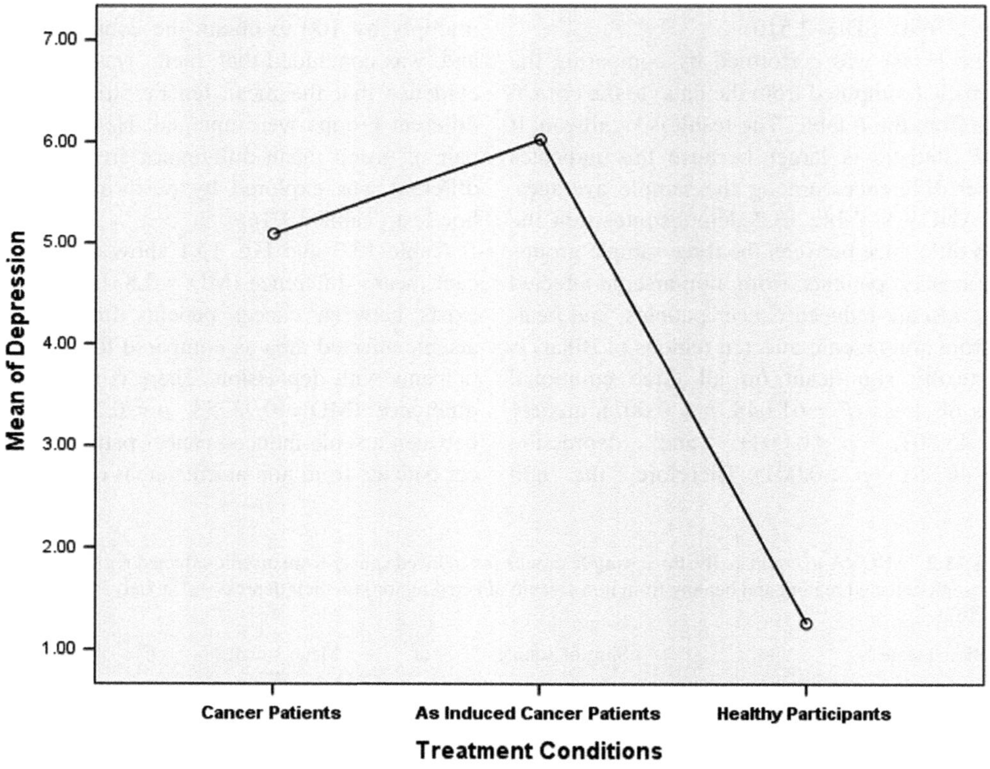

Fig. 13.1 Mean of depression on the x-axis and three treatment groups on the y-axis

A significant difference (MD = 4.76404, $p < 0.001$) exists between healthy participants and arsenic-induced cancer patients with depression. The Tukey's test isolated the statistically significant relationships.

Table 13.3 and Fig. 13.2 show that significant means difference (MD = 3.49438, $p < 0.001$; MD = 1.29213, $p < 0.035$; MD = 4.78652, $p < 0.001$) exists on anxiety level between all three possible combinations (i.e., cancer patients

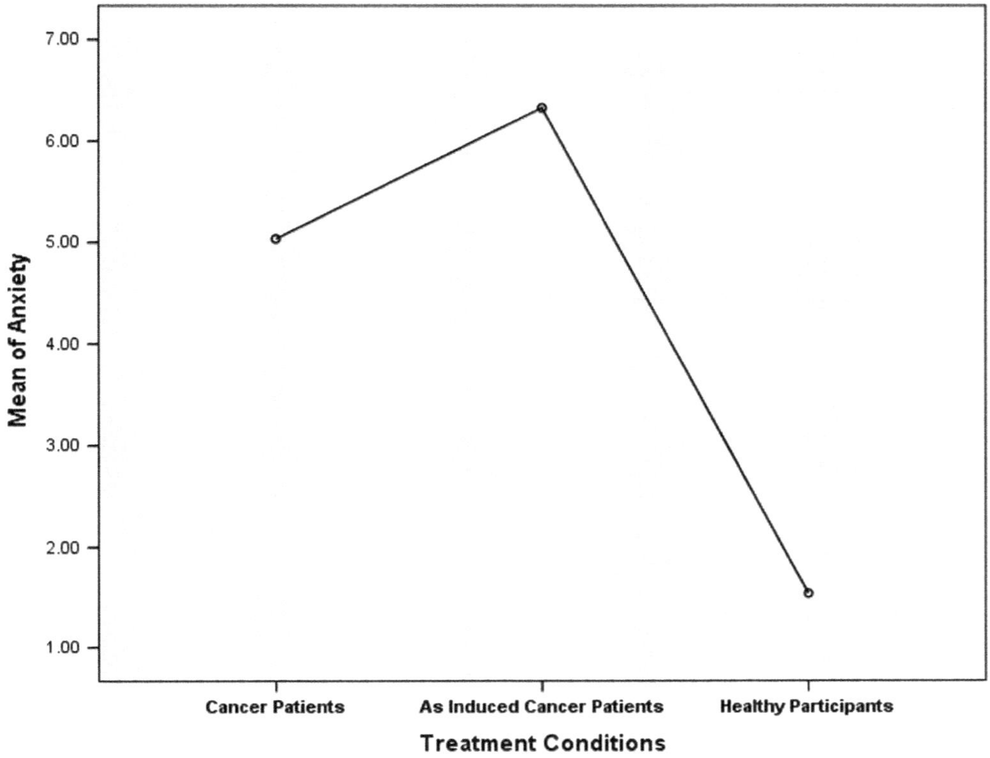

Fig. 13.2 Mean of anxiety on the *x*-axis and three treatment groups on the *y*-axis

from non-arsenic-affected area as compare to healthy participants, arsenic-induced cancer patients and cancer patients from non-arsenic area and healthy participants and arsenic-induced cancer patients).

Table 13.3 and Fig. 13.3 demonstrate that significant means difference (MD = 4.70787, $p < 0.001$; MD = 1.43820, $p < 0.037$; MD = 6.14607, $p < 0.001$) exists on stress between all three possible combinations (i.e., cancer patients from the non-arsenic-affected area as compared to healthy participants, arsenic-induced cancer patients and cancer patients from non-arsenic area and healthy participants and arsenic-induced cancer patients).

13.4 Discussion

In the era of changing climate, biodegradation, deforestation, and micro and macropollutants, humans must sustain a healthy life. Arsenic-like environmental hazards are affecting human physical and psychological well-being (Chowdhury et al. 2016). Though arsenic can cause many diseases, cancer is the most life-threatening above all (Kumar and Bharti 2021, 2023). Diagnosis of cancer can cause a tremendous emotional as well as economic burden on the patient and their caregiver (Céilleachair et al. 2012). The United Nations Sustainable Development Goals for 2030 call for action to promote and sustain health and well-being for all as its SDG Goal 3 (United Nations 2015, 2020). This is a way to a new approach for sustainable arsenic mitigation by 2030, where exploring and understanding the arsenic menace holistically and sustainably is the need of the hour (Bundschuh et al. 2022). This study aimed to explore the groundwater arsenic contamination and arsenic-induced cancer prevalent in the middle Gangetic plain of Bihar from the psychological perspective and assessed the emotional states (viz. stress, anxiety, and depression) of arsenic-induced cancer patients

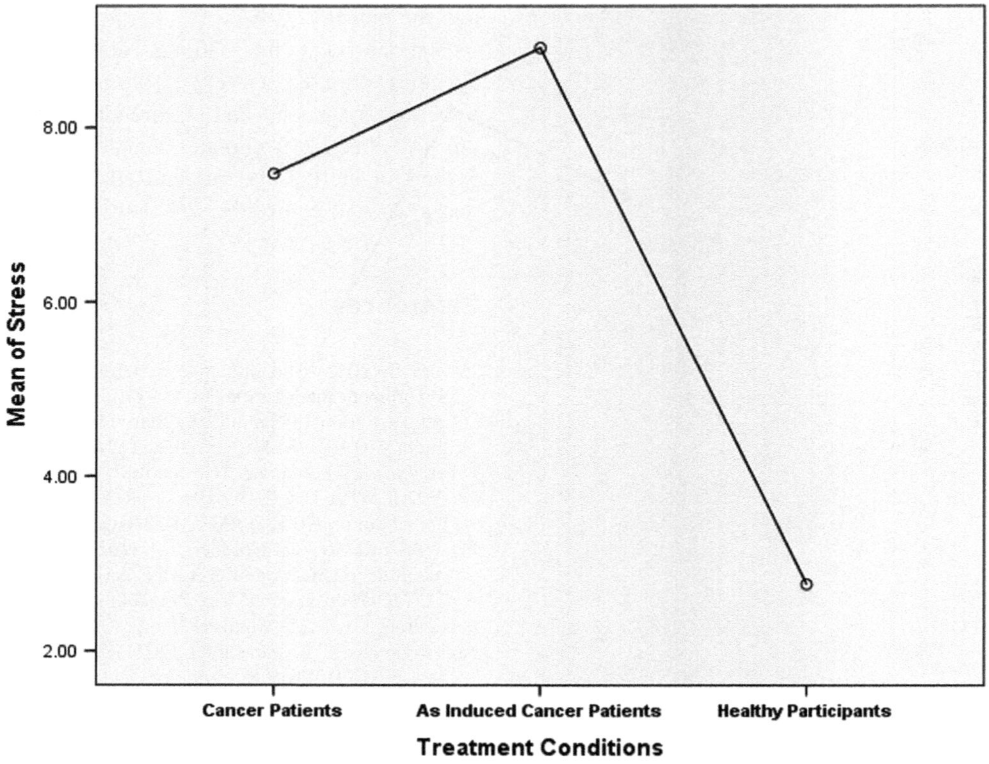

Fig. 13.3 Mean of stress on the *x*-axis and three treatment groups on the *y*-axis

residing in the affected region. The result gives a clear picture that arsenic-induced cancer patients and cancer patients from the non-arsenic-affected area have not differed significantly in depression, but arsenic-induced cancer patients and cancer patients from the non-arsenic area have scored higher on depression as compared to their healthy counterparts. Hence, it can be said that depression may be pervasive more or less equally in both types of cancer patients (i.e., arsenic-induced cancer patients and cancer patients from non-arsenic areas). However, both groups of cancer patients may face higher levels of depressive symptoms like low mood, feelings of hopelessness, significant weight loss or weight gain, etc., as compared to their healthy counterparts. In line with the current study (Céilleachair et al. 2012), found that in diagnosed cancer patients symptoms of depression and anxiety are more prevalent as compared to their healthy counterparts. Diagnosis of cancer itself may be the one reason behind their depressive symptoms because the diagnosis of cancer in patients itself arises in traumatic situations (Sumalla et al. 2009).

Arsenic-induced cancer patients from arsenic-affected middle Gangetic plain of Bihar scored higher on anxiety as compared to the other two groups, i.e., cancer patients from the non-arsenic-affected area and healthy people from the non-arsenic area of Bihar. This shows that arsenic-induced cancer patients from arsenic-affected regions may face more feelings of guilt, social withdrawal, sadness, suicidal preoccupation loss of interest in activity once enjoyed, worthlessness, the feeling of hopelessness, lowered self-esteem, etc. This result can be due to the reason that they live in such an environment where arsenic is present everywhere like water, corps, food, and even soil (Mandal and Suzuki 2002). They are conscious of the presence of arsenic and aware of the cause of their arsenicosis but are unable to control the environmental hazard due to poverty (Hassan et al. 2005). A person having arsenicosis symptoms may face more psychosocial problems due to the stigma associated with arsenicosis (Brinkel et al. 2009). In line with the present study, previous studies from Bangladesh and China have reported mental health complications mainly depression is more common among people exposed to arsenic (Brinkel et al. 2009). Some other studies (McDaniel et al. 1995; Raison and Miller 2003; Spiegel and Giese-Davis 2003) found that cancer patients were three to five times more likely to have severe depression than non-cancer patients.

Stress level too was also reported higher for arsenic-induced cancer cases compared to cancer patients from non-arsenic areas and healthy counterparts of Bihar. This could be understood as arsenic-induced cancer patients suffering more from several physical ailments (such as headaches, high blood pressure, chest pain, and sleep problems) as well as emotional issues (such as depression, panic attacks, and other form of anxiety and worry due to high stress). Some studies support the result of the present study as Ravindran et al. (2019) found in their study that arsenic-induced cancer patients were hopeless and distressed and their QOL was significantly lower. It could be due to the reason that patients in arsenic-affected areas may have knowledge about the existence of arsenic in their environment and often worry about their offspring's physical and psychological health but could do little to cope with its adverse consequences.

Mostly arsenic affects regions that belong to a rural area and belong to lower economic strata and are generally not aware of the adverse effect of arsenic contamination (Shaji et al. 2021). NGO's workings in these areas educate them about the presence of arsenic in the groundwater. But after knowing the consequences, they use this groundwater for drinking, food preparation, irrigation, etc., because people residing in these areas cannot afford pure drinking water. Thus, arsenic-induced cancer patients have been found to have more stress and anxiety level than those cancer patients who do not live in this belt.

However, both arsenic-induced cancer patients and cancer patients from non-arsenic-affected areas were found to be significantly higher in anxiety and stress as compared to healthy participants. Céilleachair et al. (2012) also found that diagnosed cancer patients may have high levels of depression and anxiety as compared to their healthy counterparts. Hence, it can be said that cancer patients from non-arsenic areas may feel worried or fearful about the future, uneasy, cold, worthlessness, the feeling of hopelessness, or/and lowered self-esteem, etc. One of the main reasons behind their psychological problems is a diagnosis of cancer (Céilleachair et al. 2012).

13.5 Conclusions

Exploration of the arsenic menace in the middle Indo-Gangetic plain of Bihar from a psychological perspective holds far reached advantages for the diseased community here. The findings present the psychological burden that the diseased inhabitants of this region are forced to live with and also were an attempt to understand the adverse consequences of groundwater. As contamination more holistically and sustainably. The emotional aspects of arsenic-induced cancer patients from arsenic-affected regions are much understudied topics. Arsenic-induced cancer patients have reported high levels of anxiety and stress as compared to cancer patients from non-arsenic areas in Bihar. Both, arsenic-induced cancer patients and cancer patients from non-arsenic area experience depression, anxiety, and stress as compared to a healthy person of non-arsenic-affected regions in Bihar. Such research pursuits were warranted to promote and sustain good health and well-being among this population. Findings call for immediate attention and action from policymakers and service providers to promote better psychological health at individual, societal, and systems levels.

13.6 Implications

- Brings findings first of its kind to report the adverse effect of arsenic contamination in groundwater on the psychological perspective of inhabitants of the middle Gangetic plain.

- Provides baseline findings for the adverse psychological consequences of anxiety, stress, and depression in arsenic-induced cancer caused due to exposure to high As concentrated groundwater, a much understudied topic.
- Further, calls for action from social workers, academicians, policymakers, and researchers to plan and promote psychological interventions and awareness accordingly for good health and well-being among this population (especially arsenic-induced cancer patients) afflicted by the major environmental issues of this region.

Acknowledgement The authors acknowledge the ethical committee of Mahavir Cancer Sansthan and Research Centre, Patna, for ethical approval to collect data from this esteemed institution.

References

Ahmed F, Ali I, Kousar S, Ahmed S (2022) The environmental impact of industrialization and foreign direct investment: empirical evidence from Asia-Pacific region. Environ Sci Pollut Res 1–15

Ahsan H, Perrin M, Rahman A, Parvez F, Stute M, Zheng Y, Hasnat Milton A, Brandt- Rauf P, van Geen A, Graziano J (2000) Associations between drinking water and urinary arsenic levels and skin lesions in Bangladesh. J Occup Environ Med 42:1195–1201

American Psychological Association (2010) Ethical principles of psychologists and code of conduct. Retrieved from https://www.apa.org/ethics/code/principles.pdf

Argos M, Kalra T, Pierce BL, Chen Y, Parvez F, Islam T, Ahmed A, Hasan R, Hasan K, Sarwar G, Levy D, Ahsan H (2011) A prospective study of arsenic exposure from drinking water and incidence of skin lesions in Bangladesh. Am J Epidemiol 174(2):185–194

Aspinwall LG (2004) Dealing with adversity: self-regulation, coping, adaptation, and health

Balali-Mood M, Naseri K, Tahergorabi Z, Khazdair MR, Sadeghi M (2021) Toxic mechanisms of five heavy metals: mercury, lead, chromium, cadmium, and arsenic. Front Pharmacol 227

BIS (2012) Indian standard: drinking water-specification. Bureau of Indian Standards, New Delhi

Brinkel J, Khan MH, Kraemer A (2009) A systematic review of arsenic exposure and its social and mental health effects with special reference to Bangladesh. Int J Environ Res Public Health 6(5):1609–1619

Bundschuh J, Khan Niazi N, Alam MA, Berg M, Tomaszewska IHB, Maity JP, Ok YS (2022) Global arsenic dilemma and sustainability. J Hazard Mater 436:129197. ISSN 0304-3894, https://doi.org/10.1016/j.jhazmat.2022.129197

Céilleachair ÓA, Costello L, Finn C, Timmons A, Fitzpatrick P, Kapur K, Sharp L (2012) Inter-relationships between the economic and emotional consequences of colorectal cancer for patients and their families: a qualitative study. BMC Gastroenterol 12:1–10

Chakraborti D, Mukherjee SC, Pati S, Sengupta MK, Rahman MM, Chowdhury UK, Basu GK (2003) Arsenic groundwater contamination in Middle Ganga Plain, Bihar, India: a future danger? Environ Health Perspect 111(9):1194–1201

Chatterjee M (2014) Arsenic in groundwater impacts 7 crore lives: panel. Times of India

Chattopadhyay S, Bhaumik S, Chaudhury AN, Gupta SD (2002) Arsenic induced changes in growth development and apoptosis in neonatal and adult brain cells in vivo and in tissue culture. Toxicol Lett 128(1–3): 73–84

Chowdhury AMR (2004) Arsenic crisis in Bangladesh. Sci Am 291(2):86–91

Chowdhury S, Krause-Pilatus A, Zimmermann KF (2016) Arsenic contamination of drinking water and mental health. In: DEF-discussion papers on development policy, vol 222

Garza-Lombó C, Pappa A, Panayiotidis MI, Gonsebatt ME, Franco R (2019) Arsenic induced neurotoxicity: a mechanistic appraisal. J Biol Inorg Chem 24:1305–1316

George CM, Gamble M, Slavkovich V, Levy D, Ahmed A, Ahsan H, Graziano J (2013) A cross-sectional study of the impact of blood selenium on blood and urinary arsenic concentrations in Bangladesh. Environ Health 12(1):1–7

Ghosh AK, Singh SK, Bose N, Chaudhary S (2007) Arsenic contaminated aquifers: a study of the Ganga levee zones in Bihar, India. In: Annual conference 2007. Royal Geographical Society, London. Session key: BSGSession

Ghosh AK, Singh SK, Bose N, Roy NP, Singh SK, Upadhyaya AK, Kumar S, Singh A (2009) Arsenic hotspots detected in Patna District, Bihar (India): a serious health hazard for estimated population of 5.5 lakhs, assessment of ground water resources and management. International Publishing House, New Delhi, pp 62–70 IK

Guha Mazumder DN (2003) Chronic arsenic toxicity: clinical features, epidemiology, and treatment: experience in West Bengal. J Environ Sci Health, Part A 38(1):141–163

Hassan MM, Atkins PJ, Dunn CE (2005) Social implications of arsenic poisoning in Bangladesh. Soc Sci Med 61(10):2201–2211

Hopenhayn C, Ferreccio C, Browning SR, Huang B, Peralta C, Gibb H, Hertz-Picciotto I (2003) Arsenic exposure from drinking water and birth weight. Epidemiology 593–602

Khan MMH, Sakauchi F, Sonoda T, Washio M, Mori M (2003) Magnitude of arsenic toxicity in tube-well drinking water in Bangladesh and its adverse effects on human health including cancer: evidence from a review of the literature. Asian Pac J Cancer Prev 4(1):7–14

Khan MMH, Aklimunnessa K, Ahsan N, Kabir M, Mori M (2006) Case-control study of arsenicosis in some arsenic-contaminated villages of Bangladesh

Kulka M (2016) A review of paraoxonase 1 properties and diagnostic applications. Polish J Vet Sci

Kumar SK, Bharti DA (2021) Arsenic exposure: sources, menace and risk management. In: Chitra VS, Balamurali RS (eds) Rejuvenating environmental studies: a multidisciplinary approach. Upanayana Publications, India, pp 92–96

Kumar SK, Bharti DA (2023) Depression, anxiety and stress among arsenic-induced cancer patients in indo-gangetic plains of Bihar: role of proactive coping. Community Health Equity Res Policy 43(4):381–387

Lindberg AL, Rahman M, Persson LÅ, Vahter M (2008) The risk of arsenic induced skin lesions in Bangladeshi men and women is affected by arsenic metabolism and the age at first exposure. Toxicol Appl Pharmacol 230(1):9–16

Lovibond SH, Lovibond PF (1995) Depression Anxiety Stress Scales (DASS–21, DASS–42) [Database record]. APA PsycTests. https://doi.org/10.1037/t01004-000

Mandal BK, Suzuki KT (2002) Arsenic round the world: a review. Talanta 58(1):201–235

Marshall G, Ferreccio C, Yuan Y, Bates MN, Steinmaus C, Selvin S, Smith AH (2007) Fifty-year study of lung and bladder cancer mortality in Chile related to arsenic in drinking water. J Natl Cancer Inst 99(12): 920–928

Mayer JE, Goldman RH (2016) Arsenic and skin cancer in the USA: the current evidence regarding arsenic-contaminated drinking water. Int J Dermatol 55(11): e585–e591

McDaniel JS, Musselman DL, Porter MR, Reed DA, Nemeroff CB (1995) Depression in patients with cancer: diagnosis, biology, and treatment. Arch Gen Psychiatry 52(2):89–99

Milton AH, Hasan Z, Shahidullah SM, Sharmin S, Jakariya MD, Rahman M, Smith W (2004) Association between nutritional status and arsenicosis due to chronic arsenic exposure in Bangladesh. Int J Environ Health Res 14(2):99–108

Mishra S, Dwivedi S, Kumar A, Chauhan R, Awasthi S, Mattusch J, Tripathi RD (2016) Current status of ground water arsenic contamination in India and recent advancements in removal techniques from drinking water. Int J Plant Environ 2(1 and 2):01–15

Mitra S, Chakraborty AJ, Tareq AM, Emran TB, Nainu F, Khusro A, Simal-Gandara J (2022) Impact of heavy metals on the environment and human health: Novel therapeutic insights to counter the toxicity. J King Saud Univ Sci 101865

Morales KH, Ryan L, Kuo TL, Wu MM, Chen CJ (2000) Risk of internal cancers from arsenic in drinking water. Environ Health Perspect 108(7):655–661

Mukherjee A, Sengupta MK, Hossain MA, Ahamed S, Das B, Nayak B, Chakraborti D (2006) Arsenic contamination in groundwater: a global perspective with emphasis on the Asian scenario. J Health, Populat Nutr 142–163

Rahman MM, Ng JC, Naidu R (2009) Chronic exposure of arsenic via drinking water and its adverse health impacts on humans. Environ Geochem Health 31 (1):189–200

Rahman MA, Rahman A, Khan MZK, Renzaho AM (2018) Human health risks and socio-economic perspectives of arsenic exposure in Bangladesh: a scoping review. Ecotoxicol Environ Saf 150:335–343

Raison CL, Miller AH (2003) When not enough is too much: the role of insufficient glucocorticoid signaling in the pathophysiology of stress-related disorders. Am J Psychiatry 160(9):1554–1565

Ravindran OS, Shankar A, Murthy T (2019) A comparative study on perceived stress, coping, quality of life, and hopelessness between cancer patients and survivors. Indian J Palliat Care 25(3):414

Shaji E, Santosh M, Sarath KV, Prakash P, Deepchand V, Divya BV (2021) Arsenic contamination of groundwater: a global synopsis with focus on the Indian Peninsula. Geosci Front 12(3):101079

Singh B, Prabhuappa KP, Eqbal S, Singh AR (2013) Depression, anxiety and stress scale: reliability and validity of Hindi adaptation. Int J Educ Manag Stud 3:446–449

Smith AH, Steinmaus CM (2009) Health effects of arsenic and chromium in drinking water: recent human findings. Annu Rev Public Health 30:107–122

Spiegel D, Giese-Davis J (2003) Depression and cancer: mechanisms and disease progression. Biol Psychiat 54(3):269–282

Sumalla EC, Ochoa C, Blanco I (2009) Posttraumatic growth in cancer: reality or illusion? Clin Psychol Rev 29(1):24–33

Thakur JK, Thakur KR, Ramanathan A, Kumar M, Singh SK (2011) Arsenic contamination of groundwater in Nepal—an overview. Water 3:1–20

United Nation (2015) Transforming our world: the 2030 agenda for sustainable development, resolution adopted by the general assembly on 25 September 2015, A/RES/70/1 Seventieth session, General Assembly, United Nations, New York. https://sustainabledevelopment.un.org/post2015/transformingourworld/publication

United Nation (2020) The sustainable development goals report 2020. United Nations Publications, New York, NY. https://unstats.un.org/sdgs/report/2020/The-Sustainable-Development-Goals-Report-2020.pdf

WHO (2011) World Health Organization guidelines for drinking water quality (4th ed) p 541 ISBN 9789241548151

Environmental Risk Over Indian Subcontinents: Consequences and Impacts

14

Rajani Srivastava and Sunita Singh

Abstract

India is facing continuous depletion of natural resources like air and water and threats of climate change. The influence of the environment on people's health changes with time. Considering that the environment and health relate to different situations is crucial because the world is changing. The present work aims to assess the impact of contemporary environmental challenges of air, water, and climate change on human health. The present paper emphasizes different environmental factors and human health with special emphasis on air, water, and climate change quality. With estimated data of more than four million premature mortality per year from open-air pollution and 3.8 from inside pollution, air pollution is the foremost environmental health risk in the world. Water quality issues are becoming more widespread globally as population demands increase. Surface water fulfills the drinking water needs of half of the world's population and is the primary method of disease transmission over long distances. Climate change is responsible for 2.4% of global diarrhea cases and 6% of malaria cases. Global temperatures are estimated to rise for the remainder of the century, raising health issues for everyone, especially in low-income nations. Therefore, public health progress needs to be studied to understand the connections between numerous environmental risk factors and human health. For more accurate estimation, quantitative examination of environmental toxins at all regional, national, and international levels is necessary for more precise estimation.

Keywords

Air quality · Climate change · Environmental risk · Human health · Water quality

14.1 Introduction

India ranks second in population in the world, with 1.54 billion inhabitants. India's growing population and wealth make pollution and other environmental problems worse. Threats to emerging nations like India include air, water, and climate pollution. India has several environmental issues, such as greenhouse gas

R. Srivastava
Department of Environmental Science (Environmental Technology), RGSC, Institute for Environment & Sustainable Development, Banaras Hindu University, Varanasi, India

S. Singh (✉)
Department of Geography, Central University of South Bihar, Gaya, India
e-mail: sunitasingh@cusb.ac.in

S. Talukdar et al. (eds.), *Environmental Risk and Resilience in the Changing World*, GIScience and Geo-environmental Modelling, https://doi.org/10.1007/978-3-031-62442-1_14

emissions, chemical and oil pollutants related to energy consumption, air pollution from vehicles and industrial effluents (Greenstone and Hanna 2014), inadequate sanitation, and the availability of non-potable water nationwide, management of municipal solid waste (Dube et al. 2014), and overpopulation and its impact on natural resources. The environmental system of the subcontinent is influenced by agricultural variables such as runoff of farmed chemicals, overgrazing, rapid crop cycles, slash-and-burn tactics, cutting of trees, and deforestation of timber for fuel (Greenstone and Hanna 2014).

Environmental risk indicates the potential loss due to environmental events or processes. Any possible or actual negative effect, regardless of size, duration, or frequency, has an unfavorable effect on the atmosphere and leads to environmental health risks (Prüss-Ustün and Neira 2016). The threat to human health has been invented due to climate change and has become a global issue (Kovats and Haines 2005a, b; Stern 2007; Srivastava 2013). These also harm ecosystems, biodiversity, and the sustainable use of natural resources (Srivastava and Pawan 2021; Srivastava 2022). The rise, comeback, and global transmission of infectious diseases, as well as the recurrence of other chronic diseases worldwide, including in India, are all being driven by climate instability and global warming.

With time, the environment's impact on people's health evolves. India was a largely rural nation, but it is becoming more urban. By 2050, four million more people can reside in cities, making up 53% of the country's population (UNDESA 2018). It must, therefore, overcome major challenges to encounter climate change (Singh et al. 2021; Srivastava 2022). The ongoing urbanization of metropolitan areas alters the atmospheric properties of its surface and boundary layer, which alters the land use patterns, affecting regional climate (Dasgupta 2004; Ren et al. 2008; Yang et al. 2012). Due to rising man-made pollution, excessive precipitation, greater temperatures, and more frequent heat waves, urbanization has a detrimental effect on human health (Lin et al. 2016; Shastri et al. 2015). Local climatic changes are redirected in various climatological circumstances, such as variations in temperature and rainfall. These regional climate changes result from man-made factors like greenhouse gas (GHG) emissions. Urbanization impacts the climate and increases the amount of aerosols and pollution in the atmosphere due to a rise in urban population and transportation. According to several studies (Ramachandran et al. 2012; Krishna Moorthy et al. 2013), one of the main causes of the air and water pollution in the Indian subcontinent that ultimately affects the climate of the particulate region is the growing urbanization along with rising population and industrialization (Dasgupta 2004). Thus, as described in this study, air and water pollution and climate change pose significant threats to humans and welfare in developing nations like India. These issues are all interrelated. Many previous studies focused on these environmental issues with human health, but information is scarce about recently developing health issues like COVID-19 and others due to new environmental challenges. Therefore, the present work aims to assess the current environmental problems of air, water, and climate change that affect people's health.

14.2 Environmental Threat and Human Health

An environmental threat is any environmental threat, including hazards and potential negative outcomes. A summary of several environmental dangers that influence the health of humans is presented in Fig. 14.1. Various elements, including nutrition, hygiene, social position, education, and standard of living, influence the population's health. Because of these causes, life expectancy has greatly increased during the economic changes that have shaped civilization. The effects of environmental risk factors on health are immensely complicated and diverse in terms of their clinical importance and severity. The impact of environmental deterioration on human health may range from mental health, which may be caused by noise, to cancer mortality caused by air pollution.

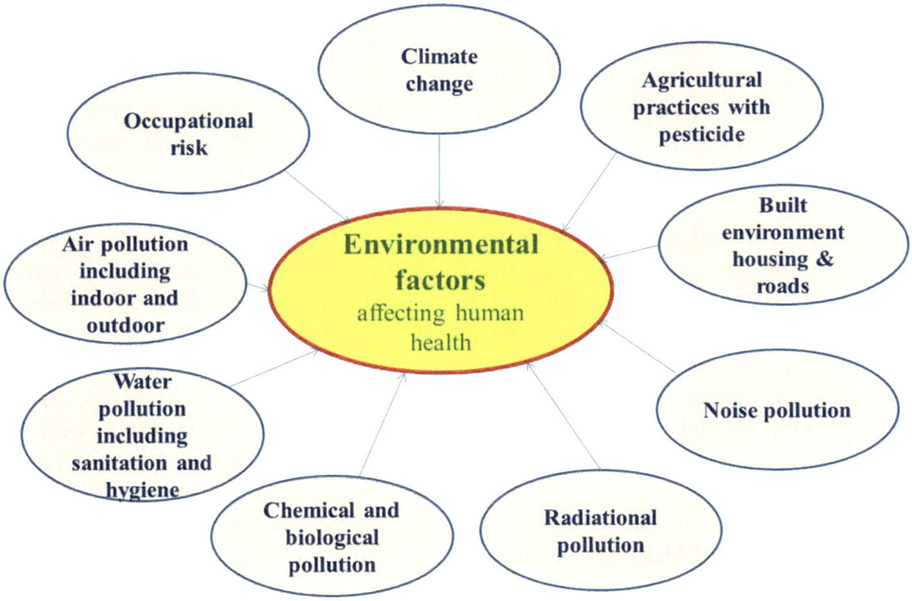

Fig. 14.1 Different environmental risk factors affecting human health

Urbanization, industrialization, mining, quarrying, and other development activities increase as population pressure rises. Many pollutants are poisoning the environment because of these activities. It not only has an impact on and degrades the environment, but it also poses health risks.

The environmental risk was predicted by the World Health Organization in 2012 to be the cause of 12.6 million deaths worldwide, or 23% of all deaths (Prüss-Ustün et al. 2019). The loss of ecosystems may raise the occurrence of cholera and malaria and the vulnerability to the emergence of new diseases (Millennium Ecosystem Assessment 2005). Rojas-Rueda has recently investigated and evaluated the effects of numerous environmental factors on human health, including air pollution, heavy metal pollution, chemical, radiation and noise pollution, cigarette smoke, and urban residential situations (Rojas-Rueda et al. 2021). They concluded that the environment has a significant impact on determining health. They recommended using this research as an additional tool to comprehend how environmental danger and human health are related.

In the past 25 years, India has experienced increased environmental risk or pollution issues, primarily because of urbanization, industrial growth, modernization of agricultural techniques, and mining and quarrying activities. Human meddling in the environment has increased, causing environmental damage that is bad for all living things. Pollution or contamination of the environment has negative impacts on living and nonliving things and people's quality of life. Other elements that affect a population's health besides environmental pollution include diet, sanitation, socioeconomic level, literacy, and way of life. Global population growth and aging exposes More people and groups to environmental risk factors. According to Prüss-Ustün et al. (2019), health-care costs around 10% of the global GDP, but financing for primary prevention and public health is quite low. Understanding threats associated with the environment is crucial for formulating decisions that will safeguard the public's health. The burden on the world's health-care systems might be lessened by funding initiatives and programs focused on reducing environmental threats.

Environmental hazards, such as pollution, and natural calamities also influence the geographic distribution of many infectious diseases. Multidisciplinary scientific research and public awareness are the best way to solve this issue. The increase in this risk should be addressed by different organizations at national and international levels, and viable solutions should also be suggested. Therefore, the goal of this chapter is to give a summary of the most current research connecting environmental risk factors and their health consequences. The present paper broadly emphasizes different environmental factors and human health with special emphasis on air quality, water quality and climate change.

14.3 Air Quality and Human Health

All age groups and sexes have reportedly been impacted by air pollution or poor air quality. According to the report, it causes over nine million deaths annually (WHO 2018). Chronic obstructive lung disease, bronchitis, dyspnea, asthma, respiratory disorders, and high rates of hospitalization (a measure of morbidity) are all directly associated with air pollution. Bronchial asthma, chronic obstructive pulmonary disease, pulmonary inefficiency, cardiovascular disease, and cardiovascular mortality (Rojas-Rueda et al. 2021). Air pollution and secondhand smoke cause 52 million lower respiratory illnesses yearly and 35% of all cases worldwide (Prüss-Ustün and Neira 2016). Additionally linked to non-communicable diseases are air pollution, chemical pollutants, and secondhand smoke, which annually cause 119 million cardiovascular disorders, 49 million cancers, and 32 million chronic respiratory ailments (Prüss-Ustün and Neira 2016).

Particulate matter (PM) is the air contaminant with the greatest number of negative effects on health. According to Chun et al. (2020), PM (2.5) contributes to the development of Parkinson's disease, chronic renal illness, chronic obstructive pulmonary disease, dementia, depression, and respiratory mortality. Suicide was also listed as a short-term effect of PM10 pollution. Pneumonia has been linked to brief exposure to PM 2.5 or PM 10 in youngsters.

Other pollutants that are present in the environment and affect human health are nitrogen dioxide (NO_2) (Chun et al. 2020), ozone (O_3) (Kim et al. 2018), household air pollution (Wu et al. 2020a, b), sulfur dioxide (SO_2) (Ioannidis 2009), carbon monoxide (CO) (Ioannidis 2009), solid fuel use (Ioannidis 2009), nitrogen oxides (Prüss-Ustün et al. 2019), desert dust (Prüss-Ustün et al. 2019), biomass burning (Prüss-Ustün et al. 2019), black carbon (Prüss-Ustün and Neira 2016), and indoor air pollution from solid fuel (Prüss-Ustün and Neira 2016). Oxides of nitrogen are associated with low birth weight and preterm birth in pregnant women (Rojas-Rueda et al. 2021). Chronic and short-term exposure to nitrogen dioxide was linked to a higher threat of pneumonia (Rojas-Rueda et al. 2021), autism spectrum disorder (Chun et al. 2020), cardiovascular mortality (Wu et al. 2020a, b), cancer mortality (Kim et al. 2018), and all-cause mortality in adults (Ioannidis 2009). Long-term O_3 exposure in adults is associated with an increase in Parkinson's disease and mortality from ischemic heart disease (IHD). The primary sources of SO_2 emissions are burning fossil fuels in factories, power plants, and other commercial establishments, as well as transportation means like trains and ships. Pregnant women exposed to SO_2 were shown to have a risk of low birth weight and gestational diabetes mellitus. Pneumonia and adult cardiac arrhythmia were associated with short-term SO_2 exposure in children. According to a study, CO increases the risk of adult cardiac arrhythmia, and pediatric pneumonia in children. The chances of premature birth during pregnancy have also been linked to CO exposure (Rojas-Rueda et al. 2021). The entire thing is integrated with climate change, and because of the risk, the consequences can be severe for human beings (Moore 2009). Most ecosystems negatively impact climate change, enhancing food insecurity, ice melting, animal extinction, and plant degradation (USGCRP 2009; Marlon et al. 2019). It has become worse in developing countries due to overpopulation, uncontrolled urbanization, and increased industrialization (Mannucci and Franchini 2017). The country with societal disparities and a lack of knowledge regarding environmentally sound management practices results in poor air quality. It is important to highlight that three hundred million people globally use the energy as mentioned above sources to fulfill their fuel wood requirement. Low earnings cause individuals to use wood or solid fuels for domestic needs, exposing them to poor-quality, polluted air within their homes. Due to the use of solid fuel for a longer period and its exposure to longer duration, women in the family have the largest risk for illness in poor or developing nations (Hashim and Boffetta 2014).

Air pollution is responsible for more than four million premature deaths yearly due to open-air pollution and 3.8 million from inside air pollution. According to WHO (2020), 90% of people breathe air with excessive pollution. Due to dangerous air pollution in India, the air quality has reached hazardous levels. The urban and peri-urban areas face pollution due to urbanization, increasing industries and a surge in motorbikes. Additionally, biomass combustion and cooking practices significantly contribute to household air pollution in India (Kankaria et al. 2014; Parajuli et al. 2016). People suffering from prerequisites related to air pollution are more susceptible to the effects of severe acute respiratory syndrome—CoV-21.

14.4 Water Quality and Human Health

When a person comes in contact with contaminated drinking or bathing water, it substantially harms its health; it may be short or long-term (Fleisher et al. 1998; Lin et al. 2016; Dwight et al. 2005). By ensuring that everyone has access to uncontaminated water and sanitary facilities, an estimated 2 million fatalities and 15% of global child mortality can be prevented. Among the most serious hazards are higher amounts of nutrients in oceanic and coastline waters caused by agricultural runoff and microbe contamination of groundwater caused by sewage mouths (Le Goffe 1995). Surface water accounts for 50% of the world's drinking water needs, yet it is the main method of long-distance disease transmission. One hundred fifty-nine million people depend on surface water that has not been treated, and it is significantly more dangerous for their health than other water sources (Paudel 2018). Water quality problems are spreading globally as population demands rise.

In certain research, active pharmaceutical compounds in the environment are linked to concerns about human noxiousness, carcinogenicity, ecologic toxicity, and antibiotic resistance (Harris et al. 2012; Sorell 2016). Additionally, it has to do with sewage discharge into water and onto land. Wet deposition is the most significant pathway to transferring metals from the atmosphere to the terrestrial surface (Dasgupta 2004; Dore et al. 2014).

The incidence of trachoma, intestinal helminth infections, and schistosomiasis are all dramatically decreased by safe drinking water (Harris et al. 2012). By changing the environment and modifying human habitation, malaria's impact on water management might decrease up to 88% (Prüss-Üstün and World Health Organization 2008). The results imply that an environmental management-focused malaria control program is very useful in decreasing the death rate and can result in long-lasting malaria prevention measures. Suppose water quality can be guaranteed up until the point of consumption. In that case, diarrhea can be significantly reduced by 28–45% depending on the kind of water supply, which is often pipeline water supply.

As persistent metallic elements in the environment bio-accumulate, bio-concentrate, and bio-amplify in the food chain, soils need to be remediated, reclaimed and restored to safeguard residents, especially children (Maddison et al. 2005; Mannucci and Franchini 2017; Liu et al. 2013). The water cycle, including surface water and wastewater, must be managed to safeguard human health from dangerous chemicals and waterborne infectious diseases (Zani et al. 2015; WHO 2017).

14.5 Climate Change and Human Health

Since 1975, the global development, resurgence, and transmission of infectious diseases, as well as the recurrence of other chronic diseases, has been largely attributed to climate instability and global warming. (McMichael et al. 2006).

The effects of climate change on health are being considered globally. Numerous factors are responsible for climate change and causing different environmental risks that can have various negative health effects (Fig. 14.2). There are three different health effects. The first is relatively immediate effects, typically brought on by weather extremes. The second one is environmental change and ecologic disruption because of climatic change. Thirdly, communities are affected by demoralizing and displacing forces such as economic upheaval caused by environmental degradation, violence, climatic change, and traumatic, infectious, nutritional, psychological, and other health effects.

India stated in 2008 that adjusting to the unavoidable effects of climate change is vital to avoid incurring enormous costs in the form of infectious diseases, medical expenses, and lost output (WHO 2009). According to estimates, 2.4% of global diarrhea and 6% of malaria cases reported in 2000 were caused by climate change (WHO 2018). According to the third report of IPCC, the world's temperature will rise throughout the rest of the century, increasing health concerns worldwide, especially in low-income countries. Additionally, it is projected that vector-borne diseases like malaria may become more common (Fankhauser 1994; Bosello et al. 2006).

The greenhouse effect is the best example showing the influence of climate change on human welfare. According to research conducted in Mexico, decreasing greenhouse gas emissions would prevent over 64,000 premature deaths over 20 years (Bosello et al. 2006). Bell et al. (2008) conducted a literature evaluation of valuation studies and evaluated the effects of greenhouse

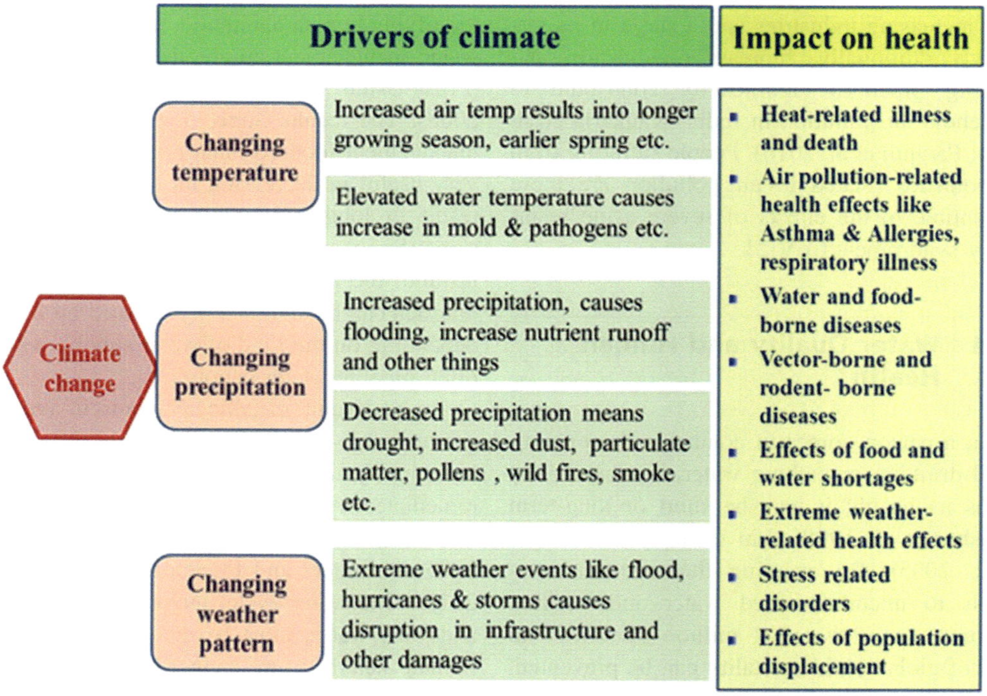

Fig. 14.2 Drivers of climate and its impact on health

gases on human health. Tseng et al. (2009) use dengue fever in Taiwan as a case study to explore the health effects of diseases linked to climate change in poor countries. After first establishing the link between climatic factors and the number of dengue fever cases, contingent valuation research was used to attempt a financial assessment. Findings from many research offer compelling proof that greenhouse gas reduction measures significantly positively affect public health.

14.6 Concluding Remarks

An unstable global health situation results from several detrimental multiplier effects caused by an environmental threat. With this risk, significant health hazards could get substantially worse over the next 50 years. Environmental and health policies should be integrated because they are closely tied together. Addressing this issue, several works have been conducted over the globe utilizing a variety of approaches to identify the health advantages of higher environmental quality.

This article reveals important values associated with longevity and health quality in both established and developing countries. It also establishes a strong link between environmental danger exposure and health issues, demonstrating the requirement for plan interventions. Improving air quality and ensuring ample quantities of uncontaminated drinking water are connected to substantial improvements in human health and well-being. There are various means that climate change can affect human health, and it is increasingly important for environmental health management to examine how it affects health at the national level.

Therefore, public health progress needs to be studied to understand the connections between numerous environmental risk factors and human health. A quantitative examination of environmental toxins at the local, national, and international levels is required for more accurate estimation.

References

Bell ML, Davis DL, Cifuentes LA, Krupnick AJ, Morgenstern RD, Thurston GD (2008) Ancillary human health benefits of improved air quality resulting from climate change mitigation. Environ Health Glob Access Sci Sour 7:41. https://doi.org/10.1186/1476-069X-7-41

Bosello F, Roson R, Tol RSJ (2006) Economy-wide estimates of the implications of climate change: human health. Ecol Econ 58(3):579–591. https://doi.org/10.1016/j.ecolecon.2005.07.032

Chun HK, Leung C, Wen SW, McDonald J, Shin HH (2020) Maternal exposure to air pollution and risk of autism in children: a systematic review and meta-analysis. Environ Pollut 256:113307. https://doi.org/10.1016/j.envpol.2019.113307

Dasgupta P (2004) Valuing health damages from water pollution in urban Delhi, India: a health production function approach. Environ Dev Econ 9(1):83–106. https://doi.org/10.1017/S1355770X03001098

Dore AJ, Hallsworth S, McDonald AG, Werner M, Kryza M, Abbot J, Fowler D (2014) Quantifying missing annual emission sources of heavy metals in the United Kingdom with an atmospheric transport model. Sci Total Environ 479–480:171–180. https://doi.org/10.1016/j.scitotenv.2014.02.001

Dube R, Nandan V, Dua S (2014) Waste incineration for urban India: Valuable contribution to sustainable MSWM or inappropriate high-tech solution affecting livelihoods and public health? Int J Environ Technol Manage 17(2/3/4):199–214. https://doi.org/10.1504/IJETM.2014.061792

Dwight RH, Fernandez LM, Baker DB, Semenza JC, Olson BH (2005) Estimating the economic burden from illnesses associated with recreational coastal water pollution—A case study in Orange County, California. J Environ Manage 76(2):95–103. https://doi.org/10.1016/j.jenvman.2004.11.017

Fankhauser S (1994) The economic costs of global warming damage: a survey. Glob Environ Chang 4(4):301–309. https://doi.org/10.1016/0959-3780(94)90030-2

Fleisher JM, Kay D, Wyer MD, Godfree AF (1998) Estimates of the severity of illnesses associated with bathing in marine recreational waters contaminated with domestic sewage. Int J Epidemiol 27(4):722–726. https://doi.org/10.1093/ije/27.4.722

Greenstone M, Hanna R (2014) Environmental regulations, air and water pollution, and infant mortality in India. Am Econ Rev 104(10):3038–3072. https://doi.org/10.1257/aer.104.10.3038

Harris SJ, Cormican M, Cummins E (2012) Antimicrobial residues and antimicrobial-resistant bacteria: Impact on the microbial environment and risk to human health—a review. Hum Ecol Risk Assess 18(4):767–809. https://doi.org/10.1080/10807039.2012.688702

Hashim D, Boffetta P (2014) Occupational and environmental exposures and cancers in developing countries.

Ann Glob Health 80(5):393–411. https://doi.org/10.1016/j.aogh.2014.10.002

Ioannidis JPA (2009) Integration of evidence from multiple meta-analyses: a primer on umbrella reviews, treatment networks and multiple treatments meta-analyses. CMAJ 181(8):488–493. https://doi.org/10.1503/cmaj.081086

Kankaria A, Nongkynrih B, Gupta SK (2014) Indoor air pollution in India: implications on health and its control. Indian J Commun Med 39(4):203–207. https://doi.org/10.4103/0970-0218.143019

Kim H-BB, Shim J-YY, Park B, Lee Y-JJ (2018) Long-term exposure to air pollutants and cancer mortality: a meta-analysis of cohort studies. Int J Environ Res Public Health 15(11):2608. https://doi.org/10.3390/ijerph15112608

Kovats RS, Haines A (2005a) Global climate change and health: Recent findings and future steps. CMAJ Can Med Assoc J 172(4):501–502. https://doi.org/10.1503/cmaj.050020

Kovats RS, Haines A (2005) Global climate change and health: Recent findings and future steps. CMAJ Can Med Assoc J 172(4):501–502. https://doi.org/10.1503/cmaj.050020

Krishna Moorthy K, Suresh Babu S, Manoj MR, Satheesh SK (2013) Buildup of aerosols over the Indian Region. Geophys Res Lett 40(5):1011–1014. https://doi.org/10.1002/grl.50165

Le Goffe P (1995) The benefits of improvements in coastal water quality: a contingent approach. J Environ Manage 45(4):305–317. https://doi.org/10.1006/jema.1995.0078

Lin H, Liu T, Xiao J, Zeng W, Li X, Guo L, Ma W (2016) Quantifying short-term and long-term health benefits of attaining ambient fine particulate pollution standards in Guangzhou, China. Atmos Environ 137:38–44. https://doi.org/10.1016/j.atmosenv.2016.04.037

Liu X, Song Q, Tang Y, Li W, Xu J, Wu J, Brookes PC (2013) Human health risk assessment of heavy metals in soil–vegetable system: a multi-medium analysis. Sci Total Environ 463–464:530–540. https://doi.org/10.1016/j.scitotenv.2013.06.064

Maddison D, Catala-Luque R, Pearce D (2005) Valuing the arsenic contamination of groundwater in Bangladesh. Environ Resource Econ 31(4):459–476. https://doi.org/10.1007/s10640-005-3364-z

Mannucci PM, Franchini M (2017) Health effects of ambient air pollution in developing countries. Int J Environ Res Public Health 14(9):1048. https://doi.org/10.3390/ijerph14091048

Marlon JR, Bloodhart B, Ballew MT, Rolfe-Redding J, Roser-Renouf C, Leiserowitz A, Maibach E (2019) How hope and doubt affect climate change mobilization. Front Commun 4:20. https://doi.org/10.3389/fcomm.2019.00020

McMichael AJ, Woodruff RE, Hales S (2006) Climate change and human health: present and future risks. Lancet 367(9513):859–869. https://doi.org/10.1016/S0140-6736(06)68079-3

Millennium Ecosystem Assessment (2005) Ecosystems and human well-being: synthesis; island press: Millennium Ecosystem Assessment series. Washington, DC

Moore FC (2009) Climate change and air pollution: exploring the synergies and potential for mitigation in industrializing countries. Sustainability 1(1):43–54. https://doi.org/10.3390/su1010043

Parajuli I, Lee H, Shrestha KR (2016) Indoor air quality and ventilation assessment of rural mountainous households of Nepal. Int J Sustain Built Environ 5(2):301–311. https://doi.org/10.1016/j.ijsbe.2016.08.003

Paudel U (2018) A review of effects of environmental change on human health. Am J Environ Sci 14(3):95–109. https://doi.org/10.3844/ajessp.2018.95.109

Prüss-Ustün A, Neira M (2016) Preventing disease through healthy environments: a global assessment of the environmental burden of disease, vol 259. World Health Organization, Geneva

Prüss-Ustün A, van Deventer E, Mudu P, Campbell-Lendrum D, Vickers C, Ivanov I, Neira M (2019) Environmental risks and non-communicable diseases. BMJ 364:l265. https://doi.org/10.1136/bmj.l265

Prüss-Üstün A, World Health Organization (2008) Safer water, better health: costs, benefits and sustainability of interventions to protect and promote health. / Prüss-Üstün A et al World Health Organization. https://apps.who.int/iris/handle/10665/43840

Ramachandran S, Kedia S, Srivastava R (2012) Aerosol optical depth trends over different regions of India. Atmos Environ 49:338–347. https://doi.org/10.1016/j.atmosenv.2011.11.017

Ren G, Zhou Y, Chu Z, Zhou J, Zhang A, Guo J, Liu X (2008) Urbanization effects on observed surface air temperature trends in North China. J Clim 21(6): 1333–1348. https://doi.org/10.1175/2007JCLI1348.1

Rojas-Rueda D, Morales-Zamora E, Alsufyani WA, Herbst CH, AlBalawi SM, Alsukait R, Alomran M (2021) Environmental risk factors and health: an umbrella review of meta-analyses. Int J Environ Res Public Health 18(2):704. https://doi.org/10.3390/ijerph18020704

Shastri H, Paul S, Ghosh S, Karmakar S (2015) Impacts of urbanization on Indian summer monsoon rainfall extremes. J Geophys Res Atmos 120(2):496–516. https://doi.org/10.1002/2014JD022061

Singh V, Singh S, Biswal A (2021) Exceedances and trends of particulate matter ($PM_{2.5}$) in five Indian megacities. Sci Total Environ 750:141461. https://doi.org/10.1016/j.scitotenv.2020.141461

Sorell TL (2016) Approaches to the development of human health toxicity values for active pharmaceutical ingredients in the environment. AAPS J 18(1):92–101. https://doi.org/10.1208/s12248-015-9818-5

Srivastava R (2013) Effect of global warming on agricultural systems. Am Eurasian J Agric Environ Sci 13(5): 677–682

Srivastava R, Pawan B (2021) Effect of climate change on Indian crop diversity: An overview. In: Krishna V

(ed) Contemporary environmental challenges in India. West, Rudra Publishers and Distributors. Karawal Nagar, New Delhi, India, pp 192–202
Srivastava R (2022) Strategies to achieve sustainable development goals (SDGs): a road map for global development. In: Srivastava R (ed). Nova Science Publishers, Hauppauge, NY. https://doi.org/10.52305/YNDL2610
Stern N (2007) The economics of climate change: the stern review. Cambridge University Press, Cambridge
Tseng WC, Chen CC, Chang CC, Chu YH (2009) Estimating the economic impacts of climate change on infectious diseases: A case study on dengue fever in Taiwan. Clim Change 92(1–2):123–140. https://doi.org/10.1007/s10584-008-9437-6
UNDESA (2018) 2018 revision of world urbanization prospects
USGCRP (2009) Global climate change impacts in the United States. In: Karl TR, Melillo JM. Peterson TC (eds) Climate change impacts by sectors: ecosystems. United States Global Change Research Program, New York, NY. Cambridge: Cambridge University Press
World Health Organization (2009) Global health risks. World Health Organization, Geneva
World Health Organization (2017) World health statistics 2017: monitoring health for the SDGs (1st ed), p 103
World Health Organization (2018) Ambient (outdoor) air pollution
World Health Organization (2019) Air pollution. Retrieved from http://www.who.int/airpollution/en/ Retrieved
World Health Organization (2020) WHO website. Retrieved from https://www.who.int/
Wu MY, Lo WC, Chao CT, Wu MS, Chiang CK (2020a) Association between air pollutants and development of chronic kidney disease: a systematic review and meta-analysis. Sci Total Environ 706:135522. https://doi.org/10.1016/j.scitotenv.2019.135522
Wu X et al (2020b) Exposure to air pollution and COVID-19 mortality in the United States. Cold Spring Harbor Lab. https://doi.org/10.1101/2020.04.05.20054502
Yang B, Zhang Y, Qian Y (2012) Simulation of urban climate with high-resolution WRF model: a case study in Nanjing, China. Asia-Pac J Atmos Sci 48(3):227–241. https://doi.org/10.1007/s13143-012-0023-5
Zani C, Restivo FM, Carcelli M, Feretti D, Pelosi G, Rogolino D, Buschini A (2015) A biotechnological approach for the development of new antifungal compounds to protect the environment and the human health. J Public Health Res 4(3):613. https://doi.org/10.4081/jphr.2015.613

Part IV

Integrated Risk Mitigation and Resilience Measures

Integrated Geospatial Approach for Mangrove Restoration and Environmental Sustainability in Mundra Coast of Gujarat, India

15

Balamurugan Guru, Pradheeps Muthulingam, R. B. Archana, and K. N. Kusuma

Abstract

Mangrove restoration is considered crucial in conserving and protecting coastal ecosystems and its habitat. Large-scale mangrove restorations were introduced to address disaster prevention across coastal habitats using selected native mangrove species. Soil characteristic study is considered an important factor in mangrove cultivation in conservation as well as developing coastal shelter belts for disaster risk reduction. In this study, we carried out a preliminary survey to assess specific physical and chemical characteristics of soil at randomly selected sites on the Mundra coast during 2014–2015. The present study was undertaken to assess the environmental status of the mangrove forest in the Mundra coast region and to study the periodical changes in the coastal habitat using high spatial resolution multi-temporal digital data. The distribution of physicochemical characteristics through the on-field determination of soil texture, soil pH, and electrical conductivity was measured. Soil Organic carbon, macronutrients (N, P, K), and micronutrients (Ca, Mg, S) are measured in the selected study sites. The results showed the even distribution of soil macro and micronutrients owing to the natural processes and human activity. Soil salinity is higher and the pH is neutral in range. The results showed that there is an increase in the natural regeneration of mangroves as well as construction activities on the Mundra coast. The information from this study was highly beneficial for the introduction of native species for mangrove regeneration, planning for conservation, and community-based management of mangroves in the study area.

Keywords

Mangrove ecosystem · Soil physicochemical parameters · Remote sensing · Restoration · Environmental sustainability

15.1 Introduction

Mangrove forest dynamics growing along the shores and estuaries contributes to the nutrient metabolism of subtropical coastal ecosystems

B. Guru (✉)
Department of Geology, School of Earth Sciences, Central University of Tamil Nadu, Thiruvarur, India
e-mail: balamurugan.guru@gmail.com

P. Muthulingam
Chennai Metropolitan Development Authority (CMDA), Egmore, Chennai 600008, India

R. B. Archana
Centre for Geoinformatics, Jamsetji Tata School of Disaster Studies, Tata Institute of Social Sciences, Deonar, Mumbai 400088, India

K. N. Kusuma
Department of Earth Sciences, Pondicherry University, Kalapet, Puducherry 605014, India

S. Talukdar et al. (eds.), *Environmental Risk and Resilience in the Changing World*, GIScience and Geo-environmental Modelling, https://doi.org/10.1007/978-3-031-62442-1_15

(Kristensen et al. 2008). Coastal ecology depends mainly on nutrient flux delivered by mangrove forests due to inadequate primary productivity. Across tropical and subtropical countries, mangroves occupy the inter-tidal zones of the sheltered coastline such as river mouth, estuaries, and lagoons (Debnath and Naskar 1999). They are widely adapted to cope with unfavorable environmental conditions for growth and regeneration, resultants of saltwater inundation, unstable soil due to tidal flow, and lack of freshwater (Elster 2000; Jagtap and Komarpant 2003; Gilman et al. 2008; Hashim et al. 2010). Mangrove forests and estuaries are the breeding and nursery grounds for numerous marine organisms including commercially important shrimps, crabs, and fish species (Nagelkerken et al. 2008). Globally, 18% of the earth's surface is a coastal zone; the mangroves cover 8% of the coastal zone are distributed across 123 countries and cover over 150,000 km^2. The mangrove ecosystem consists of 69 species in 27 genera, belonging to 20 families (Kathiresan and Bingham 2001). Comparatively, in the Asian subcontinent, mangroves cover about 5700 km^2 of the coastal area and contribute to excessive ecosystem services, thereby contributing to the economy. India accounts for 3% of global mangroves and stands fourth among the Asian countries next to Bangladesh (Singh et al. 2012). The physical and climatic factors of the central Indian Ocean region determine the influence of tides and coastal geomorphology of the country's west coast (Untavale 1986). This, in turn, influences the type and extent of coastal habitats especially mangrove forests, dispersion of coastal pollutants, seasonal patterns of sedimentation/erosion and coastal alteration, and the effect of long-term sea level changes in the coasts (Peng et al. 2016).

Mangrove ecosystem has special physicochemical characteristics, tidal currents, wind, increased temperature, murky anaerobic soil, and a rich detritus environment (Dissanayake and Chandrasekara 2014). Mangroves can survive and specialize in colonizing coastal areas, especially in tropical regions. Mangroves show better growth in the alluvial substrates with fine textured mud/silt, rich in humus but very poor in stagnant waters. Nonetheless, they find it challenging to colonize the coastal zone with waves of high energy and adapted well to sheltered shorelines. However, the importance of the mangrove ecosystem promoted the activities such as afforestation to conserve the habitat. Nutrient cycling in estuarine ecosystems is vitally managed by mangroves and the suitable nutrient requirements and information related to their tolerance to excess or insufficient nutrients could be useful in mangrove restoration and regeneration (Yates et al. 2002; Ramos e Silva et al. 2007; Rao and Babu 2017). Habitat encroachments, fragmentation, and natural hazards cause coastal depletion and are serious threats to coastal habitats. Besides, global warming causes an alarming rate of sea level increase in this century which leads to coastal erosion and uncontrolled damage to the environment. Coastal communities depend on the coastal biodiversity for their livelihoods and the loss of habitats will have economic repercussions on the livelihoods of the communities living in these sensitive areas (Gopal and Chauhan 2006). Therefore, geospatial technology is employed to understand the coastal land use and land cover changes (LULC) to develop conservation planning as well as study the historical LULC changes and are effectively used in creating different thematic maps providing the distribution of major and micronutrients in the ecosystem (Manassrisuksi and Hussin 2001; Van de Merwe et al. 2016). The changes in LULC features due to natural and anthropogenic activities caused negative environmental impacts at both global and local scales (Walker 2004; Balamurugan and Sridhara Aravind 2015). LULC is also used for monitoring and assessing of expansion speed of an area or urbanization (Carlson and Azofeifa 1999; Atasoy et al. 2006; Pielke et al. 2007; Yuan 2008). It is understood that geospatial technology approach highly successful in estimating the extent of mangrove forest distribution (Wang et al. 2004; Kirui et al. 2013) and quantitatively measuring the changes in distribution over a period of time (Balamurugan and Anubhooti 2015; Ibharim et al. 2015). Therefore, the present investigation attempted to understand the soil physicochemical properties of mangrove forests in the Mundra coast and LULC changes for proper planning for the conservation and management of the mangrove ecosystem.

15.2 Study Area

Mundra is the coastal region located between 22.77° N to 22.83° N and 69.46° E to 69.72° E along the coastline of the Kutch district in Gujarat which is considered the largest in India. It is also known to constitute 23% of the total Indian mangrove ecosystem next to the Sundarbans of West Bengal. The coastal zone of Gujarat is known for its complex habitat structure including salt marshes and mangroves (Plate 15.1‘a’, ‘b’). Our study site (Fig. 15.1) is located on the coastline of the Mundra. The most important Special Economic Zone (SEZ) is established here and known as the Mundra Port Special Economic Zone (MPSEZ) located in the Gulf of Kutch in Gujarat. The area receives its rainfall during the south–west monsoon season, however, due to weak monsoon rains and high rate of evaporation cause aridity in the study area leaving the region relatively deficient in water resources (Balamurugan and Sridhara Aravind 2015). There are 60 species of mangroves are known to exist along the Indian coastal region. Along the west coast, the Gujarat coast is home to 36 species of true mangroves (Debnath and Naskar 1999). Historical evidence confirms the practice of traditional salt production and the presence of a mangrove ecosystem (Plate 15.1c) in continuation from the Northern coast intertidal zone are lost now.

15.3 Methodology and Data Collection

Soil samples were collected through a random sampling method at six different locations (from 22.7853° N to 22.7870° N and longitude 69.4975° E to 69.4995° E) to ascertain soil chemical attributes of the location (from December 2014 to February 2015). Soil sediment samples were collected using soil sampling grabber and augers at a depth of 30 cm in polythene bags. Soil color and texture identification were done on-site and documented immediately. Later, the soils were air-dried and crushed before the soil chemical analysis in the lab. Soil pH was determined in situ and electrical conductivity (EC) was conducted using a Systronics conductivity meter. The total organic carbon, nitrogen, phosphorus, and potassium (macronutrients) and calcium, magnesium, and sulfur are determined using standard methods. Primary data for LULC analysis were collected at the study site including GPS points, and field photographs to understand the previous and existing environment of the study area. Secondary data related to information on fisherfolk and livelihoods were collected from the fisheries department and census data. Satellite images and toposheets are collected from the Indian Space Research Organization (ISRO) and Survey of India (SOI), respectively. Figure 15.2 shows the material used and the flow of methodology adopted for LULC analysis in the study.

15.3.1 Satellite Images and Thematic Maps

The satellite data were essential for the identification of land cover aspects such as the intertidal zone, and vegetation cover including existing mangrove patches, sand dunes, and urban areas along the coastal area. To ascertain a proper site for mangrove restoration, satellite images were procured for the years 2005 and 2015. The 2005 and 2015 images were captured by IRS-P6 (LISS-IV) and Resourcesat satellite (LISS-IV), respectively.

15.3.2 Creation of Spatial Soil Properties Map

The digital database was created using ArcGIS 10 software package. The soil chemical analysis such as organic carbon (OC), nitrogen (N), phosphorus (P), potassium (K), calcium (Ca),

a) Sand bars and mud flats in the study area

b) Natural Mangrove regeneration in the study area

c) Salt pans located in the region

d) Site selected for Mangrove regeneration in the inter-tidal zone

e) Developmental activities in the back scene near mangrove ecosystem in Mundra

f) Fishing net drying region

Plate 15.1 Field observations: **a** Sandbars and mudflats in the study area, **b** Natural Mangrove regeneration in the study area, **c** Salt pans located in the region, **d** Site selected for Mangrove regeneration in the intertidal zone, **e** Developmental activities in the back scene near mangrove ecosystem in Mundra, **f** Fishing net drying region

magnesium (Mg), and sulfur (S) was measured from the defined standard procedure and the interpolation of nutrient values from the different sites using Jenks natural break method as each parameter in 5 distinct classes. All seven parameters show different values concerning the site selected for testing. The soil chemical distribution maps are prepared using ArcGIS.

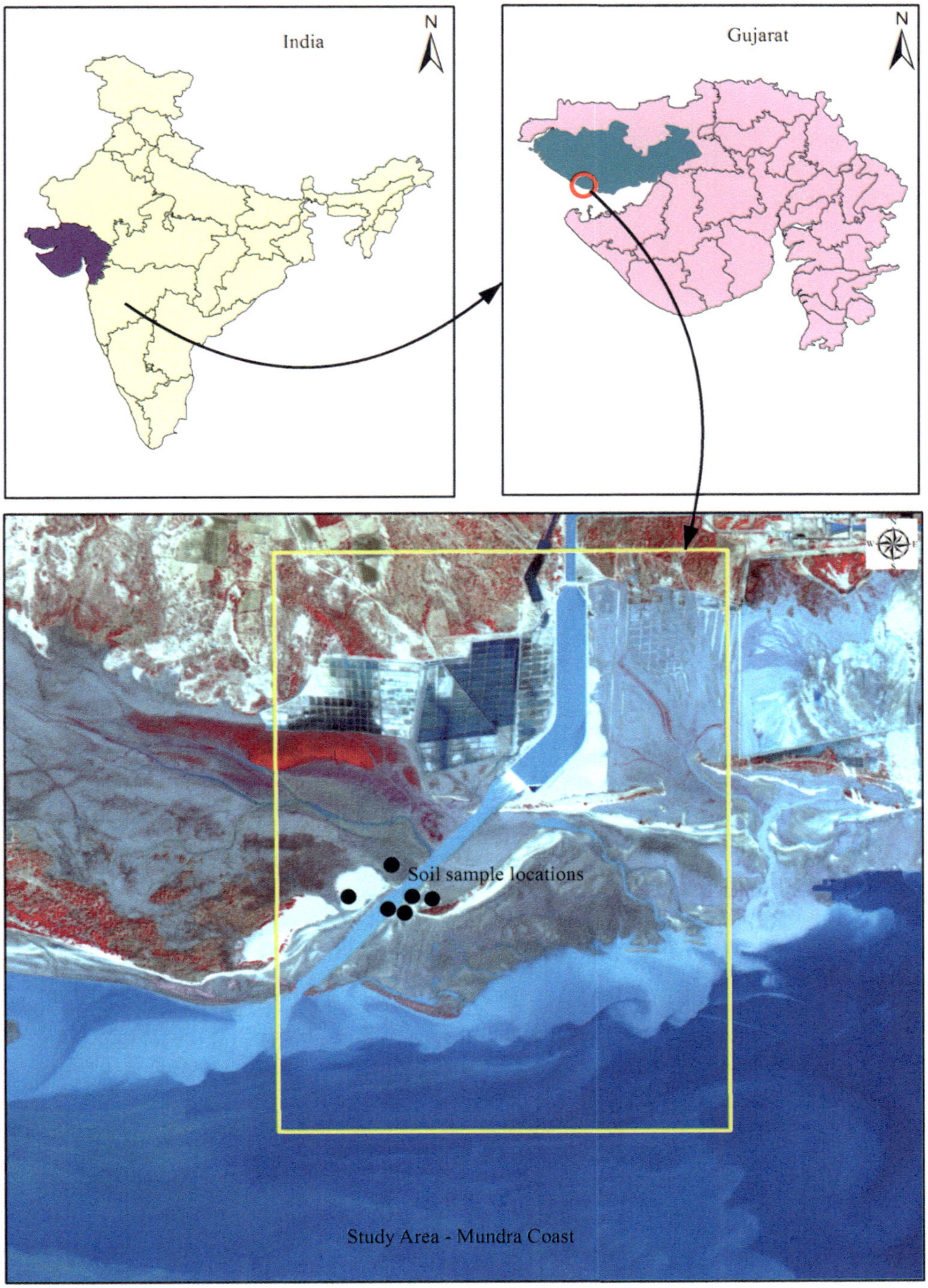

Fig. 15.1 Map showing the study area of Mundra Coast, India

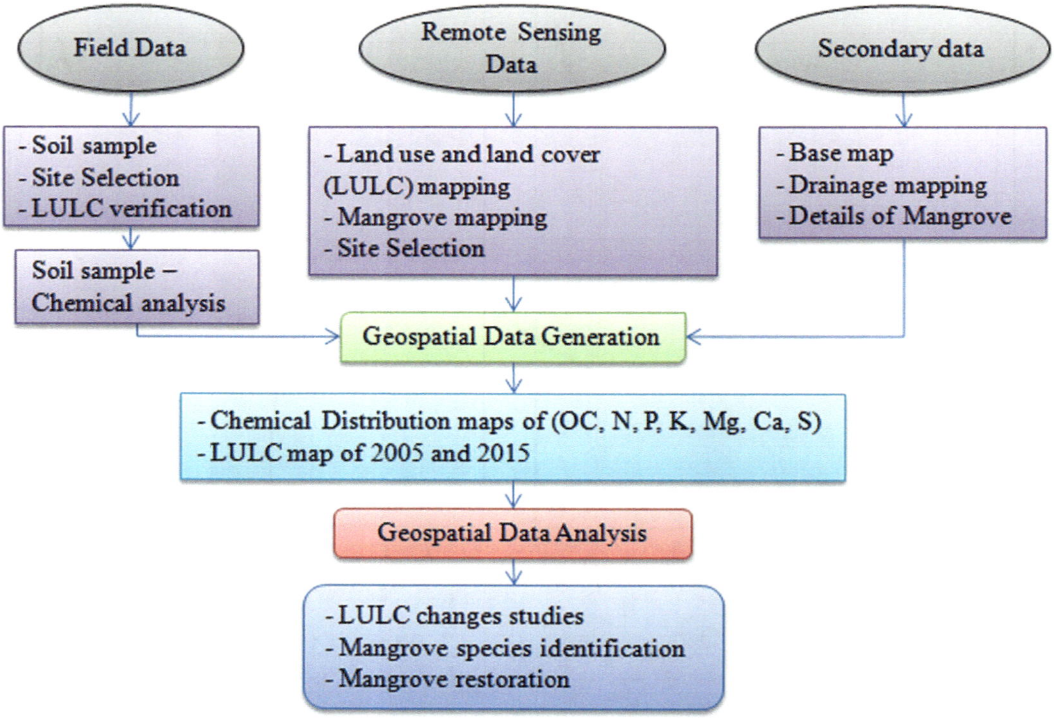

Fig. 15.2 Flowchart of material and methodology adopted in the research

15.4 Results and Discussion

15.4.1 Soil Physio-Chemical Properties

(a) **Soil Texture**

Soil properties affect the movement and retention of air, water, and solute in the soil, which subsequently influence plant growth and enhance microbial activity (Feller et al. 2010). The soil texture can be distinguished by the soil particle size distribution and is generally referred to the presence and relative proportion of sand, silt, and clay particle in the soil sample. In the present study, the soil texture was ascertained by field test method and is of the silty clay form.

(b) **Soil pH and Electrical Conductivity**

Soil pH values range from 5.16 to 7.7 and the electrical conductivity is between 5 to 12 (mS/cm) (Fig. 15.2). It is observed that the acidic pH might be due to the increased decomposition rate of organic matter by sulfur-reducing bacteria as observed by several researchers and also due to the oxidation of $FeSO_4$ and FeS to H_2SO_4. Besides, it is also attributed to the generation of CO_2 due to the biological oxidation process. The electrical conductivity (EC) values varies may be attributed to the decrease in the inflow of freshwater and the high evaporation rate in the region.

(c) **Total Organic Carbon**

Soil organic carbon assessment helps in understanding the interaction between chemicals and soils/sediment in nature. It is also considered an important factor in ecological assessment as it ensures nutrient regeneration in the soil and the productivity of the aquatic environment. Total organic carbon (TOC) from the present study ranged between 0.1 and 5.4% (Table 15.1). The major source of soil organic carbon is litter and

Table 15.1 Soil chemical characteristics of the study area

S. No.	Chemical parameters	S_1	S_2	S_3	S_4	S_5	S_6	Average value
1.	Organic carbon (%)	0.1	2.2	2.8	4.5	5.4	3.2	3.06
2.	Total nitrogen (mg/kg)	19.5	83.7	83.8	375.0	64.4	50.1	112.7
3.	Phosphorus (mg/kg)	63.9	699.0	380.0	218.0	53.4	49.4	243.9
4.	Total potassium (mg/kg)	273.0	439.0	283.0	333.3	345.0	373.0	341.0
5.	Magnesium (mg/kg)	215.0	115.0	219.0	524.0	110.0	170.0	225.5
6.	Calcium (mg/kg)	40.0	59.9	79.8	99.8	99.9	40.0	69.9
7.	Sulfur (%)	3.2	12.6	1.0	0.6	6.9	6.3	5.1

also observed to differ at various Eco-successional stages. TOC of Mundra coasts shows lower values and can be attributed to the recent regeneration status as well as poor soil nutrition.

(d) **Soil Nitrogen**

Nitrogen 'N' has a significant effect on leaf expansion and stem growth in mangroves and rich 'N' increases leaf productivity including leaf gain, leaf loss, and stem elongation. Deficiency in 'N' causes stunted growth and results in Chlorosis in Mangroves. The concentration of 'N' from the study area (19.6 mg/kg) depicts a similar distribution of N (Fig. 15.3b). The anaerobic and organic matter-rich soils of the lower intertidal zones with mangrove colonies are favorable for N fixation in tropical mangrove forests (e.g., Cusack et al. 2009). N fixation rate in association with pneumatophores and roots increases with microbial communities in the estuaries (Reef et al. 2010). It is also could be attributed to the high land runoff from rivers into the mangroves during the monsoon season.

(e) **Phosphorus**

Observation of 'P' concentration shows its availability becomes a limiting factor for the growth of mangroves due to the saline-rich ecosystem. The total phosphorous requirement for the normal growth of mangroves ranges from 240–340 mg/kg of soil. The availability of 'P' in the mangrove soils is due to the utilization of phosphates by the phytoplanktons during the summer season when the photosynthetic activity of mangroves is higher. The amount of 'P' content from our study area (Fig. 15.3c) reflects it and it accounts for an average of 243 mg/kg of soil which indicated a moderate quantity of phosphorus availability. The optimum quantity of phosphorus availability in combination with 'N' will enhance the shoot and root growth of mangroves.

(f) **Potassium**

'K' is considered an essential element for photosynthetic activity and leaf production in mangroves. Normal growth of mangroves depends on 'K' concentration ranges between 500 to 900 mg/kg in the soil. Meanwhile, the average value of 'K' from the results (Fig. 15.3d) shows a very low availability of 'K' in the study area (341 mg/kg). The relative concentration of 'Na' and 'Ca' influences the presence of potassium salts. It helps in moving the manufactured food, viz., carbohydrates (sugars) and proteins (amino acids), from leaves to roots.

(g) **Other Micronutrients**

Meanwhile, the essential micronutrients calcium, magnesium, and sulfur concentration were found to be higher (Ca = 225; Mg = 69 and S = 5) (Fig. 15.3e, f). These are the essential micronutrients vital for plant growth and play a key role in the development of cell walls, chlorophyll content in leaves, and generating extensive root systems, respectively. Comparatively, the nutrient composition in the study area was identified

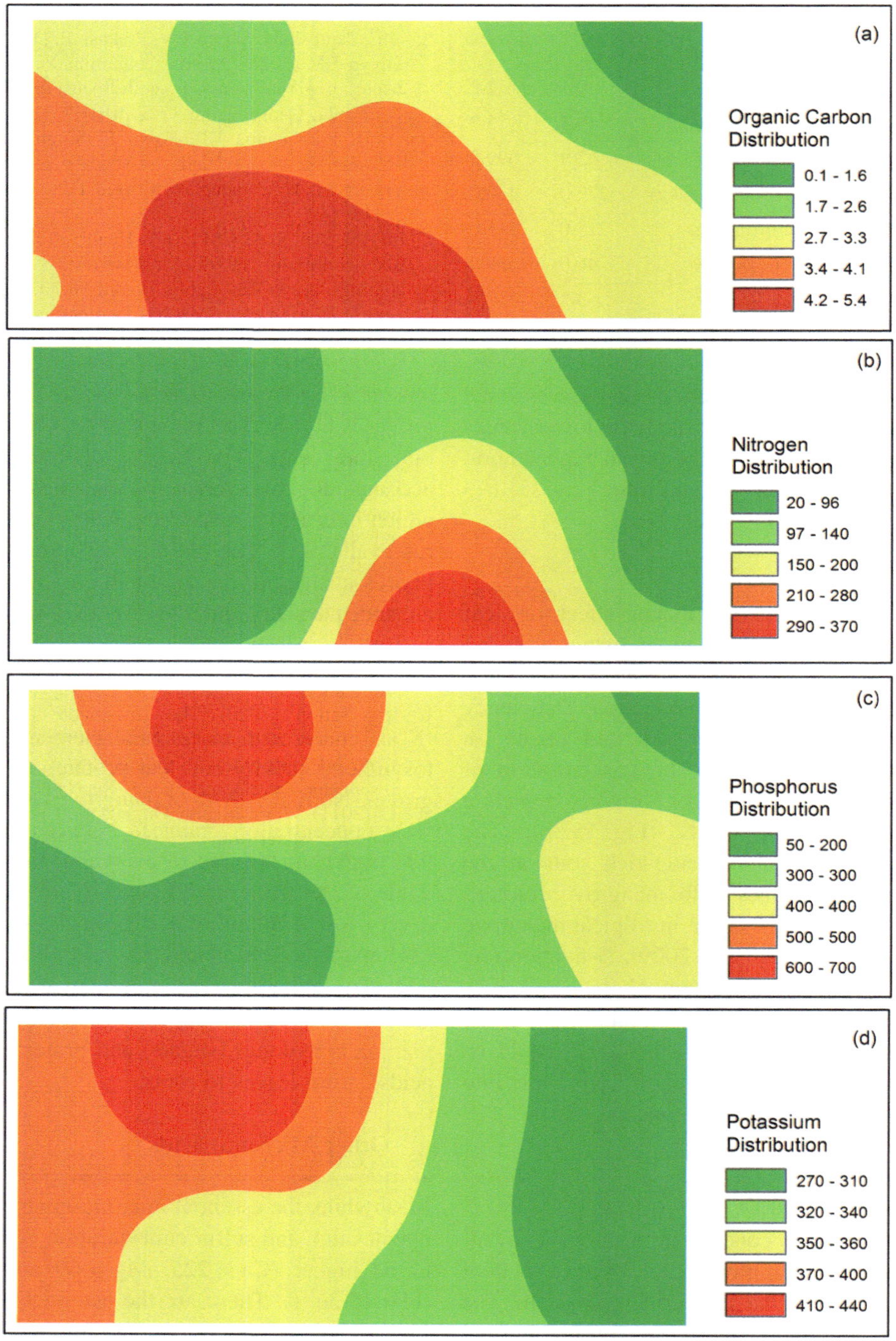

Fig. 15.3 Spatial distribution of soil nutrients **a** Organic carbon, **b** nitrogen, **c** phosphorous, **d** potassium, **e** magnesium, **f** calcium, and **g** sulfur

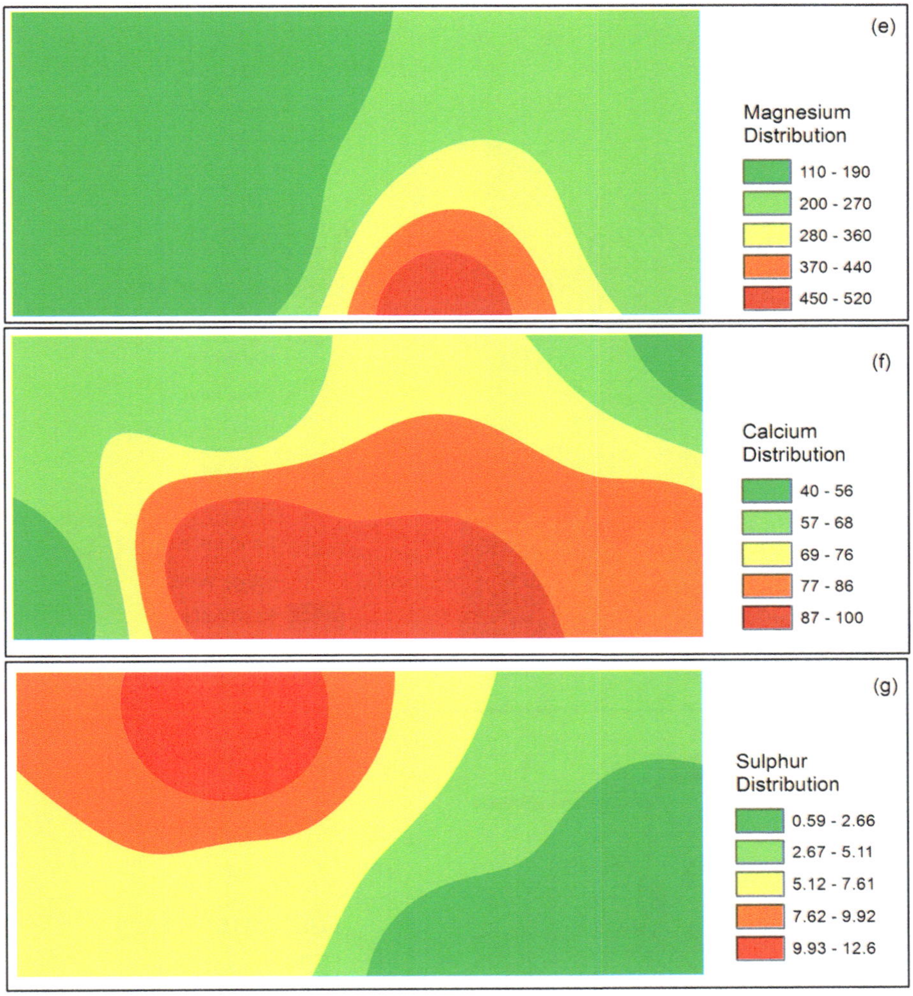

Fig. 15.3 (continued)

as similar to that of mangrove ecosystem-related studies in Sunderbans, India (Banerjee et al. 2012) and Bangladesh (Hoq et al. 2002).

15.4.2 Land Use and Land Cover (LULC) Changes

LULC map of the study area has been prepared and interpreted using ground truth investigations. LULC map of 2005 has been prepared from IRS-P6 (LISS-IV) satellite image. The primary objective of this study is to estimate the decadal changes and their transformations in LULC features. Figure 15.4a, b shows the spatial distribution of major LULC classes in the study area between 2005 and 2015 (Table 15.2). The different types of LULC features such as agricultural land, intertidal zone, mangrove, open land, open scrubland, salt pan, sand bar, sand bar with open scrub, and sandy area are experienced significant changes during the past decade in the various parts of the coastal area. Mudflats in the region seem to be reduced to 15% from 19% in 2005. Due to rapid developmental activities such as the expansion of Mundra port and heavy industrialization between 2005 and 2015, there is a rapid decline in agricultural land (53.6 to

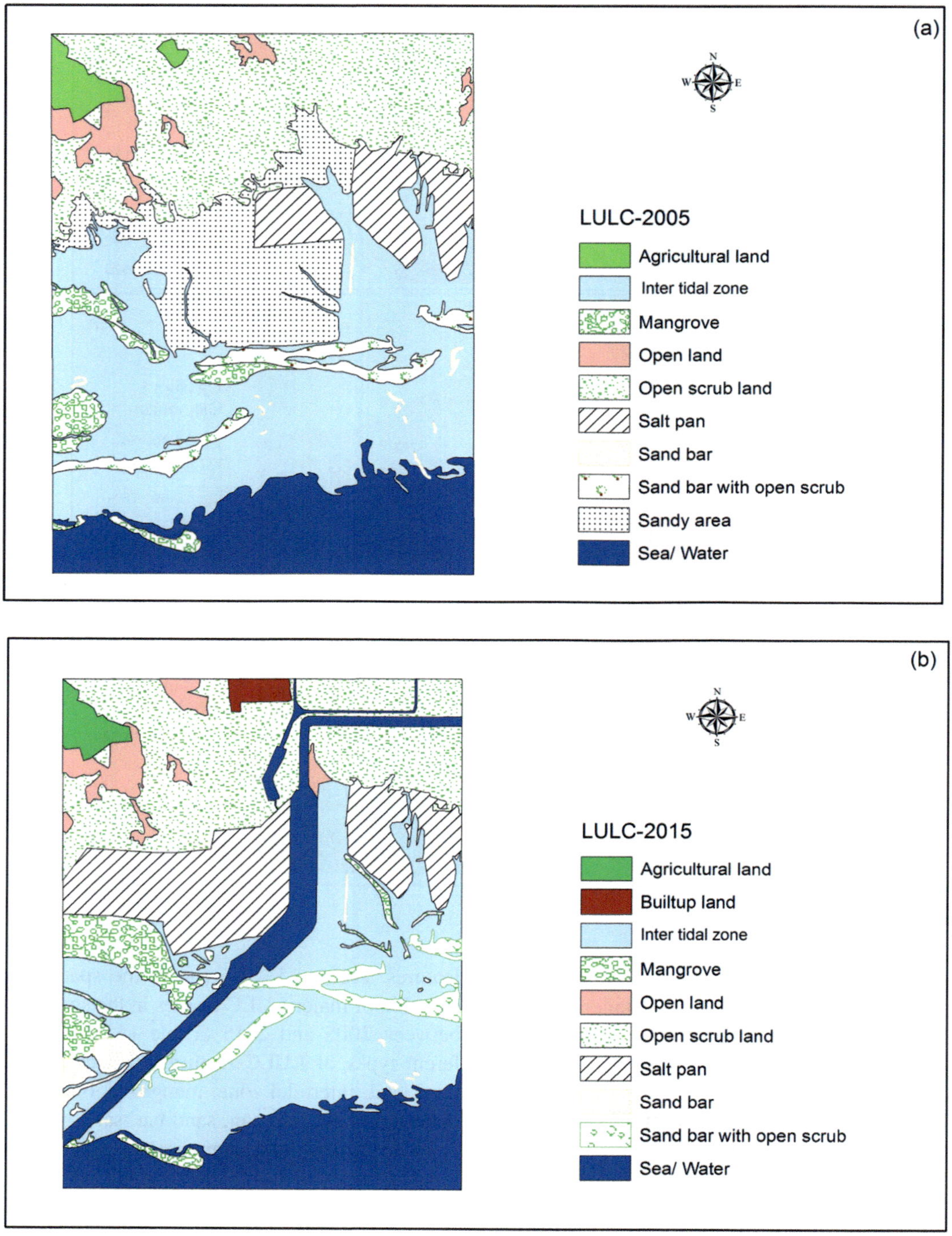

Fig. 15.4 LULC map of **a** 2005 and **b** 2015 interpreted from IRS-P6 LISS-IV and Resourcesat-2 satellite images

Table 15.2 LULC area of coverage measured from GIS map

S. No.	LULC class	2005 (in ha)	2015 (in ha)	Changes from 2005 to 2015
1	Agricultural land	31.98	27.70	− 4.28[a]
2	Built-up land	0	15.59	15.59
3	Inter tidal zone	676.22	631.49	− 44.73[a]
4	Mangrove	68.59	104.19	**35.61**
5	Open land	50.91	56.71	5.8
6	Open scrub land	469.31	444.15	− 25.16[a]
7	Salt pan	128.26	289.90	161.64
8	Sand bar	5.84	26.34	20.5
9	Sand bar with open scrub	66.42	69.74	3.32
10	Sandy area	262.63	0	− 262.63[a]
11	Sea/water	294.06	388.41	94.35
Total area		2054.22	2054.22	

[a] − ve sign indicated decreased value

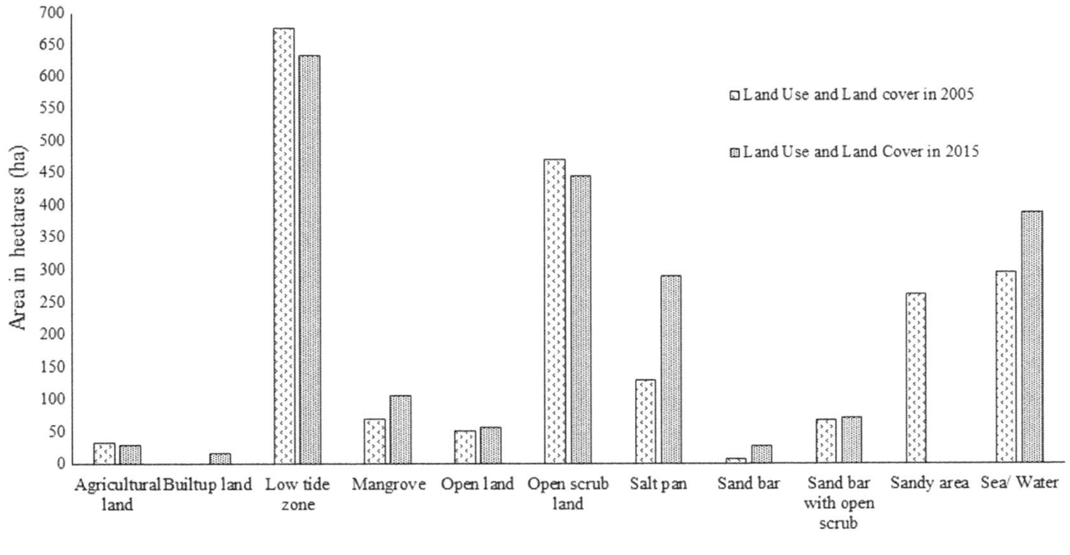

Fig. 15.5 LULC area of coverage and changes

46.4%) (Fig. 15.5). The decreasing trend in the inter-tidal zone from 51.7 to 48.3% is also observed due to the port activity and thermal power plant in the study area. On a positive sign, the mangrove area has increased from 30.7% in 2005 to 69.3% in 2015 due to the mangrove reforestation measures of the Forest department and CSR activities.

15.5 Mangrove Restoration and Sustainability

Global conservation status of Mangroves shows that there is a steep decline in mangroves in the coastal areas of India due to increase in anthropogenic pressure for fuel wood production and timber. Therefore, for recovery of mangroves proper techniques and steps are required for mangrove restoration. Mangrove planting and management is not new to South-East Asia and have a long history such as establishing mangroves in Sunderbans, India and Matang in Malaysia (Kairo et al. 2001). It is evident that such activities increased employment opportunities to local people, timber and charcoal economy, provides protection against coastal erosion, breeding grounds for fishes, firewood and building materials. The idea and initiative of mangrove restoration in the study area emerged to support the threatened livelihoods of the fisher folk (Plate 15.1f). Due to fishing ban during breeding season and onset of monsoon season forces the fisherman community to be in a difficult situation for their livelihoods in off-season. The idea of community participation in mangrove restoration such as nursery development provides an alternative livelihood opportunity and also helps in coastal management. LULC results prove the resurgence and natural regeneration of mangroves in the coastal areas of Mundra in its natural habitat. Moreover, mangroves are known for higher productivity in low nutrient environments in the tropical regions.

Various physicochemical parameters were assessed from the sediments collected at Mundra coastal region. Total soil assessment shows that there is a presence of heavy silt and clay followed by sand in all six study sites. It is also evident that the soil pH and electrical conductivity are controlled by both organic and inorganic contents of the soil environment. Such characteristic determines the selection of mangrove species be survived in an extremely dynamic environment. Based on our soil physiochemical analysis from the selected study site, six native mangrove species belongs to four different families were proposed for mangrove restoration concept for effective utilization of mangrove resurgent conditions. For example, species such as *Avicennia sp.* prefers to grow in sandy soils whereas *Rhizophora sp.* prefers humus-rich soils which are highly acidic. Similar to earlier studies, it is also observed that acidic soils with optimal pH levels are considered to be conducive to achieving a maximum growth rate of the mangrove seedlings along the coasts (Joshi and Ghose 2003). According to their natural traits of rapid and extensive growth in the seaward end of estuaries where the fresh water discharge is low (dry climatic zones), the intertidal zone of lagoons in the dry climatic zone, the following species including *Avicennia marina*, *Avicennia officinalis*, *Aegiceras corniculatum*, *Lumnitzera racemosa*, *Rhizophora mucronata* and *Pemphis acidula* were selected for restoration (Table 15.3).

These species shown a significant role in the mangrove regeneration and proven successful in India and other south east Asian countries (Kairo et al. 2001; Walker 2004; Ramesh et al. 2017). Naturally, increased soil organic matter content, enriches the availability of soil organic carbon and other nutrients (Kristensen et al. 2008). TOC is an essential element for the germination of mangroves at a rate of 21.5–53.0 mg/kg (i.e., 2.3–2.6%). Our result shows 3% of average total organic carbon availability and is considered to be optimum for the growth of mangroves as it is found in other research works conducted in the eastern coastal region of India (Sahoo et al. 2017) and Abu Dhabi (Alsumaiti and Shahid 2018). During nursery preparation method, organic nutrient supplements such as farm yard manure, green leaf manure, peat soil and bio-fertilizers are applied for seedling development and throughout transplantation schedule. Nutrient application rate followed for N:P:K at a ratio of 17:17:17 for the seedlings raised in nursery. After 60 days of transplanting, 50% of recommended dose of DAP (di-ammonium phosphate) and urea were incorporated in order to ensure perfect root establishment. In response to the artificial regeneration process the total mangrove coverage area experienced a steady increase from 2005 to 2015 (Fig. 15.4a, b). The reasons for successful regeneration and increase in mangroves is attributed to increase in constant monitoring, conservation, and reforestation efforts and community regulatory protection (Lai et al. 2015).

Table 15.3 Suitable mangrove species for the study area

S. No.	Mangrove species	Family	Community utilization	Ecosystem services	IUCN conservation status and remarks
1.	*Avicennia marina*	Avicenniaceae	Pioneer mangrove species—fuel wood, fodder for cattle	Flowers are rich source of honey and bee wax	Least concern and continuous monitoring recommended
2.	*Avicennia officinalis*	Avicenniaceae	Medicinal plant (asthma, leprosy) fuel wood	Flowers are rich source of honey	Least concern and continuous monitoring recommended
3.	*Aegiceras corniculatum*	Myrsinaceae	Wood is used for building huts and for fuel wood	Potential source for high-quality honey and bee wax	Least concern and continuous monitoring recommended
4.	*Lumnitzera racemosa*	Combretaceae	Fuel wood and bark tanning	Coastal strengthening pioneer back water species	Least concern and continuous monitoring recommended
5.	*Rhizophora mucronata*	Rhizophoraceae	Potential source of tannin and plywood, adhesive, dye bark, used as fuel wood and in medicines; leaves are used as fodder	Effective as tide breakers	Least concern and continuous monitoring recommended
6.	*Phemphis acidula*	Lythraceae	Mainly as shoreline protection, bark is used for tanning, used as ornamental plant and medicine	Coastal strengthening	Least concern and continuous monitoring recommended

15.6 Conclusion

Intensive management and monitoring of these nurseries are required for proper management of mangrove restoration process. Participatory management enables the local community to own the resources and get involved as partners for sustainable management for mangrove in the study area including the initial stage of nursery bed preparation, planting, and monitoring (Fig. 15.6). It is evident from earlier community restoration efforts taken by grass root level organization by developing community nurseries in Godavari and Krishna regions in India (Ramesh et al. 2017). The experience from nursery development insists the need for continuous monitoring action to record periodical environmental changes especially the soil dynamics (Samson and Rollon 2008), tidal hydrology and further developmental activities to get a good understanding of long-term impacts of mangrove adaptation and restoration for future conservation and management of mangroves in Mundra coast. The geospatial approach improves the precision in decision-making capability by reducing the cost and time in identifying the area for mangrove conservation. It also provides efficient data storage and retrieval facilities offering cost-free access for monitoring forest conditions (Kairo et al. 2001). LULC features show dynamic changes of various features at different rates. The preliminary results from this study can be used as baseline data to initiate larger mangrove restoration action. The Detritus food chain and rich organic nutrient environment in the mangroves improve the natural drainage system and maintain a coastal ecosystem suitable for nurseries and foraging grounds for fish species. The Mundra coast is thus capable to become

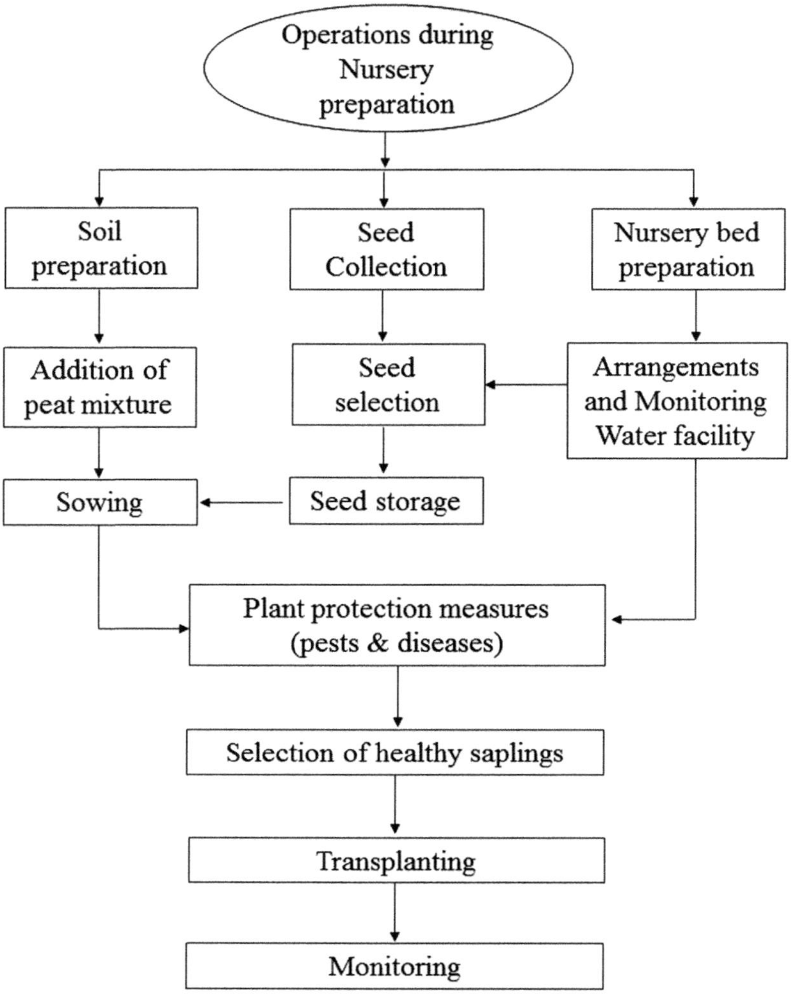

Fig. 15.6 Nursery preparation framework

the most productive mangrove forest in the Gulf of Kutch and has excellent prospects for natural coastal shelter belts.

References

Alsumaiti TS, Shahid SA (2018) Comprehensive analysis of mangrove soil in Eastern Lagoon National Park of Abu Dhabi Emirate. Int J Bus Appl Soc Sci 4(5):39–56

Atasoy M, Karsli F, Biyik C, Demir O (2006) Determining land use changes with digital photogrammetric techniques. Environ Eng Sci 23(4):712–721

Balamurugan G, Anubhooti (2015) Urban growth effects for land use changes leading to natural hazards in parts of Mumbai metro city, India. Disaster Adv 8(10):1–14

Balamurugan G, Sridhara Aravind M (2015) Land use and land cover changes in pre-and post-earthquake affected area using Geoinformatics–Western Coast of Gujarat, India. Disaster Adv 8(4):1–14

Banerjee K, Roy Chowdhury M, Sengupta K, Sett S, Mitra A (2012) Influence of anthropogenic and natural factors on the mangrove soil of Indian Sundarbans wetland. Arch Environ Sci 6:80–91

Carlson TN, Azofeifa SGA (1999) Satellite remote sensing of land use changes in and around San José, Costa Rica. Remote Sens Environ 70(3):247–256

Cusack D, Silver W, McDowell W (2009) Biological nitrogen fixation in two tropical forests: ecosystem-level patterns and effects of nitrogen fertilization. Ecosystems 12:1299–1315

Debnath HS, Naskar KR (1999) A comparative study on the mangroves and associated flora in the Ganga delta (Sunarbans) and Bay Islands (Andaman and Nicobar).

Sundarbans Mangal. Naya Prakash, Calcutta, pp 277–292
Dissanayake N, Chandrasekara U (2014) Effects of mangrove zonation and the physicochemical parameters of soil on the distribution of macrobenthic fauna in Kadolkele mangrove forest, a tropical mangrove forest in Sri Lanka. Adv Ecol
Elster C (2000) Reasons for reforestation success and failure with three mangrove species in Colombia. For Ecol Manage 131(1):201–214
Feller IC, Lovelock CE, Berger U, McKee KL, Joye SB, Ball MC (2010) Biocomplexity in mangrove ecosystems. Ann Rev Mar Sci 2:395–417
Gilman EL, Ellison J, Duke NC, Field C (2008) Threats to mangroves from climate change and adaptation options: a review. Aquat Bot 89(2):237–250
Gopal B, Chauhan M (2006) Biodiversity and its conservation in the Sundarban mangrove ecosystem. Aquat Sci-Res across Boundaries 68(3):338–354
Hashim R, Kamali B, Tamin NM, Zakaria R (2010) An integrated approach to coastal rehabilitation: Mangrove restoration in Sungai Haji Dorani, Malaysia. Estuar Coast Shelf Sci 86:118–124
Hoq ME, Islam ML, Paul HK, Ahmed SU (2002) Decomposition and seasonal changes in nutrient constituents in mangrove litter of Sundarbans mangrove, Bangladesh
Ibharim NA, Mustapha MA, Lihan T, Mazlan AG (2015) Mapping mangrove changes in the Matang mangrove forest using multi temporal satellite imageries. Ocean Coast Manag 114:64–76
Jagtap TG, Komarpant DS (2003) Evaluation of mangrove ecosystem of India for assessing its vulnerability to projected climatic changes. In: Dash SK, Rao P (eds) Assessment of climate change in India and mitigation policies. WWF, New Delhi, pp 39–51
Joshi H, Ghose M (2003) Forest structure and species distribution along soil salinity and pH gradient in mangrove swamps of the Sundarbans. Trop Ecol 44 (2):195–204
Kairo JG, Dahdouh-Guebas F, Bosire J, Koedam N (2001) Restoration and management of mangrove systems—a lesson for and from the East African region. S Afr J Bot 67(3):383–389
Kathiresan K, Bingham BL (2001) Biology of mangroves and mangrove ecosystems. Adv Mar Biol 40:81–251
Kirui KB, Kairo JG, Bosire J, Viergever KM, Rudra S, Huxham M, Briers RA (2013) Mapping of mangrove forest land cover change along the Kenya coastline using Landsat imagery. Ocean Coast Manag 83:19–24
Kristensen E, Bouillon S, Dittmar T, Marchand C (2008) Organic carbon dynamics in mangrove ecosystems: a review. Aquat Bot 89(2):201–219
Lai S, Loke LH, Hilton MJ, Bouma TJ, Todd PA (2015) The effects of urbanisation on coastal habitats and the potential for ecological engineering: a Singapore case study. Ocean Coast Manag 103:78–85
Manassrisuksi K, Hussin MWYA (2001) Assessment of a mangrove rehabilitation programme using remote sensing and GIS: a case study of Amphur Khlung, Chantaburi province, Eastern Thailand. In: Paper presented at the 22nd Asian conference on remote sensing, vol 5, p 9
Nagelkerken ISJM, Blaber SJM, Bouillon S, Green P, Haywood M, Kirton LG, Somerfield PJ (2008) The habitat function of mangroves for terrestrial and marine fauna: a review. Aquat Bot 89(2):155–185
Peng Y, Zheng M, Zheng Z, Wu G, Chen Y, Xu H, Lee SY (2016) Virtual increase or latent loss? A reassessment of mangrove populations and their conservation in Guangdong, southern China. Mar Pollut Bull 109(2):691–699
Pielke Sr RA, Adegoke J, Beltran-Przekurat A, Hiemstra CA, Lin J, Nair US, Niyogi D, Nobis TE (2007) An overview of regional land-use and land-cover impacts on rainfall. Tellus 59(B):587–601
Ramesh R, Abhilash KR, Purohit M, Krishnan P, Lakshmi A, Purvaja R, Kingsley PW (2017) Involvement of community in managing coastal wetlands in South Asia: status, issues and challenges. In: Wetland science. Springer India, pp 545–562
Ramos e Silva CA, Oliveira SR, Rêgo RDP, Mozeto AA (2007) Dynamics of phosphorus and nitrogen through litter fall and decomposition in a tropical mangrove forest. Mar Environ Res 64:524–534
Rao RK, Babu KR (2017) Seasonal variations in physicochemical characteristics of Champavathi Estuary, East Coast of India. Int J Curr Microbiol App Sci 6(6):443–447
Reef R, Feller IC, Lovelock CE (2010) Nutrition of mangroves. Tree Physiol 30(9):1148–1160
Sahoo K, Jee PK, Dhal NK, Das R (2017) Physicochemical sediment properties of mangroves of Odisha, India. J Oceanogr Mar Res 5(162):2
Samson MS, Rollon RN (2008) Growth performance of planted mangroves in the Philippines: revisiting forest management strategies. Ambio 37:234–240
Singh AK, Ansari A, Kumar D, Sarkar UK (2012) Status, biodiversity and distribution of mangroves in India: an overview. In: Uttar Pradhesh Sate Biodiversity Board. Marine biodiversity: one ocean, many worlds of life, 59–67
Untavale AO (1986) Mangroles of India. In: Mangroves of Asia and the pacific-status and management. UNDP/UNESCO Project Research and Training Pilot Program on Mangrove ecosystems. (RAS/79/002. UNOP/UNESCO. Manila)
Van de Merwe JP, Lee SY, Connolly RM, Pitt KA, Steven AD (2016) Assessing temporal and spatial trends in estuarine nutrient dynamics using a multi-species stable isotope approach. Ecol Ind 67:338–345
Walker R (2004) Theorizing Land-cover and Land-use change, the case of tropical deforestation. Int Reg Sci Rev 27(3):247–270
Wang L, Sousa WP, Gong P, Biging GS (2004) Comparison of IKONOS and QuickBird images for mapping mangrove species on the Caribbean coast of Panama. Remote Sens Environ 91:432–440

Yates EJ, Ashwath N, Midmore DJ (2002) Responses to nitrogen, phosphorus, potassium and sodium chloride by three mangrove species in pot culture. Trees-Struct Funct 16(2):120–125

Yuan F (2008) Land-cover change and environmental impact analysis in the Greater Mankato area of Minnesota using remote sensing and GIS modelling. Int J Remote Sens 29(4):1169–1184

Mapping and Visualization of Research Productivity of Scientific Literature on Agricultural Waste: A Bibliometric Analysis

16

Pasarul Islam, Enayat Bano, Syed Naushad Ahmad, and Shakil Ahmad

Abstract

The current study has used bibliometric and visualization techniques to review the literature on agricultural waste published in the Web of Science database. The study used data collected in the Scopus database and other data sources. By applying Association Search in the advanced search section of Scopus, papers from the top twenty journals were retrieved. Software packages MS Excel and VOS viewer were employed for data processing and extraction of bibliometric indicators. This is for assessing the journals' work effectiveness based on the data obtained from documents. The study analyzed the overall success of researchers dealing with agricultural waste in their research quest, efficiency, and publishing achievements. The study addressed questions related to most active countries, organizations, and authors; favorite types of sources of researchers; the collaboration of authors; the most widely used keyword and co-occurrence network in agricultural waste; and the attribution and usage of influential science. Similarly, emphasis was laid on top-ranked publications, authors per paper, degree of collaboration based on the collected data. One thousand four hundred ninety-three articles, written by 16,185 writers affiliated with 4612 institutions from around 122 countries, published these records/data review posts. The higher rate of publication, citation, and link strength is notable for the USA, India, Peoples Republic of China for agricultural waste reuse, recycling, and circular economy. The University of Tsinghua leads the group in the number of total publications. For the University of Sains Malaysia, the total highest quotations and Citations Impact were spotted as the highest. The University of Tsinghua won over the total link strength of agricultural waste.

Keywords

Agricultural waste · Scientometrics · Bibliometric · Research assessment · Review of literature

P. Islam (✉)
Indian Council of Social Science Research (ICSSR), Ministry of Education, Government of India, New Delhi, India
e-mail: islam.pasarul@gmail.com

E. Bano · S. N. Ahmad
Department of Geography, Faculty of Science, Aligarh Muslim University (Central University), Aligarh, Uttar Pradesh, India

S. Ahmad
Directorate of Library Affairs, Imam Abdulrahman Bin Faisal University, Dammam, Saudi Arabia
e-mail: shahmad@iau.edu.sa

S. Talukdar et al. (eds.), *Environmental Risk and Resilience in the Changing World*, GIScience and Geo-environmental Modelling, https://doi.org/10.1007/978-3-031-62442-1_16

16.1 Introduction

The circular economy introduced the concepts of bioeconomy, bio-based society, and green economy, which are now realigning the strategic planning of many industrial companies. Today, waste is incinerated or, better yet, another higher quality use is sought. For this reason, it is important to develop different applications for existing waste materials, considering the ecological and economic factors of wastes treatment (Väisänen et al. 2016). Sarkar et al. (2012) revealed that renewable biomass fuels such as bioethanol, biodiesel, biohydrogen, etc., are obtained from sugarcane, corn, switchgrass, algae, etc. Bioethanol production from rice straw, wheat straw, corn stalks, and sugarcane bagasse is now a topic of worldwide interest. Rice straw is the most abundant waste compared to the other primary wastes, and rice straw can potentially produce 205 billion liters of bioethanol per year, which is the highest among these four agricultural wastes mentioned. Hoseinzadeh Hesas et al. (2013) reviewed the effect of the microwave radiation process on activated carbon production from agricultural wastes. The study shows that activated carbon production by a microwave radiation process is more significant in terms of microwave radiation time, microwave power, impregnation ratio, and agent flow rate.

The agricultural and forestry wastes containing cellulose, hemicellulose, and lignin generate heat for generators, fuel gases, solid fuels, and diesel fuels by thermochemical measurement. Anaerobic digestion and ethanol fermentation is a biochemical conversion that produces methane and ethanol (Hardoyo 2013). Another important biofuel, butanol, is also made from agricultural waste, which is more efficient than ethanol. It has a higher energy content, is more miscible with diesel, is less corrosive, and has a lower vapor pressure and flash point than ethanol (Ezej and Blaschek 2010). Cellulose, the main component of lignocellulosic biomass, has attracted worldwide attention to producing greener and cleaner fuels by creating fermentable sugars that can then be converted to second-generation bioethanol (Bansal et al. 2012). Bio-fibers from agricultural residues have attracted increasing interest due to their wide application in food packaging and biomedical sciences. These eco-friendly polymers are reduced rapidly and replace petroleum-based synthetic polymers due to their safety, low production costs, and biodegradability (Mostafa et al. 2018). Hence, agricultural waste will be an essential energy source to replace non-renewable energies shortly. In 2015, world biogas production was around 57 billion cubic meters, of which Germany's biogas production exceeded 9 billion cubic meters, and Swedish biogas covered 30% of automotive gas needs. The biogas potential of agricultural waste is estimated at 26 billion cubic meters of biomethane in Europe. Therefore, the appropriate use of agricultural waste helps to solve global environmental and resource problems and brings enormous economic benefits to many countries (Ahmad et al. 2020; Ahmad and Islam 2017; Islam et al. 2020; Wei et al. 2020).

The bibliometric study is based on quantitative aspects of productions, dissemination, and use of recorded information. Pritchard first used this term in 1969 (de la Cruz-Lovera et al. 2017). Bibliometric is a discipline developed by the extensive intersection and combination of philology, information science, mathematics, and statistics. Hence, it explores other fields' systems and patterns using statistics and analysis techniques (Yu et al. 2019). Ahmad et al. (2018) reported the bibliometric tools used to measure the scientific journal's quality. In this study, bibliometric tools were used to analyze the field of agricultural waste in the construction industry. Agricultural waste refers to waste generated because of various agricultural processes. Agricultural waste, defined as the animal or plant residues or remains that cannot further converted into food or feeds, causes additional pressure on the Environment and economy. Agricultural waste has the utmost importance in a circular economic approach (Florido et al. 2019). If human civilization exists, the agricultural sector is also everlasting, which goes for its waste. In agriculture, the circular economy proposed an alternative model to the linear model, considering the ecosystem's fundamental role, its roles, and the relationship between the ecosystem and the economic system. Using agricultural wastes in the construction industry is proposed as one of the circular economy approaches (Meng et al. 2018). Meng et al. (2018) further showed that building materials within a certain green sphere have strategic market competitive advantages, leading to higher revenue and monopoly.

16.1.1 Literature Review

Wuni, Shen and Osei-Kyei have explored global research productivity on green building from 1992 to 2018. The study analyzed a total of 1147 research articles for this study from 1992 to 2018. It revealed that research on the green building was increased exponentially from 1992 to 2018. It has also been found that 44% of countries are highly engaged in green buildings' research activity, major ten broad themes on the green building identified and suggested for future research. Finally, these papers provide a new insight to the modern researcher, funding agencies, policymakers, and other professionals about the research progress in green buildings (Ahmad et al. 2018). Rahaman, Md. Safiqur, Ansari, Khadeeja M. N, Kumar, Suchetan, and Rahman, Md. Rafique conducted a scientometric study of global research productivity on coronavirus. The 25 years of data were downloaded from the Scopus database; a total of 2661 data were retrieved and analyzed with Biblioshiny, VOS viewer, and Microsoft excel. The journal "Journal of virology" is the most impactful in the field of coronavirus, and Yuen, KY, is the prolific author (75 papers). A predominantly utilized keyword is coronavirus. The University of Hong Kong is the leading producer of research papers (362 papers) and the study also shows that most of the research was published in the form of articles (68%) (Aznar-Sánchez et al. 2018).

Baudoin, Sapinho, Maddi, and Miotti analyzed research productivity in the term "Microbiota" from 1999 to 2017. The study examined 28,000 publications indexed in the web of science databases. Terms breakdown by journal subject and topic modeling. The study revealed that research increased over time; research moved from dentistry to ecology areas. The study suggested that microbiota research has opened new trends and directions for researchers and policymakers (Re Cecconi et al. 2019). Kumar Rai, Senger, and Lohiya demonstrated the research output of Astronomy in India during 1991–1995 and 2011–2015. The study assessed the progress of astronomy research in India and its Impact on the Science Citation Index from 1991 to 1995 and 2011 to 2015. The parameter studies included the growth of article publications, authorship pattern, and citation pattern during 1991–1995 and 2011–2015, citation received, international collaboration pattern, most productive author, and most preferred journals. The study also revealed the status of FORSA institutes regarding the total number of research publications and total citations received. This paper's writer belongs to one of the FORSA Institutes (Zhou et al. 2019). M. K. Singh & Tripathi examined the top ten research organizations in Biotechnology in India from 2001 to 2016. Five thousand four hundred twenty-three bibliographic records were downloaded from the Scopus database and analyzed according to the study's objectives. The study applied statistical and mathematical approaches to know the pattern of publication and types of documents. The study used different Scientometrics methods, including collaboration co-efficient, co-authorship index, and collaborative research in biotechnology. The study also used the activity index tool to know the trends in authorship patterns and institute data mapping. The USA is the most preferred country for international collaboration in the study domain (Chartered Institution of Wastes Management (CIWM) 2020). Pattanashetti & Harinarayana demonstrated research contribution in the field of Mechanical Engineering by India, Japan, and South Korea during 2000–2014. The study focused on different Scientometrics parameters, including the growth of literature, collaboration pattern, and activity index. The data was retrieved from the Web of Science database from 2000–2014. The number of papers to cite among India, Japan, and South

Korea is 20,836:197,679, 24,494:204,393, and 30,578:266,902, respectively. The study revealed that the research trends in India and South Korea are increasing while Japan's publication in Mechanical engineering is decreasing. The publication of India grew three times in the field of mechanical engineering during the time, while South Korea doubled its research publications from 2000–2014. There is an increasing trend of collaboration in all the areas of science and technology, but the rate of growth and collaboration pattern varies from one subject to another and from one country to another country (Sarmah et al. 2009). Velmurugan and Radhakrishnan examined research output on Phytochemistry in India from 1994 to 2014. The study data collected and downloaded from the web of science database indexed 1280 publications with 1216 citations, cited reference 2083; out of all these global publications, India contributed 121 research papers with 1446 citations & 473 authors with 8732 cited references over 1994–2014. The study focused on authorship pattern, international collaboration pattern, institution-wise distribution of publications, subject-wise distribution of articles, most cited papers, a ranking of core journals in phytochemistry, highly cited authors, papers, and productive keywords. The research paper tries to attempt the different parameters, including H-Index, G-Index, E-Index, HI-Index, HI-Norm, H-Annual, HM-Index, AW-Index, AWCR, and AWCRPA (Obi et al. 2016). Verma et al. (2013) claimed that agricultural wastes can be used to prepare fiber reinforced polymer composites for commercial use. They indicated that the use of natural fibers such as bagasse, coir, sisal, jute, developed composites, etc. may replace petroleum and other non-decaying materials-based products. Their low cost, easy availability, low density, acceptable specific characteristics, simplicity of separation, enhanced energy recovery, CO_2 neutrality, biodegradability, and recyclable nature were cited as potential reasons. (2004) Mannan and Ganapathy (2004) managed to produce concrete from an agricultural waste-oil palm shell with a strength requirement suitable for lightweight structural concrete. According to Väisänen et al. (2016), natural fiber–polymer composites are being more widely used in several applications because they are an environmentally friendly and cost-effective alternative to petroleum-based materials. Aprianti et al. (2015) advocated the use of agricultural wastes (such as rice huskash, palm oil fuel ash, sugar cane bagasse ash, wood waste ash, bamboo leaf ash, corn cob ash, fly ash, silica fume, and ground granulated blast slag) as supplementary cementitious materials in the production of concrete and eco-concrete. They pointed out their reliable and suitable physical and chemical properties as alternative preventive solutions promoting industrial environmental sustainability. In their study on rice husk ash, Givi et al. (2010) found that replacing cement with rice husk ash enhances concrete's compressive strength and workability while lowering its water permeability. The compressive strength and water permeability of cured concrete were enhanced by reducing the average particle size of rice husk ash, while the workability of new concrete was negatively affected. Rice husk has become a source for several silicon compounds, including silicon carbide, silica, silicon nitride, silicon tetrachloride, zeolite, and pure silicon, according to Sun and Gong (2001). According to Pode (2016), the amorphous silica-rich rice husk could be a low-cost precursor for the fabrication of value-added silica-based materials for practical applications. Soltani et al. (2015) attempted the production of advanced materials—including SiO_2, SiC, Si_3N_4, elemental Si, Mg_2Si, and more recently, active carbon, using rice husk. Rice husk ash is categorized as a highly active pozzolan by van Tuan et al. (2011) because it has a considerable amount of amorphous SiO_2 and a big surface area.

Demir (2008) recommended the use of by-products of industrial and agricultural processes (such as sawdust, tobacco residues, and grass) in an environmentally safe way as organic pore-forming agents in the brick-clay industry. Such agricultural industrial by-products may be of organic origin like polystyrene, paper sludge, coal, and coke or inorganic pore-forming agents like perlite, diatomite, calcite, pumice, and vermiculite. According to Raut et al. (2011), the various waste-create bricks have improved performance in terms of lighter density, lower thermal conductivity, and higher compressive strength, making them a cost-effective solution for green building design. Khedari et al. (2003, 2004) suggested the development of new insulating particleboards from durian (Duriozibethinus) peel and coconut (Cocos nucifera) coir as the raw material to manufacture particleboards for energy conservation of the building. Madurwar et al. (2013) affirmed the application of agro-waste as sustainable construction materials in reducing natural resource use as well as energy. According to Reis (2006), coconut fiber is an effective reinforcement for polymer concrete, boosting fracture and flexural outcomes.

16.1.2 Purpose

This study aims to analyze research output in agricultural waste literature indexed in the Scopus database between 2001 and 2020. Agricultural waste refers to waste generated because of various agricultural processes. It is to be noted that any agricultural waste that is defined as animal or plant residue and cannot be converted into food or feed adds to the environmental and economic pressures. The following objectives were identified for the present study.

- To explore the yearly growth of literature and citation trends on agricultural waste
- To identify the most relevant sources of publication
- To illustrate the most productive countries, organizations, and authors in agricultural waste research
- To know the authorship and citations patterns
- To discover the most widely cited publications
- To find the most used keywords
- To determine which country is the most active collaborator.

16.2 Research Methodology

Bibliometric methods were applied to analyze 747 research publications in agricultural waste from 2001 to 2020. The current study concentrated on various parameters of bibliometrics indicators, especially annual publication trends, relevant sources of publications, prolific authors, authorship patterns, productive organizations and countries, and keyword analysis of agricultural waste were examined in the study. Bibliometric methods were applied for the data analysis. Scientometrics and bibliometrics tools, namely BibExcel, Biblioshiny (RStudio)VOS viewer, and Gephias well as MS Excel, were used to investigate the study of annual scientific productivity, the top contributing authors, countries and institutions, most relevant sources of publications, cited documents, and frequently used keyword in agricultural waste research were carried out for this study. The bibliographic records of agricultural waste research were extracted were used SCOPUS, a multidisciplinary database that ensures the maximum number of documents for the analysis. Because of its more comprehensive coverage (Mongeon and Paul-Hus 2016; Powell and Peterson 2017; Sweileh et al. 2017; Zyoud 2016) compared to the web of science and other databases, the Scopus database was selected. The following search query was used to download the records for the study. Data were downloaded on 23rd May 2021.

(TITLE-ABS-KEY-AUTH ("Agricultural waste" OR agro-waste) AND ("Construction Material*" OR "Building Material*")) AND (LIMIT-TO (DOCTYPE, "ar") OR LIMIT-TO (DOCTYPE, "cp") OR LIMIT-TO (DOCTYPE, "re") OR LIMIT-TO (DOCTYPE, "ch") OR LIMIT-TO (DOCTYPE, "bk")) AND (LIMIT-TO (PUBYEAR, 2020) AND (LIMIT-TO (LANGUAGE, "English")).

In comparison to other search queries, authors found the maximum records for download using the "TITLE-ABS-KEY" search option using the

Boolean operator "OR" and "AND" with the limit of publication year up to 2020 and English language only. The total number of records, 747, were downloaded in Bibtex, RIS, and CSV format for the analysis. The purpose of this study was restricted to papers on "agricultural waste" that were indexed in the Scopus database, with no other databases being used for bibliographic data due to their lack of coverage. The results of the findings should therefore be a consideration given the study's limitations.

16.3 Results

In terms of total publications and citations, the findings showed a gradual increase in literature from the year 2010 onwards on agricultural waste research. There were 747 documents published in 316 different sources and obtained 12,127 citations from 2001 to 2020. The highest research publications (171) were published in the year 2020 and got 572 citations. Malaysia appeared as the top contributing country along with 164 publications and received 2,303 citations. University Tun Hussein Onn Malaysia secured the highest-ranked in agricultural waste research. Construction and Building Materials (IF = 4.419) appeared as the highest contributed source.

16.4 Data Analysis and Discussion

Table 16.1 displays the primary data related to the research topic on agricultural waste indexed in Scopus data between 2001 and 2020. The analysis reveals that 747 publications were published in 316 different sources (Journals, Books, etc.), with 12,123 citations and contributed by 2368 authors. It was also revealed that, out of 2368 authors, single-authored documents were 27 and multi-authored documents 2341, respectively. Most of the researchers were preferred to publish their research in the form of article 461 publications, followed by conference papers with 190 publications and a review with 64 publications. Book chapters were the slightest type of research on agricultural waste during the study period.

Table 16.1 Main information about data

Description	Results
Periods	2001:2020
Sources (journals, books, etc.)	316
Documents	747
Authors	2,368
Single-authored documents	27
Multi-authored documents	2,341
Total citations received by all publications	12,123
References	30,159
Average citations per document	16.23
Keywords plus (ID)	4,394
Author's keywords (DE)	1,784
Document types	
Article	461
Book chapter	32
Conference paper	190
Review	64

16.4.1 Annual Research Growth and Impact of Citation

Table 16.2 shows the publication and citation trend of agricultural waste over a twenty-years from 2001 to 2020. Total 747 documents were published on the research topic and obtained 12,127 citations. The first article on the research topic was published in 2001 and remained at a single-digit publication up to 2009 (TP ranged between 2 and 8). Double-digit research publications have been seen since 2010 (TP = 15). The research topic boom has occurred during the last 5 years, with 543 publications and 5214 citations. The most productive year was 2020 (n = 171), while the citations (n = 572) were received, followed by 2019 (n = 124) publications and (n = 731) citations, and 2018 (n = 112) publications and (n = 1262) citations. The highest number of citations (n = 1850) was recorded in 2016 for 58 publications, followed by 2015 with (n = 1373) citations for 43 publications and 2018 with (n = 1262) citations for (n = 112) publications. The calendar year 2020 also has the highest number of total cited publications (n = 108), followed by 2018 and 2019 with total cited publications (n = 90) each. The average citations per publication (C/P) and the average citations per cited publication (C/CP) were the highest in 2007 with 202.50 each, followed by 2001 with 135.67 C/P and 2003 with 189.00 C/CP. The highest h-index was appeared in 2016 with 23, followed by 2018 with 22, 2015 with 20, and 2019 with 16 (Fig. 16.1).

16.4.2 Country-Wise Research Output on Agricultural Waste

Table 16.3 lists the top 10 countries for agricultural waste research output and their citation structure. The table reveals that the top 10 countries collectively produced 747 research publications with 12,123 citations. Only the top

Table 16.2 Annual research growth and impact of citation on agricultural waste research between 2001 and 2020

Year	TP	TCP	TC	C/P	C/CP	h-index
2001	3	3	407	135.67	135.67	3
2002	3	3	148	49.33	49.33	2
2003	2	1	189	94.50	189.00	1
2004	4	4	305	76.25	76.25	2
2005	2	2	12	6.00	6.00	1
2006	6	6	451	75.17	75.17	6
2007	2	2	405	202.50	202.50	2
2008	7	7	306	43.71	43.71	5
2009	8	8	188	23.50	23.50	5
2010	15	14	442	29.47	31.57	8
2011	17	15	672	39.53	44.80	9
2012	25	22	621	24.84	28.23	12
2013	33	30	885	26.82	29.50	15
2014	34	30	509	14.97	16.97	11
2015	43	38	1373	31.93	36.13	20
2016	58	51	1850	31.90	36.27	23
2017	78	65	799	10.24	12.29	15
2018	112	90	1262	11.27	14.02	22
2019	124	90	731	5.90	8.12	16
2020	171	108	572	3.35	5.30	12

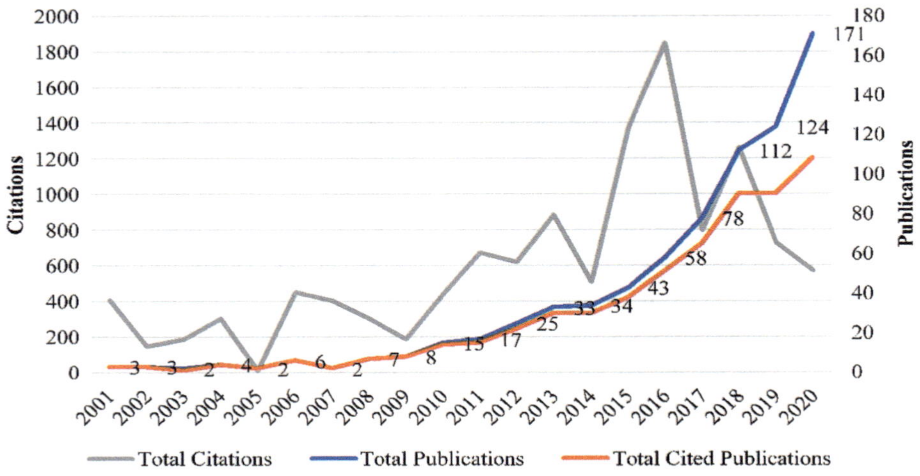

Fig. 16.1 Analysis of the overall growth trend

Table 16.3 Top ten influential countries on "agricultural waste" research

Rank	Country	TP	TCP	TC	C/P	C/CP	h-index
1	Malaysia	164	132	2303	14.04	17.45	24
2	India	133	97	2023	15.21	20.86	20
3	United States	83	74	1974	23.78	26.68	31
4	Nigeria	47	30	169	3.60	5.63	7
5	China	43	35	676	15.72	19.31	12
6	Brazil	32	31	816	25.50	26.32	14
7	Spain	28	24	351	12.54	14.63	11
8	Pakistan	25	21	363	14.52	17.29	7
9	Thailand	24	19	671	27.96	35.32	12
10	United Kingdom	24	22	567	23.63	25.77	11

three countries (Malaysia, India, and the USA) contributed more than 380 publications, Malaysia emerged as the top contributing country (n = 164) publications and received (n = 2303) citations, followed by India with (n = 133) publications and (n = 2023) citations and the United States with (n = 83) publications and (n = 1974) citations. France had the least (n = 15) productive country in the top 20 list, with (n = 190) citations. Malaysia (TCP = 132 and TC = 2303), India (TCP = 997 and TC = 2023), and the United States (TCP = 74 and TC = 1974) are among the top recipients in terms of total citations (TC) and total cited publications (TCP). Saudi Arabia had received total citations (124 for 16 publications), which are the least among the top 20 productive countries. South Korea secured the highest average citation per publication with 29.72 C/P, followed by Thailand with 27.96 C/P and Turkey with 25.95 C/P. On the other hand, South Korea recorded the maximum average citation per cited publication with 35.67 C/CP, followed by Thailand with 35.32 C/CP and the USA with 26.68 C/CP (Fig. 16.2).

16.4.3 Author Impact

The top 15 productivity analyses of the authors who contributed to agricultural waste research publications are depicted in Table 16.4. It was

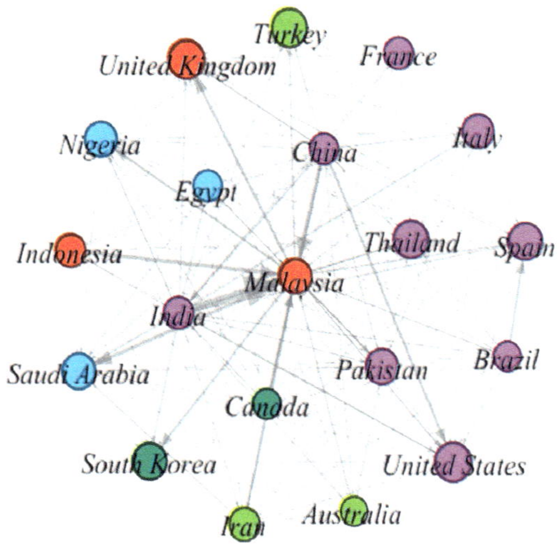

Fig. 16.2 Top 20 influential countries on agricultural waste research

clear among the top 15 authors; the contributions were varied between six to ten publications. Shafigh, P. (from the University of Malaya) contributed the highest number of research articles (ten) and received (398) citations; there were (eight) number of cited papers (NCP) out of (ten) and a citation impact of 39.80. Followed by Jumaat MZ (University of Malaya) and Mohamad N (Universiti Tun Hussein Onn Malaysia), who contributed (nine) publications each and received 280 and 26 citations, respectively, and (nine) number of cited papers (NCP) out of (nine) and (eight) number of cited documents (NCP) out of (nine) and a citation impact of 31.11 and 2.89, respectively. The h-index of Jumaat MZ was the highest (seven). Shafigh, P. received the highest number of citations (TC = 398) for ten publications, followed by Pinto J with 310 citations for his six

Table 16.4 Author impact (with total publication, total citation)

AU	AFF	CO	NP	NCP	NC	C/P	C/CP	h_index	PYS
Shafigh, P	University of Malaya	Malaysia	10	8	398	39.80	49.75	6	2010
Jumaat, M. Z	University of Malaya	Malaysia	9	9	280	31.11	31.11	7	2010
Mohamad, N	Universiti Tun Hussein Onn Malaysia	Malaysia	9	8	26	2.89	3.25	4	2014
Samad, A. A. A	Universiti Tun Hussein Onn Malaysia	Malaysia	8	7	37	4.63	5.29	5	2013
Othuman, M. A	Universiti Sains Malaysia	Malaysia	8	5	35	4.38	7.00	3	2014
Hussin, M. W	Universiti Teknologi Malaysia	Malaysia	7	6	151	21.57	25.17	4	2015
Mirza, J	Research Institute of Hydro-Quebec	Canada	7	7	270	38.57	38.57	5	2012
Thenmozhi, R	Government College of Technology	India	7	5	31	4.43	6.20	4	2014
Monz, J	Valencia Polytechnic University	Spain	7	7	159	22.71	22.71	5	2013
Ali, M	Capital University of Science and Technology	Pakistan	6	4	118	19.67	29.50	3	2015
Borrachero, M. V	Valencia Polytechnic University	Spain	6	6	123	20.50	20.50	4	2013
Ismail, M	Universiti Teknologi Malaysia	Malaysia	6	5	130	21.67	26.00	4	2015
Ismail, M. A	Curtin University Malaysia	Malaysia	6	5	68	11.33	13.60	3	2015
Pay, J	Valencia Polytechnic University	Spain	6	6	123	20.50	20.50	4	2013
Pinto, J	University of Trás-os-Montes and Alto Douro	Portugal	6	6	310	51.67	51.67	6	2011

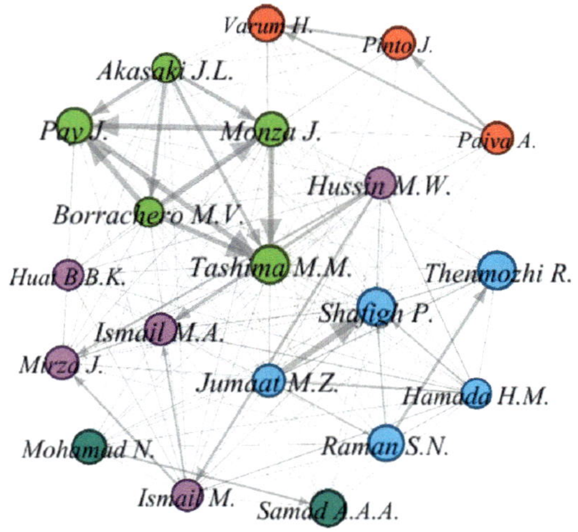

Fig. 16.3 Top 20 influential authors on agricultural waste research

publications. Pinto J is at the top regarding average citation per publication (C/P) and average citation per cited publication (C/CP) with 51.67, respectively. It was also found that more than half (8 authors) of the top 15 authors belong from Malaysia, followed by two from Spain and one from Canada, India, Pakistan, and Portugal for each. It explains that the most productive author in agriculture waste research is not the most cited (Fig. 16.3).

16.4.4 Highly Productive Organizations

Table 16.5 describes the top ten most productive organizations in agricultural waste. The top ten contributing organizations' total publications, number of cited papers, total citations, citation impact, and total cited paper impact have been mentioned in the table. It shows that the top 10 organizations collectively produced 164 research publications on agricultural waste with 2057 citations. University Tun Hussein Onn from Malaysia is the top contributing organization with (n = 27) publications and (n = 83) citations received with a citation impact of 3.07 and total cited the paper impact of 4.15, followed by University Teknologi Malaysia (n = 26) publications and (n = 333) citations received with a citation impact of 12.81 and total cited paper impact 16.65, the University of Malaya with (n = 23) publications and (n = 689) citations received with a citation impact of 29.96 and total cited paper impact 32.81.

The University of Malaysia Sarawak from Malaysia was the least productive (eight) organization among the listed organizations, and 22 citations were received with a citation impact of 2.75. A total cited the paper impact of 5.50. The University of Malaya was the most cited

Table 16.5 Top ten highly productive organizations

Affiliations	Country	TP	TCP	TC	C/P	C/CP
University Tun Hussein Onn Malaysia	Malaysia	27	20	83	3.07	4.15
University Teknologi Malaysia	Malaysia	26	20	333	12.81	16.65
University of Malaya	Malaysia	23	21	689	29.96	32.81
University Sains Malaysia	Malaysia	17	14	163	9.59	11.64
University Putra Malaysia	Malaysia	16	14	495	30.94	35.36
Landmark University	Nigeria	15	9	78	5.20	8.67
Covenant University	Nigeria	13	9	39	3.00	4.33
University Kebangsaan Malaysia	Malaysia	10	10	111	11.10	11.10
National Research Centre	Egypt	9	8	44	4.89	5.50
University Malaysia Sarawak	Malaysia	8	4	22	2.75	5.50

organization, with 689 citations received for 23 publications. The table also revealed that 7 (70%) organizations were from Malaysia, two from Nigeria, and one from Egypt. Highly contributed organizations were not highly cited.

Source Impact

Table 16.6 shows the top twenty most relevant sources of the publications of agricultural waste research. Total 747 documents analyzed were published in 316 different sources. These all publications were published in peer-reviewed journals. The top five sources were published double-digit publications on agricultural waste research, and the remaining were published single digit. *Construction and Building Materials* had the highest number of publications (81) with the number of cited papers 81, total citations of 3192, citation impact 39.41, total cited paper impact 39.41, impact factor 4.419, and h-index 29. *Journal of Cleaner Production,* the second-most important source of publication, published 28 documents with several cited papers, 28 total citations 736, citation impact of 26.29, totally cited paper impact 26.29, impact factor 7.246, and h-index 16. Other vital sources like *Energy and Buildings, Journal of Materials in Civil Engineering, Key Engineering Materials, Waste, and Biomass Valorization, and Waste Management* produced six documents each and received 242, 66, 21, 58, and 574 citations, respectively. *Bioresources, Cement and Concrete Composites, Cement and Concrete Research, Journal of Thermoplastic Composite Materials,* and *Journal Teknologi* were the minor productive sources. All sources were published four documents each and received 70, 196, 112, 27, and 25 citations, respectively. *AIP Conference Proceedings* published only seven papers and received 14 citations; the lowest citations were among the top twenty sources. *Building and Environment* had only five publications and received the highest citation impact, 186.00. *Journal of Cleaner Production* is the most impactful source for the research scholar, JIF 7.246. Most of the sources are belong from the USA (6), England (6), followed by the Netherlands (4), Switzerland (3), and (1) from Malaysia.

Table 16.6 Source impact of highly influential research journals

SO	PU	CO	NP	NCP	TC	C/P	C/CP	IF	h_index
Construction and Building Materials	Elsevier	England	81	81	3192	39.41	39.41	4.419	29
Journal of Cleaner Production	Elsevier	USA	28	28	736	26.29	26.29	7.246	16
Industrial Crops and Products	Elsevier	Netherlands	14	14	387	27.64	27.64	4.244	13
Materials	MDPI	Switzerland	14	14	189	13.50	13.50	3.057	8
Journal of Building Engineering	Elsevier	Netherlands	10	10	126	12.60	12.60	3.379	6
Materials Today: Proceedings	Elsevier	England	9	9	73	8.11	8.11	NA	5
Case Studies in Construction Materials	Elsevier	Netherlands	8	8	72	9.00	9.00	NA	4
AIP Conference Proceedings	American Institute of Physics	USA	7	7	14	2.00	2.00	NA	2
Waste Management	Elsevier	USA	6	6	574	95.67	95.67	5.448	6
Energy and Buildings	Elsevier	Switzerland	6	6	242	40.33	40.33	4.867	6

16.4.5 Authorship Pattern

The authorship patterns of agricultural waste are represented in Fig. 16.4. The number of collaborating authors was examined for each publication, which ranged from one to ten or more. And, there were only 29 articles and 1068 citations received as a single author. A total of 129 publications were contributed by two authors, with 1843 citations received. Three authors were contributed to the highest publications (181) and obtained (2509) citations. Four authors were contributed to 165 publications, and they obtained the highest citations (3343). Ten or more authors contributed only six publications, and they received 32 citations, the lowest number of contributions and citations received as shown in the figure. As a result, agricultural waste researchers are likely to collaborate with others in the field. According to the figure, the four-authorship pattern receives the highest citations, with 3343 for 165 publications, followed by three authorships with 2509 citations for 129 publications. The authors that contributed the more were not the most cited.

16.4.6 Most Highly Cited Articles

The twenty most-cited documents on agricultural waste have been presented in Table 16.7. Title of the article, the first author's name, source of publication, year of publication, the average total citation per year (TC/PY), normalized total citation (NTC) in Table 16.7. It explains the top twenty highly cited articles in the agricultural waste field. Citations for the top 20 listed publications ranged between 110 and 362 citations, with only six publications receiving more than 200 citations. The article entitled "Solid Wastes Generation in India and Their Recycling Potential in Building Materials" (2007) by Pappu A, appeared in Building and Environment was the most cited research with 362 citations, 24.133 TC/PY, and 1.788 NTC, followed by "Silicon-Based Materials from Rice Husks and Their Applications" (2001) by Sun L with 296 citations, "Potential Applications of Rice Husk Ash Waste from Rice Husk Biomass Power Plant" (2016) by Pode R with 292 citations, "Review On the Physicochemical Treatments of Rice Husk for Production of Advanced Materials" (2015) by Soltani N with 285 citations and "Development of Sustainable Construction Material Using Industrial and Agricultural Solid Waste: A Review of Waste-Create Bricks" (2011) by Raut SP with 226 citations. The article "Microstructure and Durability Properties of Cement Mortars Containing Nano-TiO_2 and Rice Husk Ash," (2016) by Mohseni E, published in Construction and Building Materials, was the least cited article on the top twenty list with 110

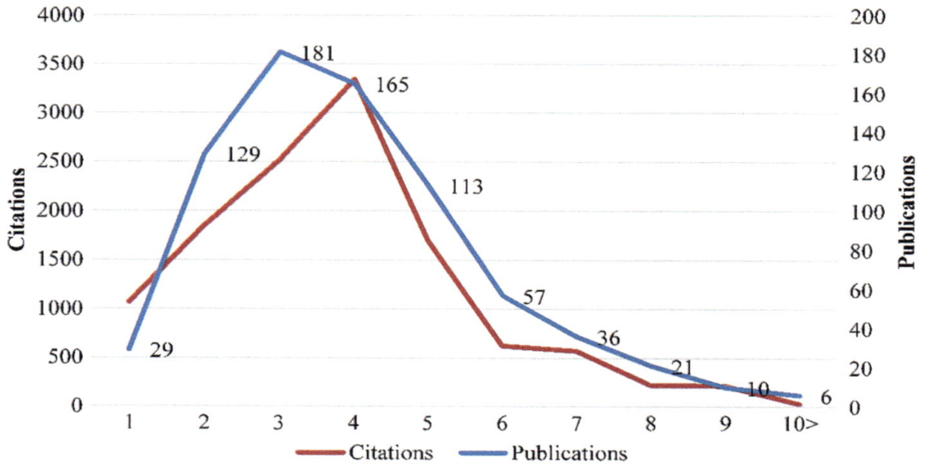

Fig. 16.4 Top ten authorship pattern of agricultural waste

Table 16.7 Top 20 most cited research papers in agricultural waste between 2001 and 2020

TC	Title	AU	SO	PY	TC/PY	NTC	GSC
362	Solid Wastes Generation in India and Their Recycling Potential in Building Materials	Pappu, A.	Building and Environment	2007	24.133	1.788	725
296	Silicon-Based Materials from Rice Husks and Their Applications	Sun, L.	Industrial and Engineering Chemistry Research	2001	14.095	2.182	494
292	Potential Applications of Rice Husk Ash Waste from Rice Husk Biomass Power Plant	Pode, R.	Renewable and Sustainable Energy Reviews	2016	48.667	9.155	420
285	Review On the Physicochemical Treatments of Rice Husk for Production of Advanced Materials	Soltani, N.	Chemical Engineering Journal	2015	40.714	8.926	407
226	Development of Sustainable Construction Material Using Industrial and Agricultural Solid Waste: A Review of Waste-Create Bricks	Raut, S. P.	Construction and Building Materials	2011	20.546	5.717	364
219	Application of Agro-Waste for Sustainable Construction Materials: A Review	Madurwar, M. V.	Construction and Building Materials	2013	24.333	8.166	350
192	Utilization of Agricultural and Forest Industry Waste and Residues in Natural Fiber–Polymer Composites: A Review	Visnen, T.	Waste Management	2016	32	6.02	282
189	Assessment of the Effects of Rice Husk Ash Particle Size on Strength, Water Permeability and Workability of Binary Blended Concrete	Givi, A. N.	Construction and Building Materials	2010	15.75	6.414	295
189	New Insulating Particleboards from Durian Peel and Coconut Coir	Khedari, J.	Building and Environment	2003	9.9474	2	249
179	Effect Of Organic Residues Addition on The Technological Properties of Clay Bricks	Demir, I.	Waste Management	2008	12.786	4.095	269
178	Fracture and Flexural Characterization of Natural Fiber-Reinforced Polymer Concrete	Reis, J. M. L.	Construction and Building Materials	2006	11.125	2.368	304
170	Supplementary Cementitious Materials Origin from Agricultural Wastes - A Review	Aprianti, E.	Construction and Building Materials	2015	24.286	5.324	266
160	The Study of Using Rice Husk Ash to Produce Ultra-High-Performance Concrete	Van Tuan, N.	Construction and Building Materials	2011	14.546	4.048	246
154	Concrete From an Agricultural Waste-Oil Palm Shell (OPS)	Mannan, M. A.	Building and Environment	2004	8.5556	2.02	278
149	New Low-Cost Insulation Particleboards from Mixture of Durian Peel and Coconut Coir	Khedari, J.	Building and Environment	2004	8.2778	1.954	243

(continued)

Table 16.7 (continued)

TC	Title	AU	SO	PY	TC/PY	NTC	GSC
146	A Huge Number of Artificial Waste Material Can Be Supplementary Cementitious Material (SCM) For Concrete Production A Review Part II	Aprianti, S. E.	Journal of Cleaner Production	2017	29.2	14.25	182
120	Coir Fiber Reinforcement and Application in Polymer Composites: A Review	Verma, D.	Journal of Materials and Environmental Science	2013	13.333	4.475	222
115	Engineering Properties of Concrete with Oil Palm Shell as Coarse Aggregate	Mannan, M. A.	Construction And Building Materials	2002	5.75	2.331	187
111	Fracture and Fatigue of Natural Fiber-Reinforced Cementitious Composites	Savastano, Jr. H.	Cement and Concrete Composites	2009	8.5385	4.723	175
110	Microstructure and Durability Properties of Cement Mortars Containing Nano-TiO_2 and Rice Husk Ash	Mohseni, E.	Construction and Building Materials	2016	18.333	3.449	143

citations. Table 16.7 also reveals that most of the cited articles were published in "Construction and Building Materials" (six articles), followed by "Building and Environment" (four articles), and Waste Management (two articles). The average total citation per year (TC/PY) and normalized total citation (NTC) was highest for the article entitle "Potential Applications of Rice Husk Ash Waste from Rice Husk Biomass Power Plant" (2016) by Pode R, appeared in Renewable and Sustainable Energy Reviews with 48.667 TC/PY and 9.155 NTC, respectively (Fig. 16.5).

16.4.7 Co-occurrence Network of Author Keywords

Figure 16.6 shows the co-occurrence network of authors' keywords. It depicts that the keyword "Agricultural Wastes" and "Mechanical Properties" have a maximum co-occurrence link of 23 for each, followed by "Compressive Strength" (22), "Lightweight Concrete" (22), and "Rice Husk Ash" (20). The keyword "Silica Fume," "Composites," "Recycling," "Bio-composites," and "Thermal Properties" have a minimum link of 5. Total link strength is highest in the case of the keyword "Compressive Strength" (128), followed by "Agricultural Wastes" (98), "Lightweight Concrete" (79), "Rice Husk Ash" (76), and Mechanical Properties" (63). The keyword "Silica Fume" has the slightest strength link with seven, followed by "Composites," "Bio-composites," and "Agricultural Residues," with 8 for each. There were 102 documents where two keywords, "Agricultural Wastes" and "Compressive Strength," occurred, but "Silica fume," "Corncob," "Composites," "Coconut shell," "Biomass" and "Agricultural waste materials" were the least appearance words with ten documents.

16.4.8 Country Collaboration Map

Figure 16.7 depicts the collaboration patterns among the top contributing countries in agricultural waste research in a table within the figure. Malaysia has the highest level of international collaboration, according to the data. Malaysia has the most collaborations with six of the top collaborating countries. Malaysia and Indonesia (9) and Spain and Brazil (9) each had the highest research collaboration, followed by Nigeria and

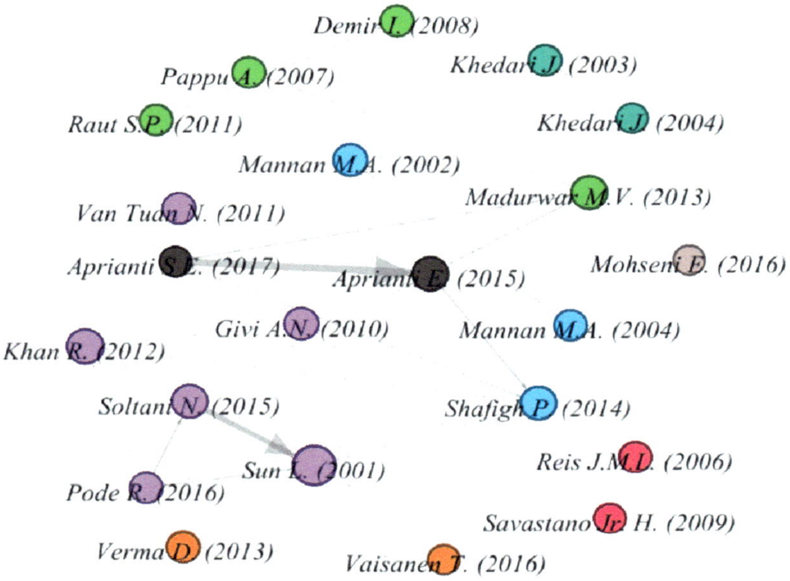

Fig. 16.5 Top document analysis

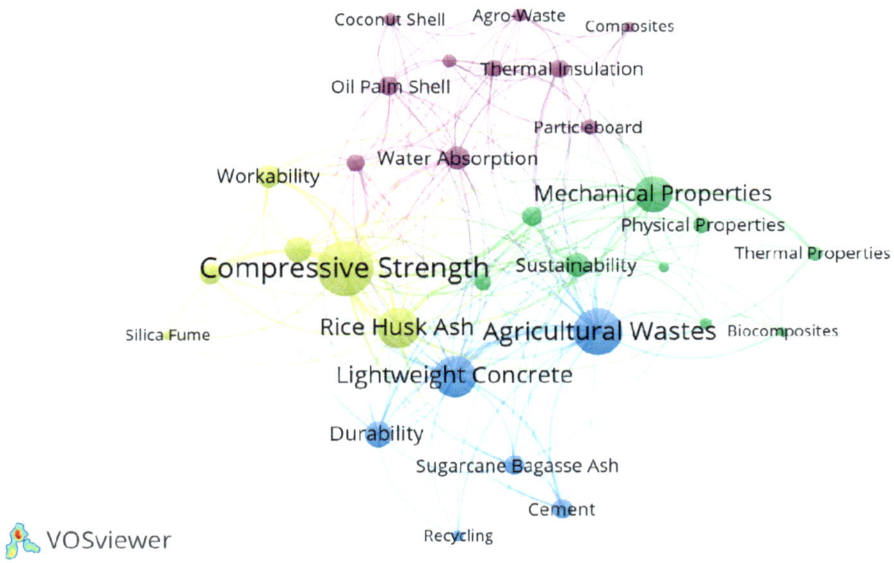

Fig. 16.6 Co-occurrence network of author keywords (minimum occurrence 10)

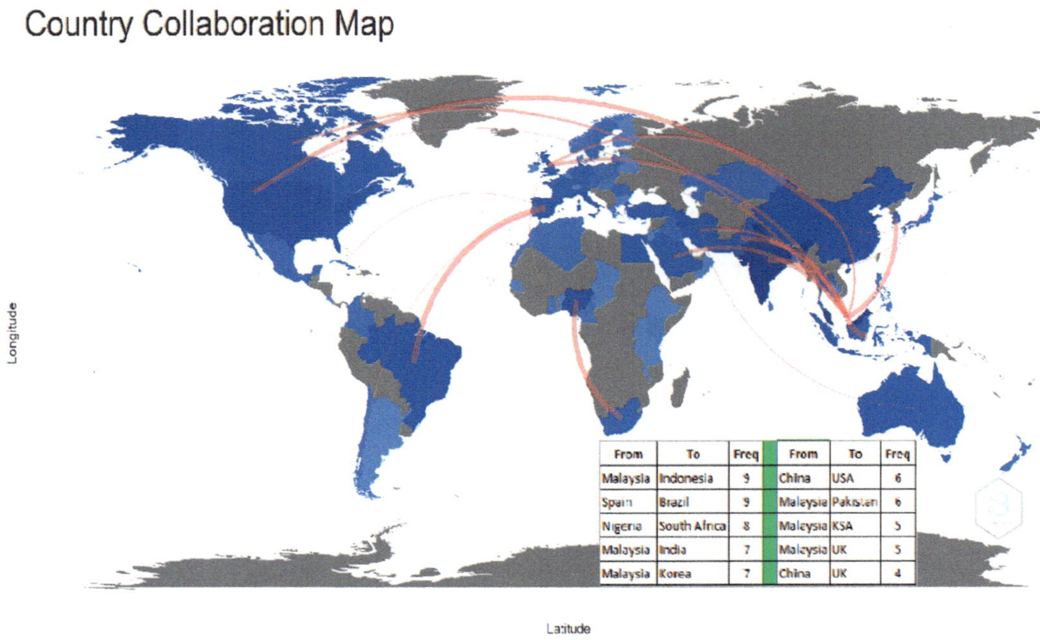

Fig. 16.7 Country collaboration map

South Africa (8). Only China and the United Kingdom had the least amount of collaboration (4). China was another significant country that collaborated on agricultural waste studies with the United States and the United Kingdom.

16.5 Conclusions

This research paper mapped the literature in agricultural waste (AW) and its citation impact between 2001 and 2020 using bibliometric methods. Network analysis was conducted using VOS viewer to analyses authors' keywords. The result of the study reveals that literature on agricultural waste has increased manifold for the last five years. The total publications and citations showed a gradual increase from the year 2010 onwards on agricultural waste research. 2020 is recorded as the most significant contributed year with 171 publications. Malaysia was a leading country in AW research, and the University Tun Hussein Onn Malaysia was a highly productive institution. In total, 747 research appeared in 316 sources, and the journal Construction and Building Materials was one of the most relevant in AW research. Malaysia with Indonesia and Spain with Brazil emerged as the top collaborating country for agricultural waste research. This study is one of the first to use bibliometric analysis to obtain insight into the AW domain, and its findings complement current research in the field. Finally, this study is mainly based on high-quality literature indexed on the Scopus database. However, the inclusion of other bibliographic databases, such as the web of science, PubMed, and Google Scholar, may have a certain degree of influence on the study's findings. The study's bibliometric findings show that agricultural waste research is a developing field with significant future research potential for agriculturalists, environmentalists, farmers, and local governments, and among others.

Acknowledgements The scholar (Dr. Pasarul Islam) is the awardee of ICSSR Post-Doctoral Fellowship. This paper is largely an outcome of the Post-Doctoral Fellowship sponsored by the Indian Council of Social Science Research (ICSSR). However, the responsibility for the facts stated, opinions expressed, and the conclusions drawn is entirely of the author.

References

Ahmad M, Islam P (2017) Agricultural modernization and rural development: a case study of Mahamaya Nagar District, Uttar Pradesh. Natl Geogr J India 63(2):17–28

Ahmad S, Sohail M, Waris A, Elginaid A, Mohammed I (2018) SCImago, eigenfactor score, and H5 index journal rank indicator: a study of journals in the area of construction and building technologies. J Libr Inf Technol 34(4):278

Ahmad M, Islam P, Siddiqui S (2020) Role of agricultural technology on socio-economic development in Hathras district, UttarPradesh. Natl Geogr J India 66 (3):222–235. https://doi.org/10.48008/ngji.1743

Aprianti E, Shafigh P, Bahri S, Farahani JN (2015) Supplementary cementitious materials origin from agricultural wastes—a review. Constr Build Mater 74:176–187. https://doi.org/10.1016/J.CONBUILDMAT.2014.10.010

Aznar-Sánchez J, García-Gómez J, Velasco-Muñoz J, Carretero-Gómez A (2018) Mining waste and its sustainable management: advances in worldwide research. Minerals 8(7):284

Bansal N, Tewari R, Soni R, Soni SK (2012) Production of cellulases from Aspergillus niger NS-2 in solid state fermentation on agricultural and kitchen waste residues. Waste Manage 32(7):1341–1346. https://doi.org/10.1016/J.WASMAN.2012.03.006

Chartered Institution of Wastes Management (CIWM) (2020) Agricultural waste. Available from: https://www.ciwm.co.uk/ciwm/knowledge/agricultural-waste.aspx

de la Cruz-Lovera C, Perea-Moreno AJ, de la Cruz-Fernández JL, Alvarez-Bermejo JA, Manzano-Agugliaro F (2017) Worldwide research on energy efficiency and sustainability in public buildings. Sustainability 9(8):1294. https://doi.org/10.3390/SU9081294

Demir I (2008) Effect of organic residues addition on the technological properties of clay bricks. Waste Manage 28(3):622–627. https://doi.org/10.1016/J.WASMAN.2007.03.019

Ezej T, Blaschek H (2010) Butanol production from lignocellulosic biomass. In: Blaschek H, Ezej T, Scheffran J (eds) Biofuels from agricultural wastes and byproducts. Wiley Blackwell

Florido C, Jacob M, Payeras M (2019) How to carry out the transition towards a more circular tourist activity in the hotel sector. The role of innovation. Admin Sci 9 (2):47. https://doi.org/10.3390/ADMSCI9020047

Givi AN, Rashid SA, Aziz FNA, Salleh MAM (2010) Assessment of the effects of rice husk ash particle size on strength, water permeability and workability of binary blended concrete. Constr Build Mater 24(11):2145–2150. https://doi.org/10.1016/J.CONBUILDMAT.2010.04.045

Hardoyo H (2013) Material utilization technology of agriculture and forestry waste. In: 2nd International conference on engineering and technology development

Hoseinzadeh Hesas R, Wan Daud WMA, Sahu JN, Arami-Niya A (2013) The effects of a microwave heating method on the production of activated carbon from agricultural waste: a review. J Anal Appl Pyrol 100:1–11. https://doi.org/10.1016/J.JAAP.2012.12.019

Islam P, Ahmad M, Bano E (2020) Exploring the adoption behaviour and determinants of agricultural modernization using path analysis in Hathras district, Uttar Pradesh. Ann Rajasthan Geogr Assoc XXXVI:1–12

Khedari J, Charoenvai S, Hirunlabh J (2003) New insulating particleboards from durian peel and coconut coir. Build Environ 38(3):435–441. https://doi.org/10.1016/S0360-1323(02)00030-6

Khedari J, Nankongnab N, Hirunlabh J, Teekasap S (2004) New low-cost insulation particleboards from mixture of durian peel and coconut coir. Build Environ 39(1):59–65. https://doi.org/10.1016/J.BUILDENV.2003.08.001

Madurwar, MV, Ralegaonkar RV, Mandavgane SA (2013) Application of agro-waste for sustainable construction materials: a review. Constr Build Mater 38:872–878. https://doi.org/10.1016/J.CONBUILDMAT.2012.09.011

Mannan MA, Ganapathy C (2004) Concrete from an agricultural waste-oil palm shell (OPS). Build Environ 39(4):441–448. https://doi.org/10.1016/J.BUILDENV.2003.10.007

Meng Q, Zhu H, Li Z, Du J, Wang X, Kim MJ (2018) How green building product decisions from customers can be transitioned to manufacturers: an agent-based model. Sustainability 10(11):3977. https://doi.org/10.3390/SU10113977

Mongeon P, Paul-Hus A (2016) The journal coverage of Web of Science and Scopus: a comparative analysis. Scientometrics 106(1):213–228. https://doi.org/10.1007/S11192-015-1765-5/FIGURES/6

Mostafa NA, Farag AA, Abo-dief HM, Tayeb AM (2018) Production of biodegradable plastic from agricultural wastes. Arab J Chem 11(4):546–553. https://doi.org/10.1016/J.ARABJC.2015.04.008

Obi FO, Ugwuishiwu BO, Nwakaire JN (2016) Agricultural waste concept, generation, utilization and management. Niger J Technol 35(4):957–964. https://doi.org/10.4314/njt.v35i4

Pode R (2016) Potential applications of rice husk ash waste from rice husk biomass power plant. Renew Sustain Energy Rev 53:1468–1485. https://doi.org/10.1016/J.RSER.2015.09.051

Powell KR, Peterson SR (2017) Coverage and quality: a comparison of Web of Science and Scopus databases for reporting faculty nursing publication metrics. Nurs Outlook 65(5):572–578. https://doi.org/10.1016/J.OUTLOOK.2017.03.004

Raut SP, Ralegaonkar RV, Mandavgane SA (2011) Development of sustainable construction material using industrial and agricultural solid waste: a review of waste-create bricks. Constr Build Mater 25(10):4037–4042. https://doi.org/10.1016/J.CONBUILDMAT.2011.04.038

Re Cecconi F, Moretti N, Dejaco MC (2019) Measuring the performance of assets: a review of the facility condition index. Int J Strateg Prop Manag 23(3):187–196. https://doi.org/10.3846/IJSPM.2019.7955

Reis JML (2006) Fracture and flexural characterization of natural fiber-reinforced polymer concrete. Constr Build Mater 20(9):673–678. https://doi.org/10.1016/J.CONBUILDMAT.2005.02.008

Sarkar N, Ghosh SK, Bannerjee S, Aikat K (2012) Bioethanol production from agricultural wastes: an overview. Renew Energy 37(1):19–27. https://doi.org/10.1016/J.RENENE.2011.06.045

Sarmah A, Copeland C, Ugwuanyi J, Aktar W, Sinha V, Rai V (2009) Agricultural wastes. In Ashworth GS, Azevedo P (eds) Agriculture issues and policies: agricultural wastes. Nova Science Publishers, Inc. http://agrifs.ir/sites/default/files/Agricultural_Wastes_0.pdf

Soltani N, Bahrami A, Pech-Canul MI, González LA (2015) Review on the physicochemical treatments of rice husk for production of advanced materials. Chem Eng J 264:899–935. https://doi.org/10.1016/J.CEJ.2014.11.056

Sun L, Gong K (2001) Silicon-based materials from rice husks and their applications. Ind Eng Chem Res 40(25):5861–5877. https://doi.org/10.1021/IE010284B

Sweileh WM, Al-Jabi SW, AbuTaha AS, Zyoud SH, Anayah FMA, Sawalha AF (2017) Bibliometric analysis of worldwide scientific literature in mobile—health: 2006–2016. BMC Med Inform Decis Mak 17(1):1–12. https://doi.org/10.1186/S12911-017-0476-7/TABLES/9

Väisänen T, Haapala A, Lappalainen R, Tomppo L (2016) Utilization of agricultural and forest industry waste and residues in natural fiber-polymer composites: a review. Waste Manage 54:62–73. https://doi.org/10.1016/J.WASMAN.2016.04.037

van Tuan N, Ye G, van Breugel K, Fraaij ALA, Bui DD (2011) The study of using rice husk ash to produce ultra high performance concrete. Constr Build Mater 25(4):2030–2035. https://doi.org/10.1016/J.CONBUILDMAT.2010.11.046

Verma D, Gope P, Verma D, Gope PC, Shandilya A, Gupta A, Maheshwari MK (2013) Coir fibre reinforcement and application in polymer composites. J Mater Environ Sci 4(2):263–276

Wei J, Liang G, Alex J, Zhang T, Ma C (2020) Research progress of energy utilization of agricultural waste in China: bibliometric analysis by Citespace. Sustainability 12(3):812. https://doi.org/10.3390/SU12030812

Yu D, Xu Z, Antuchevičienė J (2019) Bibliometric analysis of the Journal of Civil Engineering and Management between 2008 and 2018. J Civ Eng Manag 25(5):402–410. https://doi.org/10.3846/JCEM.2019.9925

Zhou W, Chen J, Huang Y (2019) Co-citation analysis and burst detection on financial bubbles with scientometrics approach. Econ Res—Ekonomska Istraživanja 32(1):2310–2328. https://doi.org/10.1080/1331677X.2019.1645716

Zyoud SH (2016) Global research trends of Middle East respiratory syndrome coronavirus: a bibliometric analysis. BMC Infect Dis 16(1):1–7. https://doi.org/10.1186/S12879-016-1600-5/TABLES/4

Unveiling the Factors Influencing Organizational Sustainability Among Indian Millennials: A Systematic Review

17

Unnikannan P. Santhosh Kumar, Chetna Jaiswal, and Narsingh Kumar

Abstract

Amidst the global momentum towards sustainability, it becomes imperative to recognize and comprehend the formidable force of the millennial generation, whose birth years span the era from 1980 to 2000, as they saturate the contemporary workforce, and in doing so, it helps to unlock their full potential in order to propel organizational sustainability to greater heights. The purpose of the study is to investigate those factors affecting millennials/Generation Y in the contemporary workplace which would contribute towards organizational sustainability. Understanding the requirements and expectations of employees, particularly those of the millennial age, is becoming increasingly important for a growing number of enterprises. To get a thorough understanding, a set of journal articles and review papers published between 2010 and 2022 were analysed for identifying modal trends and patterns of millennials contributing to organizational sustainability, organizational commitment, and productivity. The selected studies (n = 50) have been sorted out based on geographical location in which the study has been conducted (Only INDIA), relevance of the paper, and the prestige of the journal in which a particular paper has been published. The first section of this article situates our review within the context of past generational and leadership research. It then goes on to describe the review approach used and our significant findings which revealed various factors influencing millennial's job satisfaction, job performance, and employee retention which in ultimatum would boost organizational sustainability.

Keywords

Millennials · Organizational sustainability · Organizational commitment · Organizational productivity

17.1 Introduction

In the current epoch of heightened global consciousness and increasing socio-environmental complexities, the imperative of organizational sustainability has ascended to an unprecedented zenith, as it constitutes an indispensable pillar for organizations to navigate the labyrinthine terrain of corporate responsibility, cultivate resilient brand equity, fortify stakeholder trust, and forge

U. P. S. Kumar (✉) · C. Jaiswal
Department of Psychological Sciences, Central University of South Bihar, Gaya 824236, Bihar, India
e-mail: unnikannan@cusb.ac.in

N. Kumar
Department of Psychology, University of Allahabad, Allahabad, India

S. Talukdar et al. (eds.), *Environmental Risk and Resilience in the Changing World*, GIScience and Geo-environmental Modelling, https://doi.org/10.1007/978-3-031-62442-1_17

a path towards harmonious coexistence with the delicate ecosystems upon which the very fabric of our existence is inextricably interwoven.

Today's workplace conditions are usually described as the worst since the Great Depression, making it extremely difficult for candidates to find work, while companies claim that it is impossible to find suitable staff (Ludviga 2020). Because today's workforce is more varied than ever, one of the most significant ways to explain employee disparities is by generation (Mahmoud et al. 2020) where about one-fourth of the workforce will be above 55 years of age by 2025 (Bass and Valdez 2020). As of now, change is one of the prominent issues the organizations are facing and in addition, novel technologies and Global competition also add to that misery (Malik and Khera, n.d.).

The millennial generation, also known as Generation Y, has become a large demographic group with unique traits and preferences that have a big impact on many facets of society, including organizational sustainability. Strauss and Howe (1991) defined the millennial cohort as consisting of individuals born between 1982 and 2004. India is the second-largest country in terms of the population where 34% are millennials (440 million) and form approximately 47% of the total workforce. Millennials have grown up in educational, economic, social, and political environments that are unlike those of earlier generations. For example, helicopter parents, regular positive feedback and reassurance, substantial technological advancements, and political and economic turbulence have all affected this age (Thompson and Gregory 2012).

Due to these reasons, millennials tend to exhibit different attitudes, preferences, and expectations compared to their older generations (Myers and Sadaghiani 2010; Ng et al. 2010). Leaders or immediate managers of the millennials tend to play an important role in retaining and increasing the commitment level of the Gen Y employees (Thompson and Gregory 2012). As an effect of overindulging parents, millennials tend to seek constant feedback, praise, guidance, and individual attention as they have been brought up in such an environment where these were provided in abundance by their parents (Hershatter and Epstein 2010; Ng et al. 2010).

Building on the existing knowledge about the millennials and the workplace will be of advantage in future as they will be flooding the workforce within the next few years and it will be of prime importance to retain and develop them. They have also been considered as an impulsive, job-hopping, and self-centred generation while compared to the previous generation which has been put to question by various researchers stating otherwise. Therefore, this study will focus on reviewing various articles/studies and review papers published in renowned journals specifically in Asian Culture, so that we will be able to understand more about the millennial generation sharing similar cultural backgrounds at various levels.

Environmental sustainability and organizational sustainability can be achieved through a combined effort of both employees and organization. When organizational commitment of the employees, environmental commitment of the employees, and organization's commitment towards environment are inline, several behaviours and outcomes can be seen which would ensure environmental sustainability (Mesmer-Magnus et al. 2012). Organizations of different sectors like vehicle production industries, mining industries, factories, etc., have a huge impact on environmental sustainability. A sustainable organization would tackle such issues with proper code of conduct for employee behaviour, ethics, and sound organizational practices.

Organizational sustainability plays a vital role in advancing the SDG agenda by embedding sustainable practices within businesses' core operations, supply chains, and corporate strategies. It enables organizations to contribute to the broader global effort of creating a sustainable, equitable, and prosperous future for all, as envisioned by the SDGs. Employee retention and job satisfaction are directly linked to several Sustainable Development Goals (SDGs) outlined in the SDG 2030 agenda, where employee retention contributes to Goal 8 (Decent Work and Economic Growth) by promoting stable employment opportunities and reducing turnover rates.

Secondly, job satisfaction is connected to Goal 3 (Good Health and Well-being) and Goal 5 (Gender Equality). Satisfied employees often experience better physical and mental well-being, leading to improved health outcomes. Moreover, fostering a positive work environment that supports gender equality and inclusivity contributes to achieving Goal 5. Additionally, through encouraging a supportive and moral work environment, employee retention and job satisfaction support Goal 16 (Peace, Justice, and Strong Institutions). In organizations, trust, cooperation, and solid institutional structures are fostered when people feel valued and respected.

Although organizational commitment and productivity are indirectly related to the SDG 2030 goals, they contribute to sustainable development in multiple ways and one such instance is when employees are committed to their organization's goals and values (Rosati and Faria 2019), they are more likely to actively contribute to the organization's overall performance and success. This increased productivity can lead to economic growth and prosperity, which aligns with Goal 8 (Decent Work and Economic Growth). Additionally, organizations that prioritize sustainable practices and social responsibility are more likely to have high levels of dedication and productivity. The SDGs Goal 7 (Affordable and Clean Energy), Goal 12 (Responsible Consumption and Production), and Goal 17 (Partnerships for the Goals) are all aligned with these activities. They might also contribute to the community, adopt ecologically friendly production methods, promote responsible consumption, and invest in renewable energy.

As millennials are those characterized by low organizational commitment and high turnover intention (Stewart et al. 2017), it is important to look upon the factors which would promote their organizational commitment which in turn enhances organizational sustainability and productivity. Given the lack of empirical evidence to understand the millennials, this study focuses on identifying various concepts, trends, and characteristics related to millennials which will contribute in building a sustainable organization. The review reported throughout this paper addresses the following research questions:

(1) How are the millennials characterized based on their self?
(2) What work expectations do the millennials seek from their leaders?
(3) What are the different work environment factors that affect their job satisfaction, job performance, and employee retention?

By analysing a significant number of studies, we aimed to develop a set of factors that would act as a facilitator for millennial's overall development and satisfaction in an organization which would foster organizational sustainability and an inhibitor for millennial's turnover intention. This research paper was done in anticipation that it would enable researchers to study the changes in research patterns over time and provide direction for future research.

17.2 Methods

There is no one technique to perform a systematic review, as previous researchers have noted, but it is critical to explain the search process clearly and transparently (Castillo and Hallinger 2018). As that being stated, the approach followed in this paper is systematic review research where we reviewed a significant number of articles which was studied on Indian population and published between the years 2010 and 2022 from esteemed journals like SAGE publications, JSTOR, Google Scholar, Psych Articles (APA), Science Direct, Annual Reviews, Emerald Publication, and Taylor & Francis. Following the above-mentioned research questions, we also look upon various other trends followed by the millennials as well as their desired style of leadership and leader qualities.

17.2.1 Criteria for Selection

After the selection of key terms like millennial/ Gen Y, job satisfaction, job performance, career growth and employee retention, and clarifying the inclusion criteria, an extensive literature search was conducted on the desired topic from the year 2010 to 2022. The reason for including studies from the year 2010 is because of the ever-changing workplace scenario, especially in this vulnerable, uncertain, complex, and ambiguous (VUCA) environment so that the information would be of more relevance. The year 2010 and after can also be interpreted as years in which the millennials started to enter the workforce in abundance; hence the studies published after 2010 would be more empirically applicable. The studies included were only from Indian background.

17.2.2 Data Extraction and Analysis

Each article taken for this study has been analysed with the sole aim to extract as much information as possible and relevant to the research questions stated. A basic thematic analysis was performed to analyse and establish the most common features or characteristics exhibited by the millennials based on their self, leader, organization, and work environment level (Table 17.1 and Fig. 17.1).

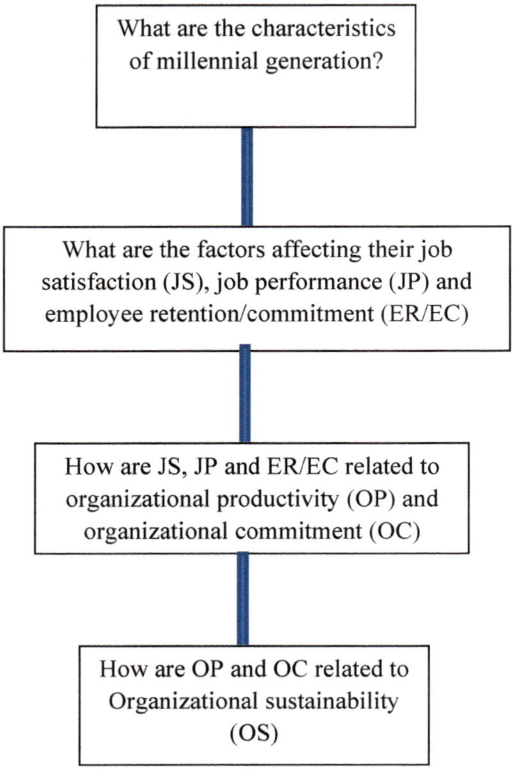

Fig. 17.1 Framework followed in the present study to reach a valid conclusion

17.2.3 Publication Volume

The literature search was purely based on the intention to know more about the millennial's characteristic, their expectations from the leader as well as the organizations and to investigate the factors that affect the job satisfaction, job performance, and employee retention. In order to make the study more valid in the contemporary setting, the study included articles published between the period of 2010 and 2022 (Fig. 17.2).

Table 17.1 Research questions and data collection

Research question	Data collected
How are the millennials characterized based on their self?	Characteristics attributed to the millennial generation
What work expectations do the millennials seek from their leaders?	Expected qualities of a leader, factors necessary for their career growth
What are the different work environment factors that affect their job satisfaction, job performance, and employee retention/commitment?	Work timing, salary, organizational norms and leadership style

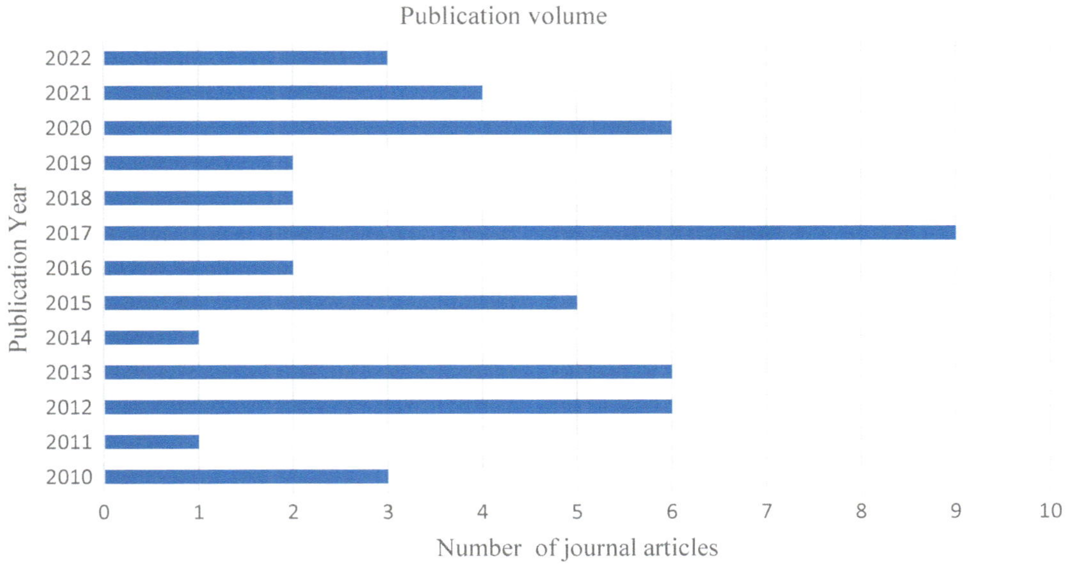

Fig. 17.2 Publication volume of articles published on the theme of millennials, 2010–2022 (*n = 50)*

17.3 Results

In this initial section, the focus is on *how are the millennials characterized based on their self?* Hence, the areas being explored are the differences between millennials and baby boomers and characteristics attributed to the millennial generation.

17.3.1 Differences Between Millennials and Baby Boomers

Before dwelling on the characteristic difference, it is important to know about the difference in the way they were brought up and in what environment did they live as well as the advantages each generation had (generation gap). They have been brought up in a period where there was a boom in globalization, information technologies, and employment outsourcing (Thangavel et al. 2021).

At work, Indian millennials have different attitudes and behaviours than previous generations, and they take an alternative approach to managing their careers that differs from standard career pathways (Saxena and Jain 2012) as they are keener on taking up contemporary careers rather than traditional ones (Sen and Hooja 2015). These distinguishing characteristics of the millennial generation represent a self-directed approach to job management and personal growth, which is comparable to that of those with protean career attitude (Gulyani and Bhatnagar 2017).

Compared to the narcissistic and judgemental nature of the baby boomers, millennials tend to be disciplined, educated, and open-minded which could be due to the circumstances which they have lived through as the baby boomers experienced a large social change and political push-back which urged them to fight against race and gender issues, whereas the millennials were brought up in an era where technology has taken over most aspects of an individual's life and they tend to be equipped with a far-reaching, global social conscience (Sharma 2019).

17.3.2 Characteristics Attributed to Millennials

Young workers in India, like their counterparts around the world, are a powerhouse of talent that can work miracles but also pose some obstacles. Therefore, in order to better serve this generation and be served by them, businesses need to gain deeper insights into this generation if they hope to be viable in the market (Malik and Khera 2014). Indian millennials, like their American counterparts, are digital natives who are far more adept at using the Internet than their parent's generation (Mitra 2011). While comparing US millennials and Indian millennials, it was found that both the populations share many common characteristics like optimism and entrepreneurial orientation (Telefonica 2013). Even with all the merits, millennials are also known for their lower loyalty (Myers and Sadaghiani 2010) and can adapt to the innovative changes in technology (Shankar et al. 2010).

Millennials are more focused on their personal growth and prefer to be in organizations that promote out-of-the-box thinking, autonomy, and freedom to express their views and to take initiatives (Prakash et al. 2021; Indhira and Shani 2016; Singh et al. 2012). For them to stay in an organization for a long period, then the above-mentioned factors must be fulfilled, and additionally the millennials should feel that there is a consistency between the assigned task and their career expectations (Motiram and Singh 2012). From all the studies being quoted above, the millennials incline towards protean career attitude, moreover the study conducted by Sen and Hooja (2015) has proved a link between psychological capital and protean career attitude as the former will affect an individual's intrinsic motivation.

Millennials tend to be self-focused and concentrate on the instant gratification of needs which is clearly associated with protean career attitude (Kaushal and Vashisht 2021). Additionally, they also seek autonomy in various activities they engage in (Srinivasan 2012).

17.3.3 What Work Expectations Do the Millennials Seek from Their Leaders?

A study conducted by Dokadia et al. (2015) discovered that Indian millennials yearn for immediate and constructive input on their personal development from their bosses. In addition, they found that Indian millennials wanted their mentor/supervisor to take an active role in holding regular and timely professional dialogues centred on their requirements, demonstrating interest in their career objectives (Naim and Lenka 2017). Mentoring also affects task performance, contextual performance, and total job performance (Chatterjee et al. 2021).

As India is collectivistic in nature, traditional culture and social ties will play an important role in mentoring relationship (Chadda and Deb 2013). Furthermore, the findings of international studies on mentoring have been verified by the few known Indian studies on mentoring. For example, research of 151 mentees from a single business in eastern India discovered that mentoring had an impact on work performance, with traditional mentoring having a higher impact than relational mentoring (Srivastava and Jomon 2013). They also seek appreciation, supportive work environment, and bright future prospects in order to be motivated at work (Bansal 2017).

A study of 64 mentors and 88 mentees from two public sector organizations found that mentoring enhanced employee job performance for over 60% of the mentees and 87% of the mentors (Buddhapriya 2017) and at the same time, it was also revealed that mentoring had its effect on self-efficacy, personal learning, communication satisfaction, and relationship quality which in turn affected job performance (Jyoti and Sharma 2017). Mentoring, team work and nature of working style have significant impact in engaging millennials (Anitha and Aruna 2016). After a rigorous review, Rao and Pavani (2015) concluded that to provide an effective leadership style for the millennials, the leader must possess certain qualities such as imagination, strategic thinking, empathy, communication skills, future-oriented, flexibility, broad-minded, knowledge, and a team builder.

17.3.4 What Are the Different Work Environment Factors that Affect Their Job Satisfaction, Job Performance, and Employee Retention/Commitment?

Before dwelling into the factors affecting the millennial's job satisfaction, job performance, and employee retention, it is important to look at those factors that employees consider principal to be received from their organizations. Fair pay, speedy promotion, and transparent organizational policies are some of the most important factors which the millennials expect to receive from the organization they are working for (Srinivasan 2012).

As millennials are prone to exhibit less loyalty and commitment, their intention to leave a job would be high and it is important to retain millennial employees as they will be flooding all the areas of an organization including the managerial positions in the coming 10 years. Another reason adding to this issue is that the millennials are thought to have high expectations for their employment and the organizations for which they work, as well as special considerations before accepting work (Arora and Kshatriya 2017). Therefore, it is important to look at those factors which would contribute to their retention.

Millennials look for opportunities for quicker professional advancement. To emotionally connect millennials, a link like "My Career" on the employee's dashboard that highlights the organization's career path and growth process may be a useful tactic (Jha et al. 2018). Those factors which the millennials consider to be of more importance when selecting a job are employment prestige, challenging job profile, job security, starting salary, and employer location (Yadav and Chaudhari 2020; Arora and Kshatriya 2017). Academic experts have found that workplace spirituality also plays an important role in employee loyalty among millennials in India (Aboobaker et al. 2020).

Job satisfaction of the millennials is also being determined by various factors like leadership behaviour, promotion criteria, interpersonal relationships, personal recognition, and growth (Koorella and Perumal 2019; Khanolkar 2013). The commitment level of Indian millennial employees towards their organization was positively effected by empowering leadership style they received (Naim 2022).

Another factor that contributes to both job satisfaction and employee retention among the millennials was found to be job mismatch (Chavadi et al. 2022). Workplace ostracism, where the employees feel avoided, tends to be negatively associated with job satisfaction which in turn decreased job performance (Bugalia et al. 2022). A challenging task, working environment, feedback, flexible schedule, and learning and growth in the organization itself are considered to be of utmost importance for the millennials to be satisfied with their job and to perform well (Sharma and Chully 2020). Yet another factor affecting the job performance of the millennials is work engagement as found by Gaikwad and Swaminathan (2020).

17.3.5 Effect of Job Satisfaction on Organizational Commitment

When employees experience high job satisfaction, they tend to be more engaged in their work. They find fulfilment and meaning in their roles, leading to increased motivation, enthusiasm, and commitment to their tasks and the organization (Garg et al. 2018). Organizational commitment is significantly influenced by job satisfaction and work engagement. Employees become deeply committed to the company when they have high levels of job satisfaction and are actively involved in their work (Biswas and Bhatnagar 2013).

Job satisfaction, which indicates contentment with one's work, encourages loyalty and produces a favourable impression of the workplace. Employee commitment and passion at work strengthen their sense of loyalty to the company and its objectives, improved employee loyalty, retention, and a stronger desire to put in time and effort towards the success of the company are results of this improved commitment, whether it be affective, continuing, or normative.

A study conducted on 396 academicians in India found a significant positive impact of job satisfaction on organizational commitment (Bashir and Gani 2020). When employees experience high levels of job satisfaction, they develop a stronger sense of attachment and loyalty to the organization. Satisfied employees perceive their work environment favourably, feel valued and respected, and have their needs met. This positive perception fosters a deep emotional connection to the organization, leading to increased affective commitment.

Despite the prevailing cultural differences, job satisfaction was found to elicit affective commitment towards organizations was found to be more in Indian samples while compared to the US samples (Kwantes 2009; Chordiya et al. 2017). Yet another study discovered a positive relationship between job satisfaction, locus of control and organizational commitment and in addition, internal locus of control moderated the relation between job satisfaction and organizational commitment (Chhabra 2013).

17.3.6 Effect of Employee Retention/Commitment on Organizational Productivity and Organizational Commitment

Employee commitment plays a vital role in positively influencing organizational productivity. When employees are committed to the organization, they demonstrate a higher level of dedication, engagement, and motivation in their work. Committed employees are more likely to invest their time and effort into their tasks, go beyond the minimum requirements, and take ownership of their responsibilities. This heightened commitment leads to increased productivity as employees are focused on achieving organizational goals and delivering high-quality results. Committed employees are also more willing to collaborate, share knowledge, and support their colleagues, fostering a positive work environment that further enhances productivity (Chanda and Goyal 2020). A study conducted by Dixit and Bhati (2012) confirms it as the results revealed a highly significant relationship between employee commitment and sustained productivity of the organization.

Apart from the direct effect, employee commitment also acts as a mediator between organizational productivity a quality of work-life (Nayak and Sahoo 2015). Committed/loyal employees are considered to be far more important than the motivated ones in order to foster organizational productivity (Varma 2017).

17.3.7 Role of Organizational Commitment and Organizational Productivity in Attaining Organizational Sustainability

In order to promote organizational sustainability, organizational commitment is essential. Employee loyalty to the company increases the likelihood that they will work there for a longer period, lowering turnover rates and related costs. With this continuity, the organization is stable and important information and expertise may be developed and retained. Additionally, more engaged, driven, and eager to go above and beyond for the success of the company are committed employees. For long-term sustainability, their dedication fosters higher levels of production, innovation, and customer pleasure (Nahak and Ellitan 2022). Independent organizational commitment towards its employees has a favourable and significant impact on economic

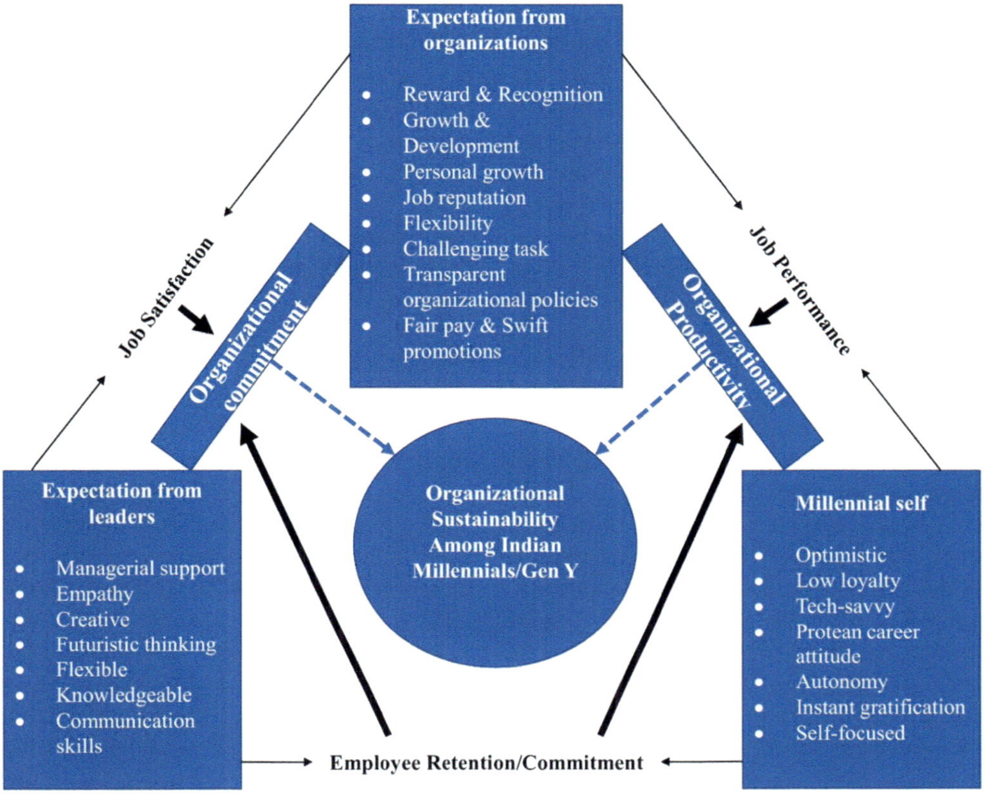

Fig. 17.3 Model based on the findings in accordance with the research questions and framework of the study

performance, environmental performance, social performance, and sustainability performance of the organization (Muslimat et al. 2020).

Organizational commitment is also found to play a mediating role between organizational justice and employee sustainability (Fatima et al. 2020). The dynamic, future-oriented, and tech-savvy nature of the millennials could boost organizational sustainability in multiple ways possible like technological innovation on sustainable manufacturing and supply chains (Assoratgoon and Kantabutra 2023). Furthermore, according to a study by Fok et al. (2021), organizational culture, the maturity of quality management, and organizational productivity are all related to an organization's overall green/sustainable attitude (Fig. 17.3).

17.4 Discussion

The above proposed three-headed model proclaims those factors affecting the millennial's job satisfaction, job performance, and employee retention in three different aspects, namely millennial self, expectation from leaders, and expectation from organizations. As the above-mentioned factors are those contributing to organizational productivity and commitment, it is evident that employee retention, reward and recognition, personal growth, managerial support, empathy from the manager, optimistic thinking, and protean career attitude are much needed for the millennials to strive towards organizational sustainability.

While looking into a wider picture, those individual factors under each of the three different aspects (millennial self, expectation from leaders and organizations) influence each other in terms of job satisfaction, job performance, and employee retention. For example, a person with optimistic thinking, technological knowledge, and protean career attitude along with an adequate amount of managerial support and recognition from the organizations tend to be more satisfied and will also affect their job satisfaction and performance.

Additionally, a leader's level of empathy, knowledge, and creative thinking would also promote the employees to take up challenging tasks, which are of prime importance for the millennials and ultimately will add on to their personal growth as well. The tech-savvy nature of the millennials would also be of advantage in the completion of challenging tasks and when the tech-savvy, futuristic, and optimistic nature of the Indian millennial self combines with creative managerial support and flexibility, personal growth, and commitment offered by the organization would pave a path leading to organizational sustainability (Yu et al. 2018; Sinha and Ghosh 2022; Jahanshahi et al. 2022).

The protean career attitude has been positively linked with the subjective and objective career success of the millennials (Kaushal and Vashisht 2021). If this is the case, then protean career attitude along with adequate mentoring and good managerial support would boost the ability of the employees to adapt and overcome the ever-changing work environments and challenging tasks easily. Low loyalty among the employees also seems to mediate the effect between leader-member exchange and organizational sustainability (Dawra et al. 2022).

Those factors expected by the millennials from organizations and leaders will also have a positive impact on their self. For example, reward and recognition and personal growth clubbed together with managerial support, knowledge, futuristic thinking, and empathy would encourage them to engage in more autonomous and challenging tasks which in turn would increase their loyalty towards their organization and leader. The flexibility on the part of the manager would greatly influence the protean career attitude of an individual as he/she enjoys freedom in managing tasks which can contribute to their better adaptation to the changing work environment (Briscoe et al. 2012). This same factor can also create a "protean paradox" in which the self-directed/self-focused nature of the millennials as well as their value orientation, serves the purpose of both the individual and the organization they work for (Hall et al. 2018). In this way, if an organization succeeds to align their needs and millennials' expectations, it would allow the organization to attain sustainability.

From the findings, it is also evident that the instant gratification characteristic of the millennial will have a huge impact on their expectations from organizations like the swift promotions they seek. As employee retention has been recognized as a major issue among the millennials, one factor contributing to it is known as workplace ostracism, where the employees feel excluded or ignored by others (Anderson and Bushman 2018) and this particular problem can be tackled by manager with good communication skill and empathy which would allow all the employees to interact and develop together, so that in the long run, turnover intentions of the millennial employees can be hindered to a large extend.

The level of employee commitment to the organization's objectives and core values is known as organizational commitment, and it significantly affects the operation and success of the organization. This dedication improves teamwork, produces a positive work atmosphere, and lowers turnover rates. Organizational productivity, on the other hand, refers to how effectively and efficiently resources are used to produce desired results. Organizations can achieve sustained growth when both dedication and productivity are strong by making the most of their resources, adapting to shifting market conditions, and continually providing value to stakeholders. In the end, a firm basis for organizational sustainability is built by a strong dedication and high productivity, guaranteeing long-term success and resilience.

17.5 Implications and Limitations

This study creates a platform for further research on the millennials and ways to improve their work engagement as well as personal growth and development. As the data collected are solely from India, it makes it more culturally valid in the Indian setting compared to those validated in western culture and this very reason also acts as a limitation as the number of accessible valid studies were less. The current findings also pave the path to the upliftment of the millennial employees who face ill-treatment since most of the higher authorities are still following the traditional method while dealing with their employees. This view must be changed and should focus on the generational differences present within the employees so that a better understanding of the millennials can be attained and can foster organizational development.

It also intends to magnify the impact of culture on the millennials and also signifies the importance of understanding what millennials expect based on their work environment. Those factors revealed in this systematic review paper can be used for various other researches on planning how to maintain millennials at work, human resource managers to organize various career development programmes based on the needs of the millennials also and ultimately how the above proposed model would enhance organizational sustainability.

17.6 Conclusion

In order to conclude, the findings about the millennials can also be viewed from various motivational theories. Considering McClelland's theory of needs, millennials can be classified into those group who prefer achievement motivation than affiliation and power as they are more task-oriented and focus on achieving greater goals with an expectation of appreciation or feedback. Based on incentive theory, millennials expect what the theory states as bonus, praise, promotion, salary hike, and most importantly paid vacation or off time as they value work-life balance and flexibility more (Gayathri and Karthikeyan 2013). Improving certain policies and norms inside an organization would enhance transparency which may lead to increased employee retention. India has also taken up a strategy known as *"Jugaad"* meaning a flexible method of problem-solving that creatively makes use of constrained resources and has a large moderating effect on knowledge management in Indian firms which would further boost employee satisfaction and organizational performance (Chatterjee et al. 2022) which then would foster organizational sustainability.

References

Aboobaker N, Edward M, Zakkariya KA (2020) Workplace spirituality and employee loyalty: an empirical investigation among millennials in India. J Asia Business Stud

Anderson CA, Bushman BJ (2018) Media violence and the general aggression model. J Soc Issues 74(2):386–413

Anitha J, Aruna M (2016) Enablers of employee engagement of Gen Y at the workplace with reference to automobile sector. Amity J Training Develop 1(1):93–108

Arora P, Kshatriya KP (2017) Millennials: the new generation with high employment expectations. Int J Adv Res Comp Sci Manage Stud 5(8):101–109

Assoratgoon W, Kantabutra S (2023) Toward a sustainability organizational culture model. J Clean Prod, 136666

Bansal N (2017) Motivation & attitude of generation y in India: an exploratory study. Indian J Indust Relat 53 (1):102–114

Bashir B, Gani A (2020) Testing the effects of job satisfaction on organizational commitment. J Manag Develop 39(4):525–542

Bass D, Valdez ED (2020) Promising approaches to workforce development in Texas. Texas Public Policy Foundation

Biswas S, Bhatnagar J (2013) Mediator analysis of employee engagement: role of perceived organizational support, PO fit, organizational commitment and job satisfaction. Vikalpa 38(1):27–40

Briscoe JP, Henagan SC, Burton JP, Murphy WM (2012) Coping with an insecure employment environment: the differing roles of protean and boundaryless career orientations. J Vocat Behav 80:308–316

Buddhapriya S (2017) Mentoring experiences: a study of Indian public sector professionals. Indian J Indust Relat 52(4):689–705

Bugalia P, Lamba N, Yadav S, Joshi M (2022) Relationship between workplace ostracism and millennials' job performance. Indian Food Process Indust. Available at SSRN 4039690

Castillo FA, Hallinger P (2018) Systematic review of research on educational leadership and management in Latin America, 1991–2017. Educat Manage Admin Leader 46(2):207–225

Chadda RK, Deb KS (2013) Indian family systems, collectivistic society and psychotherapy. Indian J Psychiat 55(Suppl 2):S299

Chanda U, Goyal P (2020) A Bayesian network model on the interlinkage between Socially Responsible HRM, employee satisfaction, employee commitment and organizational performance. J Manage Anal 7 (1):105–138

Chatterjee S, Dey AK, Chaturvedi H (2021) Effect of mentoring on job performance among Indian millennials: a quantitative study. Int J Evid Based Coach Mentor 19(1)

Chatterjee S, Chaudhuri R, Vrontis D (2022) Managing knowledge in Indian Organizations: an empirical investigation to examine the moderating role of Jugaad. J Bus Res 141:26–39

Chavadi CA, Sirothiya M et al (2022) Mediating role of job satisfaction on turnover intentions and job mismatch among millennial employees in Bengaluru. Bus Perspect Res 10(1):79–100

Chhabra B (2013) Locus of control as A mode tor in the relationship between job satisfaction and organizational commitment: a study of Indian IT professionals. Organ Markets Emerg Econ 4(08):25–41

Chordiya R, Sabharwal M, Goodman D (2017) Affective organizational commitment and job satisfaction: a cross-national comparative study. Public Admin 95 (1):178–195

Dawra S, Chand PK, Aggarwal A (2022) Leader member exchange, nepotism, and employee loyalty as the determinants of organizational sustainability in small and medium enterprises in India. Int J Sociotech Knowled Develop (IJSKD) 14(1):1–21

Dixit V, Bhati M (2012) A study about employee commitment and its impact on sustained productivity in Indian auto-component industry. European J Bus Soc Sci 1(6):34–51

Dokadia A, Rai S, Chawla D (2015) Multigenerational differences in work attributes & motivation: An empirical study. Indian J Indust Relat, 81–96

Fatima M, Izhar Z, Kazmi ZA (2020) Organizational justice and employee sustainability: the mediating role of organizational commitment. SEISENSE J Manag 3 (3):12–22

Fok LY, Morgan YC, Zee S (2021) The linkages between commitment to sustainability, organizational culture, quality management, and organizational performance. Int J Appl Manag Tech 20(1):6

Gaikwad S, Swaminathan L (2020). Impact of work engagement on job performance among millennials working in public and private banks. Int J Manage (IJM) 11(12)

Garg K, Dar IA, Mishra M (2018) Job satisfaction and work engagement: a study using private sector bank managers. Adv Dev Hum Resour 20(1):58–71

Gayathri N, Karthikeyan P (2013) Work life balance in India—a social responsibility or a competitive tool. Int Res J Bus Manag 1:103–109

Gulyani G, Bhatnagar J (2017) Mediator analysis of passion for work in Indian millennials: relationship between protean career attitude and proactive work behavior. Career Develop Int

Hall DT, Yip J, Doiron K (2018) Protean careers at work: Self-direction and values orientation in psychological success. Annu Rev Organ Psych Organ Behav 5:129–156

Hershatter A, Epstein M (2010) Millennials and the world of work: an organization and management perspective. J Bus Psychol 25(2):211–223

Indhira N, Shani N (2016) Happiness and fatigue among millennials in selected information technology organizations in India. Asian J Res Soc Sci Human 6 (7):390–399

Jahanshahi AA, Adiguzel Z, Sonmez Cakir F (2022) Managerial support for innovation as the source of corporate sustainability and innovative performance: empirical evidence from Turkey. J Public Aff 22(2): e2428

Jha N, Sareen P, Potnuru RKG (2018) Employee engagement for millennials: considering technology as an enabler. Develop Learn Organ: An Int J

Jyoti J, Sharma P (2017) Empirical investigation of a moderating and mediating variable in between mentoring and job performance: a structural model. Revista De Psicología Del Trabajo y De Las Organizaciones 33(1):55–67. https://doi.org/10.1016/j.rpto.2017.01.002

Kaushal P, Vashisht S (2021) Protean career orientation, career decision self-efficacy and career outcomes of millennial IT professionals. Manage Labour Stud 46 (3):248–262

Khanolkar RS (2013) Influence of employer branding on satisfaction and commitment of generation Y employees. IOSR J Bus Manag

Koorella G, Perumal R (2019) A study on factors affecting employee job satisfaction. IJRAR-Int J Res Anal Rev 6(1):848–854

Kwantes CT (2009) Culture, job satisfaction and organizational commitment in India and the United States. J Indian Bus Res 1(4):196–212

Ludviga I (2020) Workplace expectations versus reality: are millennials so different? In: Eurasian business perspectives. Springer, Cham, pp 39–51

Mahmoud AB, Fuxman L, Mohr I, Reisel WD, Grigoriou N (2020) We aren't your reincarnation!" workplace motivation across X, Y and Z generations. International J Manpow

Malik S, Khera SN (2014) New generation–great expectations: Exploring the work attributes of Gen Y. Global J Fin Manage 6(5):433–438

Mesmer-Magnus J, Viswesvaran C, Wiernik BM (2012) The role of commitment in bridging the gap between organizational sustainability and environmental sustainability. In: Jackson SE, Ones DS, Dilchert S (eds) Managing human resources for environmental sustainability. Jossey-Bass/Wiley, pp 155–186

Mitra K (2011) Bringing up father. Business Today, December 25

Motiram S, Singh A (2012) "How close does the apple fall to the tree?" Some evidence on intergenerational occupational mobility from India

Muslimat A, Ab Wahid H, Erlangga H (2020) Effect of organizational commitment on the sustainability performance of Indonesian industries. PalArch's J Archaeol Egypt/Egyptology 17(6):8330–8347

Myers KK, Sadaghiani K (2010) Millennials in the workplace: a communication perspective on millennials' organizational relationships and performance. J Bus Psychol 25:225–238

Nahak M, Ellitan L (2022) Organizational commitment and organizational sustainability

Naim MF (2022) How does empowering leadership contribute to organizational commitment of millennials? an Indian perspective. In: Research anthology on changing dynamics of diversity and safety in the workforce. IGI Global, pp 150–163

Naim MF, Lenka U (2017) How does mentoring contribute to Gen Y employees' intention to stay? an Indian perspective. Eur J Psychol 13(2):314

Nayak T, Sahoo CK (2015) Quality of work life and organizational performance: the mediating role of employee commitment. J Health Manag 17(3):263–273

Ng ESW, Schweitzer L, Lyons ST (2010) New generation, great expectations: a field study of the millennial generation. J Bus Psychol 25:281–292

Prakash K, Tiwari P, Jain VK (2021) Exploring workplace expectations: an empirical study on Millennials of India. World Rev Entrepreneur, Manag Sustain Develop 17(2–3):124–141

Rao KS, Pavani B (2015) Architecting leadership styles to meet millennial aspirations of organizations: a study in Indian context. In: Architecting leadership styles to meet millennial aspirations of organizations: a study in Indian context (June 1, 2015). Academicia 5(6)

Rosati F, Faria LG (2019) Addressing the SDGs in sustainability reports: the relationship with institutional factors. J Clean Prod 215:1312–1326

Saxena P, Jain R (2012) Managing career aspirations of generation y at work place. Int J Adv Res Comp Sci Soft Eng 2(7):114–118

Sen C, Hooja HR (2015) Developing Indian human resource: Role of psychological capital and protean career orientation. Indian J Positive Psychol 6(3):311

Shankar V, Venkatesh A, Hofacker C, Naik P (2010) Mobile marketing in the retailing environment: current insights and future research avenues. J Interact Mark 24(2):111–120

Sharma D (2019) How to recruit, incentivize and retain millennials. Sage Publications Pvt, New Delhi

Sharma AS, Chully AA (2020) A study on the Job expectations of the Millennial Generation in the Indian context. J Xi'an Univ Architect Tech 12

Singh P, Bhandarker A, Rai S (2012) Millennials and the workplace: challenges for architecting the organizations of tomorrow. SAGE Publications India, New Delhi

Sinha S, Ghosh K (2022) Organizational sustainability and performance improvement in microfinance institutions (MFIs): managerial insights of what, why and how. Social Respons J 18(2):240–265

Srinivasan V (2012) Multi generations in the workforce: building collaboration. IIMB Manag Rev 24(1):48–66

Srivastava S, Jomon MG (2013) Mentoring & performance: implications for business organizations. Indian J Indust Relat 48(4):711–725

Stewart JS, Oliver EG, Cravens KS, Oishi S (2017) Managing millennials: embracing generational differences. Bus Horiz 60(1):45–54

Strauss W, Howe N (1991) Generations: the history of America's future, 1584 to 2069, vol 538. Quill, New York

Telefonica (2013) Connected, yet divided: telefonica survey of the millennial generation reveals digital natives are optimistic about their individual futures, June 4. Accessed at http://blog.digital.telefonica.com/?press-release=telefonica-millennial-survey-findings

Thangavel P, Pathak P, Chandra B (2021) Millennials and Generation Z: a generational cohort analysis of Indian consumers. Benchmarking: An Int J

Thompson C, Gregory JB (2012) Managing millennials: a framework for improving attraction, motivation, and retention. Psychologist-Manager J 15(4):237–246

Varma C (2017) Importance of employee motivation & job satisfaction for organizational performance. Int J Soc Sci Interdiscipl Res 6(2)

Yadav R, Chaudhari DS (2020) Work values influencing career choice in non-life insurance PSUs: a study of Indian millennials. Bimaquest 20(2). Retrieved from http://bimaquest.niapune.org.in/index.php/bimaquest/article/view/73

Yu MC, Mai Q, Tsai SB, Dai Y (2018) An empirical study on the organizational trust, employee-organization relationship and innovative behavior from the integrated perspective of social exchange and organizational sustainability. Sustainability 10(3):864

Environmental Application of Digital Twins: A Review

18

Yamini Sahu and Vaibhav Kumar Upadhyay

Abstract

A digital twin serves as a virtual replica of a real-life object, faithfully mirroring its behavior and status in a digital environment throughout its lifespan. Initially gaining popularity in manufacturing and industrial processes, digital twin technology has now transcended industries and emerged as a valuable asset in diverse sectors, including environmental management and sustainability. This cutting-edge technology acts as a potent tool for facilitating simulations and real-time data synchronization between physical entities and their virtual counterparts, making it highly applicable to a wide array of environmental applications. The research papers presented in the list delve into various aspects of digital twin technology, shedding light on its multifaceted roles and contributions. Topics covered encompass sustainable urban road planning, where digital twins prove instrumental in optimizing urban infrastructure and creating eco-friendly transportation systems. Additionally, the studies explore city digital twin technologies, illuminating their potential for promoting sustainability in urban design and development and fostering citizen-centric smart cities that prioritize the needs of their inhabitants. Insights into the integration of digital twins with federated learning in smart cities underscore the importance of data-driven decision-making and resource optimization. Disaster management emerges as another critical domain, where digital twins play a pivotal role in enhancing emergency response strategies through the integration of artificial and human intelligence. The significance of digital twins extends to airport environments, enabling resilient disaster management responses to safeguard travelers and infrastructure. Furthermore, novel digital twin architectures, coupled with synchronization-enhanced deep learning, empower early flood risk predictions, bolstering city climate resilience planning. The application of digital twins in agriculture and railway bridge maintenance addresses essential challenges and opportunities for enhancing environmental sustainability and resource management.

Keyword

Digital twin · Sustainable urban planning · Climate resilience planning · Environmental sustainability

Y. Sahu (✉) · V. K. Upadhyay
Department of Data Science and Engineering, IISER, Bhopal, India
e-mail: yamini21@iiserb.ac.in

S. Talukdar et al. (eds.), *Environmental Risk and Resilience in the Changing World*, GIScience and Geo-environmental Modelling, https://doi.org/10.1007/978-3-031-62442-1_18

18.1 Introduction

In the era of the digital age, various concepts and technologies are reshaping the built environment, making construction, homes, and cities smarter to cater to users' convenience. Digital technologies serve as powerful tools, driving productivity and performance enhancements in projects and the built environment. (Manzoor et al. 2021). These technologies comprise the Internet of Things (IoT), Building Information Modeling (BIM), augmented reality/ virtual reality (AR/VR), mixed reality (MR), artificial intelligence (AI), and digital twin (DT). DT is among the leading technologies that can digitally depict physical entities and establish a bidirectional connection between them, enabling functions such as information exchange, visualization, monitoring, and management (El Saddik 2018; Madni et al. 2019). DT integrates multiple technologies, including BIM and IoT, to generate digital entities and facilitate the transfer of information from the physical building to the digital model. The capabilities of DT vary based on the integrated technologies and the specific purposes it is intended to fulfill.

The DT is recognized as a computer-generated replica of a physical system that simulates various scenarios by continuously synchronizing real-time data from the field. The primary essence of DT closely relates to simulation techniques and the corresponding software utilized (Phanden et al. 2021). On the other hand, DT revolves around establishing a connection between a physical system and its virtual representation through bidirectional communication, which may or may not involve human participation (2023).

DT has been applied in several fields, such as health care, manufacturing, aerospace, agriculture, urban planning, environmental sustainability, and smart city (Phanden et al. 2021; Tzachor et al. 2022). DTs have emerged as versatile technology with a broad spectrum of applications across diverse industries. In manufacturing (Kritzinger et al. 2018), they create virtual replicas of physical assets and production lines, facilitating real-time monitoring and predictive maintenance. In the aerospace (Phanden et al. 2021) and aviation sectors, DTs help analyze aircraft systems and optimize fuel efficiency for safer and more efficient flights. Smart cities (Hämäläinen 2020) leverage DTs to create virtual models of entire urban areas, aiding in urban planning and resource management. Health care (Alazab et al. 2022) utilizes DTs for personalized medicine, preoperative planning, and drug development, enhancing patient care. Energy and utilities (Clausen et al. 2021) benefit from DTs to optimize power generation and improve grid resilience. Construction and architecture (Opoku et al. 2021) rely on DTs for efficient planning and design. In agriculture (Pylianidis et al. 2021), DTs enable precision farming techniques for increased crop yields. Defense (de Hoz Diego et al. 2022) applications employ DTs for mission planning and training simulations. The retail industry (Freitag et al. 2021) utilizes DTs to personalize customer experiences. With its versatility and impact on various sectors, DT technology continues to drive advancements in efficiency, sustainability, and overall performance.

The utilization of DTs in various fields serves a range of objectives tailored to specific industries and applications. One common goal is optimization and efficiency, as DTs offer real-time data and insights to streamline processes, minimize downtime, and enhance overall performance. Another key objective is predictive maintenance, where DTs monitor physical assets to detect potential issues in advance, reducing maintenance costs and unplanned disruptions. Moreover, DTs empower decision-makers with virtual representations, facilitating scenario simulations and informed choices. In fields like health care and manufacturing, digital twins enable personalized solutions, treatments, and products by creating virtual models based on individual characteristics.

18.2 Application of Digital Twins

DTs have found a multitude of environmental applications, playing a pivotal role in promoting sustainable practices and enhancing the management of natural resources. These applications span a wide spectrum of fields, each contributing to environmental conservation and protection. Sustainable urban planning is facilitated by DT, enabling city planners to create virtual replicas of urban areas and optimize infrastructure development for energy efficiency and eco-friendliness. Water resource management benefits from DT, monitoring and managing water systems to make informed decisions for water conservation and equitable distribution. Technology extends its reach to fostering energy efficiency in buildings and industries by simulating energy consumption patterns and identifying areas for improvement, thereby reducing carbon footprints. In the fight against climate change, DTs are instrumental in assessing the impact on ecosystems and human settlements, empowering policymakers to develop strategies for mitigation and adaptation. Waste management processes benefit from DT, optimizing waste collection and recycling through predictive models that reduce environmental impact and enhance resource recovery. By analyzing air quality data from sensors and satellite imagery, DT enables authorities to visualize air pollution patterns and identify pollution sources for effective interventions and improved public health. Finally, DT contributes to natural disaster preparedness, simulating potential scenarios like floods, and wildfires allowing for better emergency response and minimizing the impact on communities and the environment.

18.2.1 Sustainable Urban Planning Using DT

The application of digital twin technology in Sustainable Urban Planning presents a transformative approach to address the complex challenges of modern cities. By creating virtual replicas of urban environments, DTs enable urban planners and policymakers to harness the power of data and simulations for more informed decision-making. Real-time data from various sources, such as IoT devices and sensors, can be integrated into the virtual model, providing a comprehensive and up-to-date understanding of the city's dynamics. DT finds applications in various fields to promote sustainability, including smart city development and sustainable road planning, among others.

The importance of sustainable urban road planning and the challenges posed by urbanization, such as traffic congestion, limited space for construction, and environmental issues. It emphasizes the need to consider various demands and constraints while proposing functional, economical, people-friendly, and eco-friendly road schemes. The use of multi-criteria decision-making (MCDM) and Geographical Information Systems (GIS) in urban road planning is mentioned, along with the increasing importance of DT in providing solutions to interpret and analyze data from the physical world. The study's objective is to present a sustainable urban road planning approach using DT, MCDM, and GIS to assist planners in considering multiple factors for both new road construction and existing road widening. The factors considered in the study include building demolition, traffic congestion, driving route selection habits, air quality, and noise.

Smarting the city is a highly intricate process, primarily due to the inherent complexity of urban environments. Unlike automated systems, cities are living entities that constantly evolve through variations in physical structures, economic activities, political dynamics, social interactions, and ecological systems. The concept of smart cities has evolved from merely controlling physical growth to encompassing a wide array of elements, including social and knowledge infrastructures. While some critics view it as ambiguous and question its ability to address urban and social challenges like inequality, social polarization, and gentrification, substantial research indicates the prospective benefits of smart cities in terms of sustainability, efficiency, optimization, collaboration, and innovation.

The ongoing global challenge posed by the COVID-19 pandemic has amplified the urgency for smart city initiatives and digitalization of services. With critical domains such as education, transportation, and entertainment severely affected, the relevance and significance of smarting the city have intensified (Fig. 18.1 and Table 18.1).

DTs enable urban planners to model and visualize the impacts of proposed changes, such as new infrastructure projects, land use plans, or transportation systems, before implementing them in the real world. This simulation-based approach helps identify potential issues and optimize designs for maximum efficiency and minimal environmental impact. Additionally, DTs support data-driven decision-making by integrating information from various sources, allowing planners to consider social, economic, and environmental factors simultaneously.

Moreover, DTs promote citizen engagement and participatory planning. By sharing DT models with the public, urban planners can involve citizens in the decision-making process, gather feedback, and address concerns, leading to more inclusive and sustainable urban development.

18.2.2 Environmental Sustainable Using DT

Environmental sustainability using DTs refers to the application of DT technology to support and enhance environmentally friendly practices, resource efficiency, and conservation efforts. DT, which are virtual representation of physical entities, are employed to model and simulate various environmental aspects, such as energy consumption, waste management, air quality, and water usage, in real time or near real time (Fig. 18.2).

Using DTs for environmental sustainability allows organizations and cities to monitor, analyze, and optimize their environmental performance. For example, in the context of buildings, DTs can help monitor energy usage, identify energy inefficiencies, and propose energy-saving strategies. In urban areas, DTs can simulate the impact of new infrastructure projects on the environment, optimize transportation systems to reduce emissions and congestion, and plan for sustainable land use (Table 18.2).

The above mentions various DT frameworks and architectures developed for specific applications, such as global wildfire prediction, early flood risk predictions for climate resilience planning, and disaster management systems. These DTs utilize reduced-order modeling, deep learning, and similarity-based hybrid modeling techniques to enhance their capabilities. Additionally, there is a specific mention of airport digital twins focused on disaster management response.

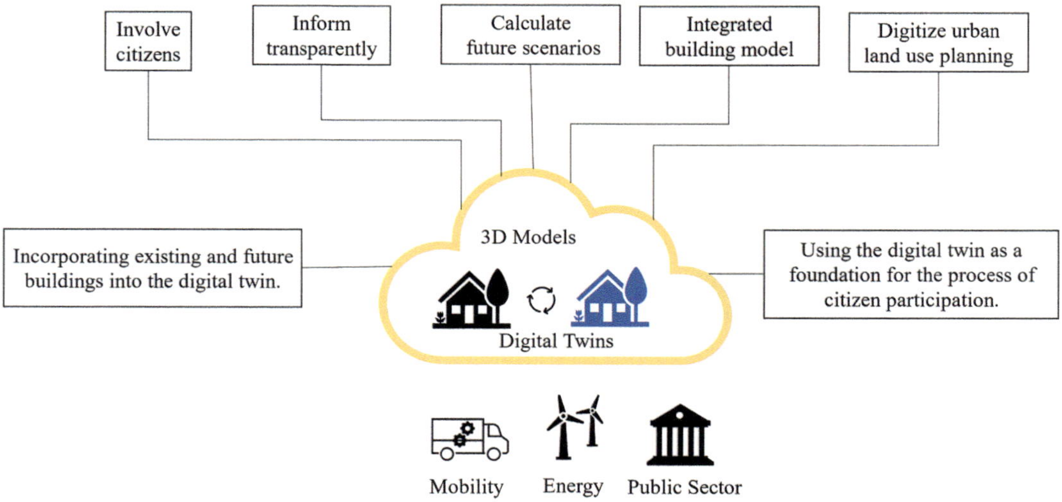

Fig. 18.1 DT in the field of sustainable urban planning

Table 18.1 Review on DT in sustainable urban planning

Authors	Summary
Jiang et al. (2022)	It integrates digital twin technology, multi-criteria decision-making (MCDM), and geographic information system (GIS) to consider factors like building demolition, traffic congestion, air quality, and noise, resulting in functional and eco-friendly road schemes for new construction and existing road widening to alleviate traffic congestion
Shahat et al. (2021)	This paper focuses on the complexity of smart city development and the emergence of DT technology. It highlights the potential benefits of DTs in enhancing city management and sustainability, particularly in the context of the ongoing COVID-19 pandemic. The paper aims to clarify the challenges and requirements of creating a fully mirrored city DT, proposing a research agenda to maximize its utilization and provide valuable insights for informed decision-making by politicians and city managers
Xia et al. (2022)	The analysis emphasizes the attention received by GIS and BIM integration but acknowledges challenges related to professional disconnect and fragmented composition. Future research should concentrate on smart city planning, updating, and management, as well as ontology-based integration platforms and collaborative management of urban rail transportation engineering
Dembski et al. (2020)	The development of an urban DT for the town of Herrenberg in Germany is highlighted, which allows seamless visualization across different scales and data categories in virtual and augmented reality. The DT involves the integration of various urban data, including social data collected from citizens, fostering collaboration and participation in urban planning and decision-making processes. While the DT is an abstraction of reality, it offers great potential for scenario development and testing, contributing to consensus-building and better-informed decision-making among diverse stakeholders
Pandya et al. (2023)	The paper discusses the integration of federated learning (FL) with smart city applications to address issues like data leakage and privacy concerns. It provides an overview of current and future developments of FL for smart cities, including its applications in smart transportation, health care, governance, disaster management, industries, and UAV monitoring. The paper also outlines research challenges and prospects for the future development of FL in smart cities
Deng et al. (2021)	The Digital Twin City (DTC) is emerging as a transformative model for building smart cities, revolutionizing city governance structures and propelling urban development. Numerous important cities worldwide are already embracing plans for constructing DT cities, driven by the rapid advancements in digital twin technologies. This paper presents a DT pattern that incorporates self-perceiving, self-determining, self-organizing, self-executing, and adaptive capabilities to facilitate urban operation and maintenance. Through surveying, IoT perception, collaborative computing, simulation, and deep learning, the DTC platform enables efficient management, historical event tracking, and exploration of new frontiers. The development of DTCs is paving the way for innovative city management paradigms and will likely shape future research in this domain
Ford and Wolf (2020)	Utilizing digital twins for smart cities (SCDT) involves integrating sensing and simulation across various infrastructure systems for effective community management. Disaster management within communities serves as a crucial foundation for advancing SCDT. This study introduces and examines a conceptual model of SCDT tailored for disaster management, addressing potential threats to its development. Emphasizing information loops over individual components emerges as a vital aspect for future SCDT progress. Key contributions encompass support for SCDT development in disaster management, a conceptual SCDT model for disaster management, and a discussion of essential considerations during SCDT development and deployment for disaster management

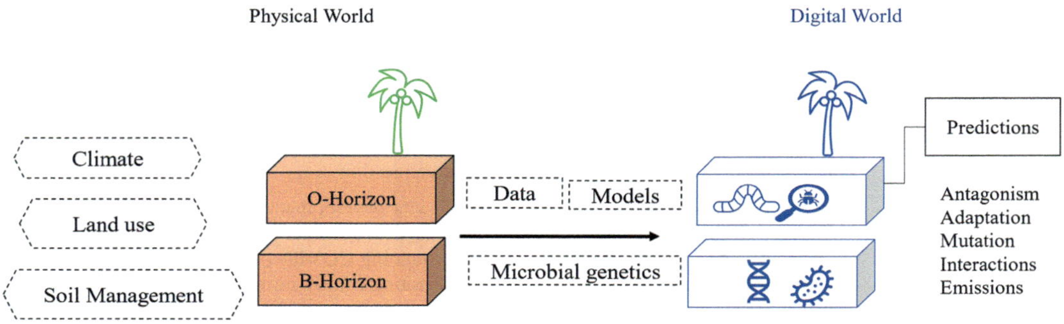

Fig. 18.2 DT in the field of environment management

Table 18.2 Review on DT in environment management

Authors	Summary
Fan et al. (2021)	This paper outlines a vision for a Disaster City DT paradigm, aiming to converge crisis informatics and information technology in disaster management. It proposes integrating artificial intelligence (AI) algorithms to enhance situation assessment, decision-making, and coordination among stakeholders, providing increased visibility into complex disaster management and humanitarian actions. The paradigm comprises four main components: multi-data sensing for data collection, data integration and analytics, multi-actor game-theoretic decision-making, and dynamic network analysis
Agapaki et al. (2022)	Airports encounter a range of hazards and challenges, including natural disasters and cybersecurity attacks, necessitating operational decisions amid unpredictable circumstances. We present the idea of a foundational twin, acting as a robust data platform that integrates diverse data sources and facilitates interaction among various twins. Our emphasis is on supplying data sources and metrics for each foundational twin, particularly focusing on the environmental airport twin designed for major US airports
Yun et al. (2022)	This study addresses the need for accurate disaster monitoring and prediction services in disaster management systems. It proposes a DT architecture with a similarity-based hybrid modeling scheme to enhance prediction accuracy. The hybrid model combines physics-based and data-driven disaster models to compensate for errors and reduce data dependency. Evaluations in wildfire scenarios demonstrate that the DT significantly decreases prediction errors compared to existing schemes, offering more reliable disaster predictions to optimize the allocation of disaster response resources
Ghaith et al. (2022)	Floods have long been costly hazards worldwide, and their frequency and impacts are expected to worsen with climate change, particularly in cities. DT technologies offer decision-makers valuable tools to assess city resilience against projected floods. However, developing DTs for flood predictions is challenging due to time-consuming and uncertain processes involved in creating hydrologic and hydraulic models. This study presents a flood prediction methodology (FPM) that integrates synchronization analysis and deep learning to directly simulate rainfall-flood relationships, bypassing the computationally expensive models. The FPM serves as the core of data-driven DTs that provide rapid early warnings and vulnerability information, aiding city resilience planning under various climate change scenarios using real-time sensor data. The City of Calgary is used as a demonstration case
Huang et al. (2022)	The Earth System Digital Twin (ESDT) is a dynamic digital representation of Earth systems, integrating models and observation data with AI and visualization tools. It allows users to explore the current state of the Earth system, predict future conditions, and simulate scenarios under different assumptions. NASA's IDEAS project aims to establish an extensible architecture for developing DTs of our physical environment for Earth Science. IDEAS collaborates with the SCO's FloodDAM project and CNES to focus on floods detection, prediction, and impacts, validating and demonstrating the IDEAS architecture

(continued)

Table 18.2 (continued)

Authors	Summary
Hyeong-su et al. (2019)	This paper introduces a novel wildfire DT framework aimed at predicting wildfire spreads more accurately. Wildfire simulations can help reduce damage, but current simulations have limitations due to the complexity of dynamically changing environmental parameters. The proposed framework integrates an interactive wildfire spread simulator with sensing data from an IoT server to create a DT for wildfire prediction
Zhong et al. (2023)	This paper discusses the use of a deep learning-based DT fire model, called JULES-INFERNO, to improve the efficiency of global wildfire predictions. The model integrates reduced-order modeling (ROM) and deep learning prediction networks to handle large amounts of data and reduce computational costs. Latent data assimilation (LA) is applied to adjust prediction results and ensure accuracy and stability. The proposed DT achieves accurate surrogate predictions and runs significantly faster than the original JULES-INFERNO model, making it a promising tool for wildfire prediction without the need for high-performance computing (HPC) clusters
Pylianidis et al. (2021)	This paper explores the adoption of DT in the agriculture industry. It emphasizes the transformative potential of DTs, offering enhanced control and management of complex systems through integrated technologies. The study conducts a mixed-method investigation, analyzing 28 DT use cases in agriculture from 2017 to 2020, comparing them with other disciplines. The findings highlight the benefits and potential of digital twins in agriculture and propose an application-based roadmap for wider adoption, outlining examples of increasing complexity
Yun et al. (2022)	This paper discusses the potential of the DT in agriculture by integrating high-fidelity modeling and bidirectional data streams, enabling real-time virtual representation. By overcoming decision-making limitations and enhancing automation in agriculture, the DT offers a unique opportunity for digitization in the industry
Purcell et al. (2023)	This paper discusses the potential of DT technology in addressing interconnected challenges like food security, land degradation, climate change, and population growth for sustainable agriculture. DTs can provide increased information availability to empower stakeholders in pursuing sustainable production methods. However, the review also emphasizes the need to assess and mitigate potential negative technical and social-ecological effects of DT implementation. The study proposes a high-level roadmap to support the open and sustainable development of DTs in agriculture, outlining necessary milestones to achieve sustainable objectives
Kaewunruen et al. (2023)	To optimize railway construction and maintenance, engineers require innovative DTs that enable visualization, information sharing, and condition monitoring during operation. Building Information Modeling (BIM) is an effective approach for creating and managing a 3D model with digital information, facilitating project management, monitoring, and operation throughout the asset's entire life cycle. The application of BIM can enhance cost management, reduce project delivery time, and improve the sustainability and resilience of railway bridges. In this study, a novel DT was developed using BIM integration for the Minnamurra Railway Bridge (MRB) in Australia, showcasing its extended application in enhancing operation, maintenance, and asset management. The DT was also utilized to determine greenhouse gas (GHG) emissions and cost consumption, promoting virtual collaboration and value co-creation among stakeholders, resulting in reduced costs and GHG emissions

18.3 Conclusion

The application of DT in environmental settings has proven to be a transformative and versatile approach with far-reaching benefits. Across various domains, including smart cities, disaster management, climate change, agriculture, and environmental sustainability, DTs have showcased their potential to revolutionize how we address complex environmental challenges.

In smart cities, DTs offer a dynamic and interactive platform to optimize urban planning, infrastructure management, and resource allocation. They empower city officials and stakeholders to make data-driven decisions, leading to more efficient, resilient, and sustainable urban environments. In disaster management, DTs play a crucial role in early warning systems, situational awareness, and resource allocation during floods and wildfires. By providing real-time virtual representations, they enable effective response strategies, minimizing damage and enhancing public safety. When it comes to climate change, DTs facilitate precise impact assessments, enabling policymakers and researchers to devise effective mitigation and adaptation strategies. They help identify trends, model scenarios, and support evidence-based policies to address this global challenge. In agriculture, DTs empower farmers with tools for precision farming, optimizing resource utilization, increasing productivity, and promoting sustainable agricultural practices. They contribute to food security while minimizing the environmental footprint of farming activities.

Moreover, DTs are instrumental in achieving environmental sustainability goals. By integrating data from various sources and employing advanced modeling techniques, they provide valuable insights into natural resource management, conservation efforts, and ecosystem health.

References

Agapaki E (2022) Airport digital twins for resilient disaster management response. In: Simos DE, Rasskazova VA, Archetti F et al (eds) Learning and intelligent optimization. Springer International Publishing, Cham, pp 467–486

Alazab M, Khan LU, Koppu S, et al (2022) Digital twins for healthcare 4.0—recent advances, architecture, and open challenges. IEEE Cons Elect Mag, 1–8. https://doi.org/10.1109/MCE.2022.3208986

Al-Sehrawy R, Kumar B, Watson R (2023) The pluralism of digital twins for urban management: bridging theory and practice. J Urban Manag 12:16–32. https://doi.org/10.1016/j.jum.2023.01.002

Clausen A, Arendt K, Johansen A et al (2021) A digital twin framework for improving energy efficiency and occupant comfort in public and commercial buildings. Energy Inform 4:40. https://doi.org/10.1186/s42162-021-00153-9

de Hoz Diego JD, Temperekidis A, Katsaros P, Konstantinou C (2022) An IoT digital twin for cybersecurity defence based on runtime verification. In: Margaria T, Steffen B (eds) Leveraging applications of formal methods, verification and validation. Verification principles. Springer International Publishing, Cham, pp 556–574

Dembski F, Wössner U, Letzgus M et al (2020) Urban digital twins for smart cities and citizens: the case study of Herrenberg, Germany. Sustain 12:2307. https://doi.org/10.3390/su12062307

Deng T, Zhang K, Shen Z-J (2021) A systematic review of a digital twin city: a new pattern of urban governance toward smart cities. J Manag Sci Eng 6:125–134. https://doi.org/10.1016/j.jmse.2021.03.003

El Saddik A (2018) Digital twins: the convergence of multimedia technologies. IEEE Multimed 25:87–92. https://doi.org/10.1109/MMUL.2018.023121167

Fan C, Zhang C, Yahja A, Mostafavi A (2021) Disaster city digital twin: a vision for integrating artificial and human intelligence for disaster management. Int J Inf Manage 56:102049. https://doi.org/10.1016/j.ijinfomgt.2019.102049

Ford DN, Wolf CM (2020) Smart cities with digital twin systems for disaster management. J Manag Eng 36:04020027. https://doi.org/10.1061/(ASCE)ME.1943-5479.0000779

Freitag M, Kotzab H, Megow N (eds) (2021) Dynamics in logistics: twenty-five years of interdisciplinary logistics research in Bremen, Germany. Springer International Publishing, Germany

Ghaith M, Yosri A, El-Dakhakhni W (2022) Synchronization-enhanced deep learning early flood risk predictions: the core of data-driven city digital twins for climate resilience planning. Water 14:3619. https://doi.org/10.3390/w14223619

Hämäläinen M (2020) Smart city development with digital twin technology. University of Maribor

Huang T, David C, Oadia C et al (2022) An earth system digital twin for flood prediction and analysis. In: IGARSS 2022–2022 IEEE International Geoscience and Remote Sensing Symposium, pp 4735–4738

Hyeong-su K, Jin-Woo K, Yun S, Kim W-T (2019) A novel wildfire digital-twin framework using interactive wildfire spread simulator. In: 2019 Eleventh International Conference on Ubiquitous and Future Networks (ICUFN), pp 636–638

Jiang F, Ma L, Broyd T et al (2022) Digital twin enabled sustainable urban road planning. Sustain Cities Soc 78:103645. https://doi.org/10.1016/j.scs.2021.103645

Kaewunruen S, AbdelHadi M, Kongpuang M et al (2023) Digital twins for managing railway bridge maintenance, resilience, and climate change adaptation. Sensors 23:252. https://doi.org/10.3390/s23010252

Kritzinger W, Karner M, Traar G et al (2018) Digital twin in manufacturing: a categorical literature review and classification. IFAC-PapersOnLine 51:1016–1022. https://doi.org/10.1016/j.ifacol.2018.08.474

Madni AM, Madni CC, Lucero SD (2019) Leveraging digital twin technology in model-based systems engineering. Systems 7:7. https://doi.org/10.3390/systems7010007

Manzoor B, Othman I, Pomares JC (2021) Digital technologies in the architecture, engineering and construction (AEC) industry—a bibliometric—qualitative literature review of research activities. Int J Environ Res Public Health 18:6135. https://doi.org/10.3390/ijerph18116135

Opoku D-GJ, Perera S, Osei-Kyei R, Rashidi M (2021) Digital twin application in the construction industry: a literature review. J Build Eng 40:102726. https://doi.org/10.1016/j.jobe.2021.102726

Pandya S, Srivastava G, Jhaveri R et al (2023) Federated learning for smart cities: a comprehensive survey. Sustainable Energy Technol Assess 55:102987. https://doi.org/10.1016/j.seta.2022.102987

Phanden RK, Sharma P, Dubey A (2021) A review on simulation in digital twin for aerospace, manufacturing and robotics. Mat Today: Proceed 38:174–178. https://doi.org/10.1016/j.matpr.2020.06.446

Purcell W, Neubauer T, Mallinger K (2023) Digital Twins in agriculture: challenges and opportunities for environmental sustainability. Current Opinion Environ Sustain 61:101252. https://doi.org/10.1016/j.cosust.2022.101252

Pylianidis C, Osinga S, Athanasiadis IN (2021) Introducing digital twins to agriculture. Comput Electron Agric 184:105942. https://doi.org/10.1016/j.compag.2020.105942

Shahat E, Hyun CT, Yeom C (2021) City digital twin potentials: a review and research agenda. Sustain 13:3386. https://doi.org/10.3390/su13063386

Tzachor A, Sabri S, Richards CE et al (2022) Potential and limitations of digital twins to achieve the sustainable development goals. Nat Sustain 5:822–829. https://doi.org/10.1038/s41893-022-00923-7

Xia H, Liu Z, Efremochkina M et al (2022) Study on city digital twin technologies for sustainable smart city design: a review and bibliometric analysis of geographic information system and building information modeling integration. Sustain Cities Soc 84:104009. https://doi.org/10.1016/j.scs.2022.104009

Yun S-J, Kwon J-W, Kim W-T (2022) A novel digital twin architecture with similarity-based hybrid modeling for supporting dependable disaster management systems. Sensors 22:4774. https://doi.org/10.3390/s22134774

Zhong C, Cheng S, Kasoar M, Arcucci R (2023) Reduced-order digital twin and latent data assimilation for global wildfire prediction. Nat Hazard 23:1755–1768. https://doi.org/10.5194/nhess-23-1755-2023

Analysis of Environmental Sustainability Through Corporate Social Responsibility

19

Pradip Ram and Roshan Kumar

Abstract

The study focuses on the identification of various efforts taken by Indian public sector steel companies in different dimensions of environmental protection. This study is based on the content analysis of the annual reports and sustainability reports of companies for the identification of various dimensions of their environmental sustainability. Moreover, various standards have been identified that guide the companies for effective sustainability practices and reporting. The study has found that Steel Authority of India Limited, RINL, and NMDC are the companies making more efforts to achieve environmental sustainability through their CSR practices.

Keywords

Corporate · Social responsibility · Sustainability · Steel sector · Public sector

P. Ram (✉) · R. Kumar
Department of Commerce and Business Studies, Central University of South Bihar, Gaya, India
e-mail: pradipram@cusb.ac.in

R. Kumar
e-mail: roshankumar02071996@gmail.com

19.1 Introduction

Environmental sustainability is a critical issue facing society, and corporate social responsibility (CSR) is one way that businesses can help address this challenge. Environmental sustainability is the use and management of natural resources in a responsible way so that it meets the needs of the present without hampering future generations. On the other hand, CSR is the idea that businesses benefit society as a whole beyond just maximizing profits for shareholders.

Numerous studies have shown that CSR can positively impact the environment. For example, a study by Saeed et al. (2018) found that CSR initiatives can lead to reductions in greenhouse gas emissions, energy efficiency improvements, and waste reduction. Similarly, a study by Lee and Lim (2017) found that CSR practices can help businesses reduce their environmental impact and improve their environmental performance. Corporate social responsibility (CSR) and environmental sustainability are two concepts that are closely related. Companies that take CSR seriously often focus on reducing their environmental impact and promoting sustainability. According to a United Nations Global Compact study, "environmental sustainability is a core element of CSR and is increasingly becoming a strategic imperative for businesses that want to create long-term value for their stakeholders" (UNGC 2019).

S. Talukdar et al. (eds.), *Environmental Risk and Resilience in the Changing World*, GIScience and Geo-environmental Modelling, https://doi.org/10.1007/978-3-031-62442-1_19

Companies that are committed to environmental sustainability can achieve a range of benefits, including reduced costs, enhanced reputation, and increased customer loyalty (Eccles and Serafeim 2013; Gao et al. 2015). Moreover, environmental sustainability initiatives can help companies comply with regulations, mitigate risks, and improve their overall business performance (KPMG 2019). There are many ways in which companies can promote environmental sustainability as part of their CSR strategy. For example, they can reduce their carbon footprint, promote energy efficiency, use renewable energy sources, minimize waste, and promote sustainable sourcing practices (UNGC 2019).

Sustainability reporting is a process of value reporting in which economic, environmental, and social performance is publicly communicated by an organization. It is a form of value reporting that aims to provide a comprehensive and impartial view of the sustainability performance of the reporting entity, taking into account both positive and negative contributions. However, sustainability reporting does not take into account the time aspect or the interconnectedness of different sustainability dimensions (Lozano 2011). Overall, environmental sustainability is an important issue facing society, and CSR is one way businesses can address this challenge. By implementing environmentally sustainable practices and investing in initiatives that reduce their environmental footprint, companies can contribute to a more sustainable future for both society and their business.

Several studies have examined the relationship between CSR and environmental sustainability. For instance, Zaman and Ali (2018) found that CSR activities can positively influence environmental sustainability by reducing waste, promoting eco-efficiency, and encouraging environmental innovations. Jabbour et al. (2019) found that companies with a strong commitment to CSR are more likely to adopt environmentally friendly practices such as reducing greenhouse gas emissions, conserving energy, and using renewable energy sources. However, some scholars argue that CSR practices may not always lead to environmental sustainability. Banerjee (2007) argues that CSR activities can be used as a form of greenwashing, where companies use CSR initiatives to enhance their reputation without addressing the root causes of environmental problems. Boiral and Gendron (2011) suggest that CSR activities may not be effective in promoting environmental sustainability if they are not integrated into the organization's core business strategy. Stakeholder pressure can influence the adoption of CSR practices that promote environmental sustainability. Montiel and Delgado-Ceballos (2016) found that companies more exposed to stakeholder pressure are more likely to adopt CSR practices promoting environmental sustainability. Other studies show that sustainability reporting is just an artificial and idealized representation that is disconnected from reality to some extent. Boiral (2013) disclosed the extent to which sustainability reporting can be viewed as a simulacrum used to camouflage real, sustainable development problems and project an idealized view of the firms' situations.

More importantly, how CSR contributes to the development of sustainability in a recent study, Mishra (2020) developed to understand the compatibility of CSR contribution to the identified activities under sustainable development goals and also studied the pattern based on the CSR contribution of several preferred Indian companies. Similarly, Lyon and Maxwell (2007) say market drivers of CSR will likely continue to grow in importance. Environmental issues that are complex require expensive remedies or require change across multiple firms, such as global warming, and political pressure is likely to remain a critical influence on CSR activities. According to Godha and Jain (2015), the applicability of GRI reporting in India is yet at a nascent stage. The cement, mining, and metal sectors are the most compliant among the companies reporting to GRI, with banking and financial services and the healthcare and pharmaceutical sectors being the least compliant.

The research focused on threefold objectives:

- Analyzing the content of environmental sustainability of public steel companies.
- Measurement and analysis of the activities conducted for environmentally sustainable behaviour of companies.
- Analyzing the extent of the content disclosed in the annual report of the company.

19.2 Methodology

The total number of public companies that fall under the steel sector is eight, and all companies' annual reports and sustainability reports have been used for the purpose of analysis.

Eight public sector steel companies have been selected as the sample unit for the study purpose, and these are SAIL, NMDC, RINL, MOIL, MSTC, KIOCL, MECON, and FSNL.

The methodology of the content analysis has been applied for the purpose of calculating the quantum of disclosure of mining companies in their annual report.

Content Analysis: The content analysis technique is defined as "A systematic research method for analyzing textual information in a standardized way that allows evaluators to make inferences about the information" (Krippendorf 1980). It involves codifying qualitative and quantitative information into predefined categories in order to derive patterns in the presentation and reporting information. Content analysis employs a three-step process (Raman 2006).

In the *first step*, an appropriate document is chosen. An appropriate document may be any written material like a book, magazine, annual report, manuscript, newspaper, etc., where the content is found related to the study. For this study, annual reports have been selected as the appropriate document. The *second step* is to determine the unit for measuring contents. The unit of measurement may be words, sentences, paragraphs, and pages. In this study a number of pages have been used as a unit of measurement. The *third step* involves the identification of themes or categories into which blocks of content can be classified. In this study, four themes, i.e. community development, human resources, physical resources, and environment contribution and product and service contribution, have been identified as major categories of the classification.

19.3 Results and Discussion

Table 19.1 is the analysis of environmental disclosure guidelines of eight public steel companies that reveal that SAIL, NMDC, and MOIL disclose their environmental issues as per GRI, ISO, CSR, and BRR guidelines, except for CREP. So, it has been ranked first among all eight companies. Next, RINL is the second-highest guidelines-following company, which

Table 19.1 Environmental guidelines disclosure of companies

Guidelines company ↓	GRI	ISO	CSR	CREP	BRR (NVG principle)	Rank
SAIL	✓	✓	✓	–	✓	1
NMDC	✓	✓	✓	–	✓	1
RINL		✓	✓	✓	–	2
MOIL	✓	✓	✓	–	✓	1
MSTC	–	✓	✓	–	✓	2
KIOCL	–	✓	✓	–	✓	2
MECON	–	✓	✓	–	–	3
FSNL	–	✓	✓	–	–	3

Sources Compiled from sustainability, environment, and annual report

Table 19.2 Certification number, principles, and standards of different companies

Company	GRI Standard 2016	ISO certificate	BRR (NVG principle)
SAIL	103-1,2,3; 301-1; 302-1 303-1,2,3; 304-1; 305-4 306-1,2,3	14,001	Principle-2,6
NMDC	103-1,2,3; 301-1,2,3;302-1,2,3,4,5 303-1,2,3;304-2,3,4;305 1,2,3,4,5,6,7;306-1,2,3,4,5	14,001	Principle-2,6
RINL	Nil	14,001	Nil
MOIL	Nil	14,001	Principle-2,6
MSTC	Nil	14,001	Principle-2,6
KIOCL	Nil	14,001	Principle-2,6
MECON	Nil	NIL	Nil
FSNL	Nil	14,001	Nil

Sources Compiled from Sustainability, Environment, and Annual Report

follows ISO, CSR, and CREP except for GRI and BRR, while MSTC and KIOCL are also the seconds which follow ISO, CSR, and BRR except for GRI and CREP. Finally, the third among the eight companies is MECON and FSNL, which follow only ISO and CSR.

The content analysis has been done in Table 19.2 of the eight companies, revealing that there are standards, certification numbers, and principles that are followed by the company. The company, as SAIL and NMDC, follows all three guidelines, i.e. GRI standard 2016, the ISO certificate, and BRR (NVG Principle). In GRI standard 2016, environmental topics followed by GRI 103 Management approach, 103-1 Explanation of the material topic and its boundary, 103-2 The management approach and its components, 103-3 Evaluation of the management approach in which GRI 301: explain Materials 2016, 301-1 Materials used by weight or volume, 301-2 Recycled input materials used, and 301-3 Reclaimed products and their packaging materials. GRI 302: explain Energy 2016, 302-1 Energy consumption within the organization, 302-2 Energy consumption outside of the organization; 302-3 Energy intensity 302-4 Reduction of energy consumption, and 302-5 Reductions in energy requirements of products and services. GRI 304: explain Water 2016, 304-1 Operational sites owned, leased, managed in, or adjacent to, protected areas and areas of high biodiversity value outside protected areas, 304-2 Significant impacts of activities, products, and services on biodiversity, 304-3 Habitats protected or restored, and 304-4 IUCN Red List species and national conservation list species with habitats in areas affected by operations. GRI 305: explain Emissions 2016, 305-1 Direct (Scope 1) GHG emissions, 305-2 Energy indirect (Scope 2) GHG emissions, 305-3 Other indirect (Scope 3) GHG emissions, 305-4 GHG emissions intensity, 305-5 Reduction of GHG emissions, 305-6 Emissions of ozone-depleting substances (ODS), and 305-7 Nitrogen oxides (NOX), sulphur oxides (SOX), and other significant air emissions. GRI 306: explain Effluents and Waste 2016, 306-1 Water discharge by quality and destination, 306-2 Waste by type and disposal method, 306- 3 Significant spills, 306-4 Transport of hazardous waste, and 306-5 Water bodies affected by water discharges and/or runoff. However, ISO certificate 14,001 explains to reduce their environmental footprint by adopting an effective environmental management system (EMS), and BRR (NVG Principles) explains Principles 2, which supports that businesses should provide goods and services in a manner that is sustainable and safe and Principles-6 is businesses should respect and make efforts to protect and restore the environment. MOIL, MSTC, and KIOCL follow only two guidelines: ISO Certificate 14,001 and BRR (NVG

Principles 2 and 6). RINL is a company that follows only the ISO 14001 certification. MECON doesn't follow any of the guidelines.

- NMDC Sustainability Report-2018

Analyzing content in Table 19.3 on environmental disclosure reveals that SAIL, RINL, NMDC, KIOCL, MOIL, MSTC, MECON, and FSNL have 312, 307, 330, 152, 145, 183, 136, and 108 total pages in their annual report, respectively, out of which the pages for environmental disclosure are 16.5, 4, 15.5, 10.5, 14, 2.5, 0.5, and 1 which is equivalent to the percentage of 5.29%, 1.30%, 4.70%, 6.90%, 9.65%, 0.01%, 0.37%, and 0.92% for SAIL, RINL, NMDC, KIOCL, MOIL, MSTC, MECON, and FSNL, respectively.

Table 19.4 is based on the different percentages of the contents within the dimension of environment that is issued by the companies in their annual report. This analysis reveals that the RINL is the highest content disclosure company in terms of environmental issues. It has the highest content regarding waste-related water and effluents, emissions, and waste-related issues. Apart from the RINL, NMDC, SAIL, KIOCL, MOIL, MSTC, MECON, and FSNL have content disclosure of 17%, 16%, 12.67%, 16.23%, 7.64%, 5.13%, and 7.4%, respectively.

The analysis of Table 19.5 reveals that all companies except the MSTC have some kind of contribution to the generation of renewable energy in their organization. SAIL has three dimensions of renewable energy generation, i.e. solar, wind, and hydraulic, whereas all other companies have only two dimensions of renewable energy generation, that is only solar and wind.

Table 19.6 represents the initiatives taken by the companies to keep the environment cool and free of air pollution. The company plants large numbers of saplings in the organizational area and in the nearby community. Here, some companies are not involved in the planting of saplings, so the reporting was not found in their annual report during the time period.

Table 19.7 gives the comparative analysis of seven public sector steel companies and reveals that a maximum of three guidelines are followed by SAIL, MOIL, and NMDC, whereas only two guidelines are followed by MSTC and KIOCL. The least guidelines are followed by RINL and MECON, i.e. 1 and 0. In the annual report disclosure, the highest percentage is covered by KIOCL, MOIL, and SAIL for environmental issues, and it is 6.90%, 9.65%, and 5.29%, respectively. For a more comprehensive

Table 19.3 Content analysis of environmental disclosure of different companies

Company	SAIL	RINL	NMDC	KIOCL	MOIL	MSTC	MECON	FSNL
Annual report 2021–2022	Number of pages in annual report							
No. of pages for environmental disclosure	16.5	4	15.5	10.5	14	2.5	0.5	1
Total no. of pages in the annual report	312	307	330	152	145	183	136	108
Percentage (%)	5.29	1.30	4.70	6.90	9.65	0.01	0.37	0.92
Total no. of pages in environmental and sustainability report	100	46[a], 47[a]	104[a]	–	Nil	Nil	Nil	Nil
Percentage (%)	100	100	100	Nil	Nil	Nil	Nil	Nil

Sources Compiled from sustainability, environment, and annual report
Note [a] Six-monthly compliance report in respect of conditions stipulated while issuing environmental clearance Period: Apr 21 to Sep 21 and Period: Oct 21 to Mar 22

Table 19.4 Content analysis of environmental aspects in different areas

Company aspect ↓	SAIL		RINL		NMDC		KIOCL		MOIL		MSTC		MECON		FSNL	
Materials	1[a]	1%	2[a]	2.15%	1[a]	0.96%	4.25[a]	2.80%	2[a]	1.37%	5.5[a]	3.00%	1.5[a]	1.10%	1[a]	0.93%
Energy	4[aa]	4%	8[a]	8.60%	3[a]	2.88%	4.25[a]	2.80%	7.5[a]	5.17%	4[a]	2.18%	3.5[a]	2.57%	2[a]	1.85%
Water and effluents	3[a]	3%	33[a]	35.48%	4[a]	3.85%	5[a]	3.29%	4.5[a]	3.10%	1.5[a]	0.81%	1[a]	0.73%	0.5[a]	0.46%
Biodiversity	2[a]	2%	7[a]	7.52%	3[a]	2.88%	1[a]	0.66%	1[a]	0.69%	0[a]	0%	0[a]	0%	0.5[a]	0.46%
Emission	3[a]	3%	25.5[a]	27.41%	3[a]	2.88%	2.75[a]	1.81%	3[a]	2.10%	1[a]	0.55%	1[a]	0.73%	0[a]	0%
Waste	3[a]	3%	19[a]	20.43%	4[a]	3.85%	2[a]	1.31%	5.5[a]	3.80%	2[a]	1.10%	0[a]	0%	4[a]	3.70%
Total value	16[a]	16%	94.5[a]	101.59%	18[a]	17.3%	19.25[a]	12.67%	23.5[a]	16.23%	14[a]	7.64%	7[a]	5.13%	8[a]	7.4%

Sources Compiled from sustainability, environment, and annual report
[a] Number of pages disclosed in annual report

Table 19.5 Renewable sources of energy initiation by companies

Company	Renewable sources of energy
SAIL	1 MW State Electricity Board at RSP 2 MW rooftop solar units at the BSL plant 0.360 MW rooftop solar power units on guest house and hospital building of BSP and ISP, respectively 10 MW hydel power plant at Mandira Dam of RSP 6.195 MW rooftop solar units on different buildings of SAIL plants
RINL	5 MW solar power PV plant has generated 3406 MWH of solar energy during the year 2021–2022
NMDC	1 MW capacity grid-connected rooftop solar (RTS) power plants are installed in various production units of NMDC 425 kWp roof top solar power plant in Bacheli Complex, Dantewada
KIOCL	5 MW solar power plant at Tumkur Dist., Karnataka 1.3MWp (1 MW AC) ground-mounted solar power plant Wind energy (non-solar) is purchased through the non-captive wheeling route for utilization in the Pellet plant. 66 lakh units of wind energy is used for FY 2021–2022
MOIL	Wind farms of 4.8 MW and 15.2 MW are situated at Nagda Hills and Ratedi Hills, respectively 5.0 MW solar power plant in Maharashtra and a 5.5 MW capacity solar plant in Madhya Pradesh
MSTC	NIL
FSNL	15 KW solar power unit at its corporate office
MECON	140 kW of solar energy power generating units have been installed, of which 60 kWp solar energy power units have been commissioned in FY 2021–2022

Sources Compiled from sustainability, environment, and annual report

disclosure of the environmental concern, three companies named SAIL, RINL, and NMDC have their sustainability report along with the annual report and disclosure in the annual report. All the companies are concerned with 5 to 6 environmental areas for sustainable development, and these are emission, air, water, waste management, biodiversity, and renewable energy.

Table 19.6 Plantation saplings were driven by different companies

Company	Plantation
SAIL	3.28 lakh saplings
RINL	7881 number of saplings planted from Oct'2021 to Mar 2022 10,618 saplings from Apr 2021 to Sep 2021
NMDC	113,650 saplings
KIOCL	200 saplings
MOIL	22.08 lakhs total saplings planted
MSTC	Nil
FSNL	Nil
MECON	Nil

Sources Compiled from sustainability, environment, and annual report

Table 19.7 Comparative analysis of the companies

Company	SAIL	RINL	NMDC	MOIL	MSTC	KIOCL	MECON	FSNL
No. of guidelines followed	3	1	3	3	2	2	0	1
% of disclosure in the annual report	5.29	1.30	4.70	9.65	0.01	6.90	0.37	0.92
% of disclosure in the environmental and sustainability report	100%	100%	100%	Nil	Nil	Nil	Nil	Nil
No. of content in environmental areas	6	6	6	6	5	6	4	5
Control on emission	19.555 MW	5 MW	1.425 MW	6.3 MW	30.5 MW	NIL	140KW	15KW
Plantation drive	3.28 lakh saplings	18,499 saplings	113,650 saplings	22.08 lakh sapling	200 saplings	Nil	Nil	Nil
Waste management	95% utilization	90.12% utilization	2883.901 (nos.) generated	Nil	Nil	Nil	Nil	Nil

Sources Compiled from sustainability, environment, and annual report

19.4 Conclusion

The extensive content analysis of the annual report of Indian public sector steel companies finds that companies are concerned about environmental degradation and are doing their best to protect against this degradation. The air pollution is treated by the plantation of saplings and the water pollution is treated by the effective waste management of hazardous elements generated during production.

Companies are concerned with the regeneration of renewable energy in the form of solar energy and wind energy and the development of hydraulic power plants. The companies have many ways to disclose the aspects of environmental concern, but many companies follow different guidelines or standards such as GRI standard 2016, ISO certification, BRR's National

Voluntary Guidelines on Social, Environmental and Economic Responsibilities of Business (NVGs), released by the Ministry of Corporate Affairs in July 2011, and corporate responsibility for environmental protection. For the effective representation of CSR activity, some companies have separate sustainability reports along with their annual report, and this shows the environmental and sustainability concerns of the companies. It can be concluded that most of the public sector companies are concerned about the environment and its degradation and are doing positive work for the same.

References

Banerjee SB (2007) Corporate social responsibility: the good, the bad and the ugly. Crit Sociol 33(1):51–79

Boiral O (2013) Sustainability reports as simulacra? a counter-account of A and A+ GRI reports. Account, Audit Account J 26(7):1036–1071. https://doi.org/10.1108/AAAJ-04-2012-00998

Boiral O, Gendron Y (2011) Sustainability reports as simulacra? a counter-account of A and A+ GRI reports. Account, Audit & Accountab J 24(8):1037–1070

Eccles RG, Serafeim G (2013) The performance frontier: innovating for a sustainable strategy. Harv Bus Rev 91 (5):50–60

Gao J, Greenberg R, Wong-On-Wing B, Wong M (2015) Corporate social responsibility and stock price crash risk. J Bank Finance 50:479–493

Godha A, Jain P (2015) Sustainability reporting trend in Indian companies as per GRI framework: a comparative study. South Asian J Bus Manage Cases 4(1):62–73. https://doi.org/10.1177/2277977915574040 http://bmc.sagepub.com

Jabbour CJC, Teixeira AA, Jabbour ABLS (2019) Environmental management and firms' competitiveness in Brazil: the moderating role of innovation. J Clean Prod 207:863–874

KPMG (2019) Sustainable development goals: are you ready? Retrieved from https://home.kpmg/content/dam/kpmg/xx/pdf/2019/07/sdg-report.pdf

Krippendorff K (1980) Content analysis: an introduction to its methodology, London: Sage

Lee SH, Lim SH (2017) The impact of CSR practices on corporate environmental performance. Sustainability 9 (9):1624

Lozano R (2011) The state of sustainability reporting in universities. Int J Sustain High Educ 12(1):67–78

Lyon TP, Maxwell JW (2007) Corporate social responsibility and the environment: a theoretical perspective

Mishra L (2020). Corporate social responsibility and sustainable development goals: a study of Indian companies J Public Affairs. https://doi.org/10.1002/pa.2147

Montiel I, Delgado-Ceballos J (2016) Corporate social responsibility and stakeholder engagement: a review and critique. J Bus Ethics 138(3):429–443

Raman SR (2006) Corporate social reporting in India—A view from the top. Glob Bus Rev 7:313–324

Saeed S, Saeed S, Akhtar S (2018) Corporate social responsibility and environmental sustainability: an empirical investigation in the Pakistani context. J Clean Prod 194:421–430

United Nations Global Compact (UNGC) (2019) Global corporate sustainability report 2019. Retrieved from https://www.unglobalcompact.org/library/5876

Zaman K, Ali MA (2018) Corporate social responsibility and environmental sustainability practices in Bangladeshi organizations: a study on selected private organizations. J Clean Prod 204:80–89

20 Rethinking Environmental Risk and Resilience: Embracing Geospatial and AI Innovations for a Changing World

Swapan Talukdar, Atiqur Rahman, Somnath Bera, G. V. Ramana, and Atish Prashar

Abstract

The Earth's ecosystems are facing unprecedented challenges due to anthropogenic activities and climate change, threatening natural environments worldwide. Loss and degradation of ecosystems have significant implications for environmental risk and resilience, impacting both local communities and global biodiversity and climate systems. Recent scientific research has made valuable contributions to understanding environmental risks and developing resilience strategies. Remote sensing and GIS techniques have been instrumental in monitoring environmental risks in various regions. Moreover, artificial intelligence (AI) has emerged as a powerful tool to address environmental challenges effectively, with applications in agriculture, ecological monitoring, and climate change solutions. The concept of nature-based solutions (NBS) has gained attention, promoting the integration of nature into urban planning to address environmental challenges. Despite the progress made through AI and NBS, challenges such as data accessibility, ethical considerations, and local ecological factors remain. This article synthesizes recent scientific contributions on environmental risk and resilience, exploring the growing role of AI and NBS in addressing challenges. It emphasizes the potential of AI and NBS in mitigating risks and enhancing resilience while acknowledging the need for comprehensive exploration and integration for sustainable environmental solutions. Collaborative action and global commitment are essential to face the complex environmental issues and pave the way for a more sustainable tomorrow.

Keywords

Environmental risk · Resilience · Geospatial · AI innovations · Changing world · Nature-based solutions (NBS)

S. Talukdar
Department of Geography, Asutosh College, University of Calcutta, Kolkata, India

A. Rahman
Department of Geography, Faculty of Natural Sciences, Jamia Millia Islamia, New Delhi 110025, India

S. Bera (✉)
Department of Geography, Central University of South Bihar, Gaya 824236, India
e-mail: somnath@cusb.ac.in

G. V. Ramana
Department of Civil Engineering, Indian Institute of Technology Delhi, New Delhi 110016, India

A. Prashar
Department of Mass Communication and Media, Central University of South Bihar, Gaya 824236, India

S. Talukdar et al. (eds.), *Environmental Risk and Resilience in the Changing World*, GIScience and Geo-environmental Modelling, https://doi.org/10.1007/978-3-031-62442-1_20

20.1 Introduction

The delicate balance of the Earth's ecosystems is facing unprecedented challenges due to anthropogenic activities and climate change. As urbanization and development expand rapidly worldwide, natural environments, including wetlands, coastal areas, and riverine landscapes are under increasing threat (Li et al. 2010; Wu et al. 2017; Allouche et al. 2021). The loss and degradation of these ecosystems have significant implications for environmental risk and resilience, impacting not only local communities but also global biodiversity and climate systems.

Recent scientific research has been instrumental in understanding the environmental risks and developing strategies for building resilience in the face of these challenges. Studies focusing on urban communities in Indonesia (Hawken and Sunindijo 2018), wetland ecological risk assessment in China's coastal zones (Xiong et al. n.d.), and historical wetland degradation evaluation (Hu et al. 2020) have provided valuable insights into the complexity of environmental issues. Utilizing remote sensing and GIS techniques, researchers have successfully monitored environmental risks in various regions, including Egypt (Elfadaly et al. 2018), Vietnam (Quang et al. 2021), and Mexico (Scullion et al. 2011). Furthermore, the integration of remote sensing, GIS, and geophysical-based approaches has enhanced the understanding of coastal environmental hazards (Youssef et al. 2021) and riverine landscape dynamics (Leuven and Poudevigne 2002).

As technology continues to advance, artificial intelligence (AI) has emerged as a powerful tool to address environmental challenges effectively. AI has proven to be instrumental in various fields, and its application in environmental research is gaining momentum. Recent studies have showcased the potential of AI to improve the resilience of agriculture production systems (Jung et al. 2021) and to monitor ecological spatial–temporal changes (Zhu et al. 2021). Chen et al. (2023) present a comprehensive review of AI-based solutions for climate change, underlining the growing relevance of AI in tackling environmental problems.

The concept of nature-based solutions (NBS) has garnered attention in recent years, promoting the integration of nature into urban and landscape planning to address social, economic, and environmental challenges (Lafortezza et al. 2018). NBS offer promising approaches to mitigate environmental risks and enhance resilience by restoring ecosystems and their functions (Seddon et al. 2020). Eggermont et al. (2015) emphasize the influence of NBS in environmental management and research in Europe, while Liu et al. (2022a, b) highlight their role in accommodating urban expansion through ecosystem services integration.

Despite the progress made through AI and NBS, there is still a pressing need for a deeper understanding of their capabilities and limitations. Integrating AI into environmental research has the potential to enhance data analysis, prediction, and decision-making processes (Nishant et al. 2020). However, there are challenges that must be addressed, such as data accessibility, ethical considerations, and the potential for bias in AI algorithms (Chen et al. 2023).

Similarly, while NBS offer innovative approaches, the development and implementation of these solutions require careful consideration of local ecological, social, and economic factors (Faivre et al. 2017). Assessing the effectiveness of NBS and ensuring their long-term sustainability are crucial elements in their successful deployment (Debele et al. 2019).

This article aims to synthesize recent scientific contributions on environmental risk and resilience, exploring the growing role of AI and NBS in addressing environmental challenges. By critically evaluating the existing research and exploring new avenues for future investigations, this article seeks to contribute to the ongoing efforts to protect and restore Earth's ecosystems.

Therefore, the integration of AI and NBS presents exciting opportunities to mitigate environmental risks and enhance resilience. The knowledge gained from recent studies utilizing remote sensing and GIS has deepened our understanding of the complexities of environmental challenges. However, there is still much to be explored in the application of AI and the implementation of NBS to achieve sustainable environmental solutions. By harnessing the potential of AI and the restorative power of nature, we can pave the way for a more resilient and harmonious coexistence with the natural world.

20.2 Recent Trends in Environmental Risk and Resilience Research and Limitations

Environmental risk and resilience research have become increasingly important in the context of global challenges such as climate change, pollution, and ecosystem degradation. Recent advancements in remote sensing technologies, AI, and data analysis have significantly enhanced our ability to assess and monitor environmental risks, as well as to develop adaptive strategies for building resilience. This article explores recent trends in environmental risk and resilience research, highlighting key studies and their limitations.

Over the past decade, the scientific community has made notable progress in understanding the complex interactions between environmental processes and human activities. Jha et al. (2023) employed a machine learning approach to assess the implications of climate risk factors on agriculture in India, providing valuable insights for climate adaptation strategies. Additionally, Jiang (2021) used linear regression analysis to analyze the air quality index in Shanghai, demonstrating the potential of data analysis in understanding urban air pollution dynamics.

Remote sensing has emerged as a powerful tool for environmental monitoring and assessment. Studies like those by Ju et al. (2022) and Kacic and Kuenzer (2022) have utilized remote sensing to monitor air pollution and biodiversity in urban and forest environments, respectively. Furthermore, Jung et al. (2021) highlighted the potential of remote sensing and AI as tools to improve the resilience of agriculture production systems.

Understanding the impacts of climate change on human health is crucial, and studies such as Kim et al. (2014) have reviewed the consequences of global climate change on human health, emphasizing the importance of addressing environmental risks in healthcare planning and policymaking. Additionally, Khan et al. (2021) and Khan and Amelie (2015) have assessed climate change readiness and ecosystem-based adaptation in Seychelles and South Asia, respectively.

Machine learning algorithms, such as support vector machines and random forests, have been increasingly utilized in environmental risk assessment. Khan et al. (2021) applied support vector machines to predict heatwaves over Pakistan, offering insights into climate change impacts. Madani et al. (2022) utilized random forests and logistic regression for predicting groundwater contamination based on ammonia concentration, highlighting the potential of AI in assessing water quality risks.

Despite the progress made in environmental risk research, there are still some limitations. One notable challenge lies in integrating local and external knowledge to develop effective ecosystem-based adaptation strategies (Mercer et al. 2012). Additionally, assessing the impact of urban growth on biodiversity remains an ongoing research gap (McDonald et al. 2020).

Monitoring environmental quality in urban areas is also complex due to data integration issues. Studies have shown that the combination of remote sensing and census data can help overcome some of these challenges (Musse et al. 2018). Furthermore, incorporating disruptive technologies into disaster risk management is an area that requires further exploration (Munawar et al. 2022).

To overcome the limitations of current research and enhance environmental risk and resilience efforts, nature-based solutions and advanced AI technologies present promising avenues. Nature-based solutions, such as drone-based mapping for landslide measurements

(Koutalakis et al. 2021), play a pivotal role in assessing and addressing environmental challenges, fostering sustainable land and water management (Mahrad et al. 2020).

Advanced AI, including machine learning and XAI, offers innovative ways to analyze complex environmental data and improve prediction models (Madani et al. 2022). For example, Naderpour et al. (2021) utilized spatial deep neural networks for forest fire risk prediction, demonstrating the potential of AI in developing precise and reliable risk assessments. Moreover, the integration of AI with Internet of Things (IoT) platforms facilitates anomaly detection in environmental systems, aiding in climate control and management (Liu et al. 2020).

Therefore, recent trends in environmental risk and resilience research have embraced advancements in machine learning, remote sensing, and AI technologies. However, there are still challenges to overcome, including the integration of local knowledge, data limitations, and the need for comprehensive monitoring and detection techniques. To address these limitations, nature-based solutions and advanced AI technologies hold significant promise. Embracing these solutions can pave the way for more effective environmental risk assessment and resilience strategies, ultimately contributing to sustainable environmental management and conservation.

20.3 AI and XAI Innovations: Revolutionizing Environmental Risk Assessment and Prediction

AI and explainable artificial intelligence (XAI) innovations have revolutionized environmental risk assessment and prediction, offering transformative solutions to build environmental resilience and tackle various risks associated with human activities and climate change. By harnessing the power of machine learning and deep learning algorithms, these technologies have shown great promise in addressing complex environmental challenges.

One area where AI and XAI have demonstrated their potential is in sustainable agriculture. Binas et al. (2019) showcased the use of reinforcement learning for sustainable agriculture, enabling more efficient and environmentally friendly farming practices. AI-powered agricultural systems can optimize resource use, reduce waste, and enhance crop yields, contributing to food security and environmental sustainability.

Environmental risk assessment and prediction are essential for identifying and mitigating potential hazards. Machine learning models have been utilized to explore the mobility and environmental assessment of toxic elements in mining-associated solid wastes (Qi et al. 2023). By predicting the distribution and potential impacts of contaminants, these models support better decision-making for waste management and remediation efforts.

In the context of water quality, machine learning-based source analyses have been used to assess trace metals in surface water (Li et al. 2022a, b). Understanding the sources of pollutants is crucial for effective pollution control and environmental protection. AI-powered analyses provide valuable insights for targeted and efficient measures to mitigate environmental risks.

AI and XAI have also made significant contributions to environmental conservation and biodiversity preservation. Deep learning techniques have been applied to image classification for plant diseases, aiding in early detection and management of crop diseases (Negi et al. 2021). Furthermore, AI has been utilized to quantify carbon emissions in machine learning processes, allowing for carbon-conscious decision-making during model training and development (Lacoste et al. 2019).

The application of AI and XAI is not limited to terrestrial ecosystems but extends to marine environments as well. These technologies have been harnessed for marine oil spill detection, prediction, and vulnerability assessment (Shamsudeen 2020; Yekeen and Balogun 2020). Timely detection and accurate prediction of oil spills enable swift response actions, reducing their adverse impacts on marine ecosystems and coastal communities.

Furthermore, AI and XAI have found relevance in addressing public health challenges, such as predicting the spread of infectious diseases. For instance, Abdulkareem et al. (2021) demonstrated how machine learning models can predict COVID-19 based on environmental factors, contributing to early detection and response planning. This application showcases the potential of AI in supporting public health interventions.

Moreover, the aviation industry has leveraged AI and machine learning for environmental impact analysis, aiding in fuel consumption optimization and emission reduction (Gao and Mavris 2022). By applying these technologies to aviation operations, the industry can contribute to more sustainable air travel and reduce its carbon footprint.

The predictive capabilities of AI and XAI extend to natural disasters, such as wildfires and snow avalanches. These technologies have been employed for wildfire susceptibility mapping and prediction, enabling proactive measures for wildfire management and prevention (Abdollahi and Pradhan 2023). Similarly, XAI-based machine learning algorithms have been used for snow avalanche susceptibility mapping, enhancing risk assessment in mountainous regions prone to avalanches (Iban and Bilgilioglu 2023).

To fully harness the potential of AI and XAI in building environmental resilience, the integration of these technologies into decision-making processes is essential. Researchers have proposed novel frameworks for spatiotemporal prediction of environmental data using deep learning (Amato et al. 2020). Such approaches facilitate evidence-based decision-making, allowing policymakers and stakeholders to make informed choices to enhance environmental resilience.

Therefore, AI and XAI innovations have transformed environmental risk assessment and prediction, offering transformative solutions to build environmental resilience. These technologies have demonstrated their potential in various fields, including sustainable agriculture, pollution control, biodiversity conservation, public health, and disaster risk management. The integration of AI and XAI into decision-making processes can lead to more sustainable and resilient environmental practices, helping society tackle the challenges of a changing world.

20.4 Nature-Based Solutions: Pioneering Sustainable Pathways for Environmental Risk Management

Nature-based solutions have emerged as a pioneering approach to building environmental resilience and tackling various risks associated with urbanization and climate change. NBS refer to interventions that utilize natural elements, processes, and ecosystem services to address environmental challenges effectively (Sowińska-Świerkosz and García 2022). This scientific review explores the role of NBS in enhancing environmental resilience and examines how these solutions can contribute to mitigating risks and promoting sustainable pathways for urban development.

Urbanization and climate change pose significant threats to ecosystems and human well-being. In response, NBS have gained increasing attention as a valuable strategy for adapting to and mitigating the impacts of environmental risks. Frantzeskaki et al. (2019) emphasized the importance of linking science, policy, and practice communities to facilitate evidence-based decision-making for implementing NBS. Furthermore, Escobedo et al. (2019) highlighted the multiple benefits of urban forests and green infrastructure as nature-based solutions, serving as essential components of resilient cities.

One of the critical features of NBS is their ability to integrate natural elements into urban landscapes, thereby providing multifunctional benefits for both people and ecosystems (Pontee et al. 2016). Girardin et al. (2021) underscored that NBS have the potential to contribute to global cooling if timely action is taken. These solutions can mitigate urban heat island effects and reduce the impacts of extreme weather events, making cities more resilient to climate change (Bush and Doyon 2019). For instance,

Song (2022) demonstrated how NBS, such as green roofs and permeable pavements, are integral components of China's sponge city initiative, which aims to enhance flood resilience.

Moreover, NBS offer a promising avenue for citizen engagement in environmental governance and decision-making processes (Kiss et al. 2022). By involving citizens in the planning and implementation of NBS projects, local communities can develop a sense of ownership and responsibility, leading to more sustainable outcomes (Kolokotsa et al. 2020). However, to fully capitalize on the potential of citizen participation, it is essential to address structural barriers and ensure inclusivity in governance (Dorst et al. 2022).

Integrated valuation of NBS is crucial for understanding their full range of benefits. Liquete et al. (2016) highlighted the hidden benefits of NBS for water pollution control, which go beyond conventional economic assessments. By considering the broader spectrum of ecosystem services provided by NBS, decision-makers can make more informed choices that balance environmental, social, and economic aspects of sustainability (Lafortezza et al. 2018). This holistic approach also enhances the chances of success and acceptance of NBS projects by stakeholders (Artmann and Sartison 2018).

To fully realize the potential of NBS, transitional pathways are needed to facilitate their adoption and mainstreaming (Davies and Lafortezza 2019). Policymakers must create conducive regulatory frameworks and foster collaboration among various stakeholders, including local communities, scientists, and practitioners (Davies et al. 2021). Bridging the gap between research and practice is essential for successful implementation.

Assessing the effectiveness of NBS requires a transformative change in the way they are planned and managed (Palomo et al. 2021). Traditional approaches that focus solely on technological solutions may not fully capitalize on the synergies between human and natural systems. By adopting a transformative perspective, cities can reconfigure themselves to become more adaptive, resilient, and sustainable (Somarakis et al. 2019).

The evidence for the effectiveness of NBS is not limited to a specific region, as exemplified by Acreman et al. (2021) in their study of NBS addressing water issues in Africa. These solutions offer promising pathways for addressing environmental risks in diverse contexts, provided they are contextually adapted and co-designed with local communities (Frantzeskaki 2019).

Nature-based solutions have emerged as a powerful and transformative approach to building environmental resilience and managing risks in the face of urbanization and climate change. By integrating natural elements and ecosystem services into urban landscapes, NBS offer multifunctional benefits that enhance the adaptability and sustainability of cities. Citizen participation, integrated valuation, and transformative thinking are essential components of successful NBS projects. To fully realize the potential of NBS, a shift in planning and governance approaches is necessary, ensuring the engagement of diverse stakeholders in a collaborative and inclusive manner. As the world grapples with the challenges of climate change and urbanization, the adoption of NBS provides a promising pathway to a more resilient and sustainable future.

20.5 Synergy and Integration: Where Nature Meets AI for Enhanced Resilience

The synergy and integration of nature-based solutions (NBS) with AI and XAI innovations offer a powerful approach to enhance environmental resilience and tackle various risks in the face of climate change and human activities. Nature-based solutions, as demonstrated in various studies, encompass a wide range of practices that utilize natural processes and ecosystems to address environmental challenges. For instance, green roofs have been identified as a nature-based solution to improve urban sustainability, offering benefits such as stormwater management, temperature regulation, and biodiversity promotion (Mihalakakou et al. 2023). Furthermore, NBS play a vital role in global climate adaptation, as highlighted by Diep and McPhearson (2022), showcasing their potential to enhance ecosystem resilience and provide multiple co-benefits for both nature and society.

AI and XAI technologies, on the other hand, have shown great promise in revolutionizing environmental risk assessment and prediction. The use of machine learning and deep learning algorithms has facilitated the exploration of complex environmental data, enabling the identification of patterns, trends, and potential risks. AI-powered models have been employed for diverse applications, including the detection of marine oil spills, wildfire spread prediction, and assessment of soil erosion susceptibility (Shamsudeen 2020; Abdollahi and Pradhan 2023; Alqadhi et al. 2023). Moreover, explainable AI has emerged as a crucial factor in building trust and transparency, providing insights into the decision-making process of AI models and enabling effective integration into environmental management practices (Javed et al. 2023).

The integration of NBS with AI and XAI opens up new avenues for enhancing environmental resilience by optimizing the potential of both approaches. A hybrid expert system that combines NBS case-based knowledge with AI techniques has been proposed, allowing for more efficient decision-making and adaptive management in environmental challenges (Sarabi et al. 2022). The synergistic approach enables the fusion of domain-specific knowledge and advanced data-driven analysis, thereby strengthening the resilience of ecosystems and communities.

One prominent area where synergy and integration have proven valuable is in climate change mitigation and adaptation strategies. Nature-based and technology-based solutions can complement each other to achieve sustainable blue growth and marine biodiversity conservation (Mustafa et al. 2019). By leveraging AI to analyze complex ecological interactions and predict environmental changes, decision-makers can enhance the design and implementation of NBS interventions.

The integration of NBS and AI in smart cities is another promising application. A survey by Ruckelshaus et al. (2020) identified new data technologies as instrumental in assessing and managing risk in coastal zones. AI-driven data analytics and predictive modeling can provide valuable insights to optimize the planning and implementation of NBS for coastal resilience, such as natural coastal buffers and wetland restoration.

Furthermore, the application of AI and XAI to evaluate the effectiveness of NBS is crucial for evidence-based decision-making. Type-II fuzzy approaches with explainable AI have been used to select nature-based leisure travel destinations amid the COVID-19 pandemic (Lin and Chen 2022). The combination of these methodologies enables a comprehensive evaluation of risk and benefits associated with NBS, aiding in the identification of suitable interventions to build resilience against environmental hazards.

The synthesis of NBS and AI can also drive advancements in disaster risk reduction and response strategies. For example, in the context of flood risk reduction, Liu et al. (2023) assessed the suitability of small dams as NBS in the Solomon Islands, using AI techniques to identify suitable locations. The integration of NBS and AI allows for more effective and context-specific approaches to disaster management and resilience building.

To achieve the full potential of synergy and integration, the collaboration among environmental scientists, engineers, urban planners, and AI experts is essential. The cooperation of multidisciplinary teams can lead to innovative solutions that address complex environmental challenges more effectively. By harnessing the power of AI to process and interpret large-scale environmental data, NBS can be optimized and adapted to meet the diverse and dynamic needs of communities and ecosystems.

Therefore, the synergy and integration of nature-based solutions with AI and XAI innovations offer a transformative approach to building environmental resilience and addressing various environmental risks. This integrated approach enables the optimization of NBS interventions, enhances climate change adaptation, improves disaster risk reduction strategies, and supports evidence-based decision-making for sustainable environmental management. The collaboration between diverse disciplines and the continuous development of advanced AI technologies will be critical to harnessing the full potential of this synergy in creating a more resilient and sustainable future.

20.6 Beyond Boundaries: Collaborative Action for a Sustainable Tomorrow

We promote the paramount importance of collaborative action and a united global commitment to address the complex issues of environmental risk and resilience. It is a call to arms for governments, policymakers, scientists, businesses, and communities to come together and work as one in the pursuit of a sustainable tomorrow. We recognize that environmental challenges know no boundaries; they transcend national borders and affect us all. Therefore, we advocate for knowledge-sharing, capacity-building, and international cooperation to amplify the impact of our collective efforts.

Collaboration is the key to unlocking solutions to the pressing environmental issues we face. By pooling our expertise, resources, and innovative ideas, we can create a powerful force for positive change. Governments must collaborate on policies and regulations that promote sustainable practices and protect natural ecosystems. Policymakers must listen to scientific experts and integrate evidence-based solutions into their decision-making processes. Scientists must engage in cross-disciplinary collaboration to understand the complex interplay between human activities and the environment fully. Businesses must adopt sustainable practices and invest in green technologies to reduce their environmental footprint.

Moreover, communities play a crucial role in building resilience and adapting to environmental challenges. Local knowledge and indigenous wisdom are invaluable resources that can inform sustainable practices and strengthen community resilience. We must empower communities to be active participants in the design and implementation of environmental initiatives, ensuring that their voices are heard and their needs are addressed.

Therefore, the call for collaborative action and a global commitment to sustainability goes beyond rhetoric; it is a rallying cry for meaningful change. It is a reminder that we share a common destiny and must work together to protect the planet for current and future generations. By uniting our efforts and resources, we can overcome environmental risks, build resilience, and create a sustainable tomorrow for all. Let us rise to the occasion, break down barriers, and collaborate wholeheartedly for the betterment of our planet and the well-being of all its inhabitants.

References

Abdollahi A, Pradhan B (2023) Explainable artificial intelligence (XAI) for interpreting the contributing factors feed into the wildfire susceptibility prediction model. Sci Total Environ 879:163004

Abdulkareem AB, Sani NS, Sahran S, Alyessari ZAA, Adam A, Abd Rahman AH, Abdulkarem AB (2021) Predicting COVID-19 based on environmental factors with machine learning. Intell Autom Soft Comput 28:305–320

Acreman M, Smith A, Charters L, Tickner D, Opperman J, Acreman S, Edwards F, Sayers P, Chivava F (2021) Evidence for the effectiveness of nature-based solutions to water issues in Africa. Environ Res Lett 16 (6):063007

Allouche KF, Gad A, Negm AM (2021) Conclusions and recommendations for "environmental remote sensing and GIS in Tunisia. Environ Remote Sens GIS in Tunisia, 351–362

Alqadhi S, Mallick J, Talukdar S, Alkahtani M (2023) An artificial intelligence-based assessment of soil erosion probability indices and contributing factors in the Abha-Khamis watershed, Saudi Arabia. Front Ecol Evol 11:1189184

Amato F, Guignard F, Robert S, Kanevski M (2020) A novel framework for spatio-temporal prediction of environmental data using deep learning. Sci Rep 10 (1):22243

Artmann M, Sartison K (2018) The role of urban agriculture as a nature-based solution: a review for developing a systemic assessment framework. Sustainability 10(6):1937

Binas J, Luginbuehl L, Bengio Y (2019) Reinforcement learning for sustainable agriculture. In: ICML 2019 Workshop Climate Change: How Can AI Help, June

Bush J, Doyon A (2019) Building urban resilience with nature-based solutions: How can urban planning contribute? Cities 95:102483

Chen L, Chen Z, Zhang Y, Liu Y, Osman AI, Farghali M, Hua J, Al-Fatesh A, Ihara I, Rooney DW, Yap PS (2023) Artificial intelligence-based solutions for climate change: a review. Environ Chem Lett, 1–33

Davies C, Lafortezza R (2019) Transitional path to the adoption of nature-based solutions. Land Use Policy 80:406–409

Davies C, Chen WY, Sanesi G, Lafortezza R (2021) The European Union roadmap for implementing nature-based solutions: a review. Environ Sci Policy 121:49–67

Debele SE, Kumar P, Sahani J, Marti-Cardona B, Mickovski SB, Leo LS, Porcù F, Bertini F, Montesi D, Vojinovic Z, Di Sabatino S (2019) Nature-based solutions for hydro-meteorological hazards: revised concepts, classification schemes and databases. Environ Res 179:108799

Diep L, McPhearson T (2022) Nature-based solutions for global climate adaptation. Nature 606(7915)

Dorst H, van der Jagt A, Toxopeus H, Tozer L, Raven R, Runhaar H (2022) What's behind the barriers? uncovering structural conditions working against urban nature-based solutions. Landsc Urban Plan 220:104335

Eggermont H, Balian E, Azevedo JMN, Beumer V, Brodin T, Claudet J, Fady B, Grube M, Keune H, Lamarque P, Reuter K (2015) Nature-based solutions: new influence for environmental management and research in Europe. GAIA-Ecol Perspect Sci Soc 24 (4):243–248

Elfadaly A, Attia W, Lasaponara R (2018) Monitoring the environmental risks around Medinet Habu and Ramesseum Temple at West Luxor, Egypt, using remote sensing and GIS techniques. J Archaeol Method Theory 25:587–610

Escobedo FJ, Giannico V, Jim CY, Sanesi G, Lafortezza R (2019) Urban forests, ecosystem services, green infrastructure and nature-based solutions: Nexus or evolving metaphors? Urban Forest Urban Green 37:3–12

Faivre N, Fritz M, Freitas T, De Boissezon B, Vandewoestijne S (2017) Nature-Based Solutions in the EU: innovating with nature to address social, economic and environmental challenges. Environ Res 159:509–518

Frantzeskaki N (2019) Seven lessons for planning nature-based solutions in cities. Environ Sci Policy 93:101–111

Frantzeskaki N, McPhearson T, Collier MJ, Kendal D, Bulkeley H, Dumitru A, Walsh C, Noble K, Van Wyk E, Ordóñez C, Oke C (2019) Nature-based solutions for urban climate change adaptation: linking science, policy, and practice communities for evidence-based decision-making. Bioscience 69 (6):455–466

Gao Z, Mavris DN (2022) Statistics and machine learning in aviation environmental impact analysis: a survey of recent progress. Aerospace 9(12):750

Girardin CA, Jenkins S, Seddon N, Allen M, Lewis SL, Wheeler CE, Griscom BW, Malhi Y (2021) Nature-based solutions can help cool the planet—if we act now. Nature 593(7858):191–194

Hawken S, Sunindijo RY (2018) City of Kampung: risk and resilience in the urban communities of Surabaya, Indonesia. Int J Build Pathol Adapt 36(5):543–568

Hu T, Liu J, Zheng G, Zhang D, Huang K (2020) Evaluation of historical and future wetland degradation using remote sensing imagery and land use modeling. Land Degrad Dev 31(1):65–80

Iban MC, Bilgilioglu SS (2023) Snow avalanche susceptibility mapping using novel tree-based machine learning algorithms (XGBoost, NGBoost, and LightGBM) with eXplainable Artificial Intelligence (XAI) approach. Stoch Env Res Risk Assess 37 (6):2243–2270

Javed AR, Ahmed W, Pandya S, Maddikunta PKR, Alazab M, Gadekallu TR (2023) A survey of explainable artificial intelligence for smart cities. Electronics 12(4):1020

Jha P, Chinngaihlian S, Upreti P, Handa A (2023) A machine learning approach to assess implications of climate risk factors on agriculture: the Indian case. Clim Risk Manag 41:100523. https://doi.org/10.1016/j.crm.2023.100523

Jiang W (2021) The data analysis of Shanghai air quality index based on linear regression analysis. J Phys: Conf Series

Ju T, Liang Z, Liu W, Li B, Huang R, Geng T (2022) Monitoring of air pollution by remote sensing in Lanzhou City from 2010 to 2019. Water Air Soil Pollut 233(9):359

Jung J, Maeda M, Chang A, Bhandari M, Ashapure A, Landivar-Bowles J (2021) The potential of remote sensing and artificial intelligence as tools to improve the resilience of agriculture production systems. Curr Opin Biotechnol 70:15–22

Kacic P, Kuenzer C (2022) Forest biodiversity monitoring based on remotely sensed spectral diversity—a review. Remote Sens 14(21):5363

Khan A, Amelie V (2015) Assessing climate change readiness in Seychelles: implications for ecosystem-based adaptation mainstreaming and marine spatial planning. Reg Environ Change 15:721–733

Khan N, Shahid S, Bin IT, Behlil F (2021) Prediction of heat waves over Pakistan using support vector machine algorithm in the context of climate change. Stoch Environ Res Risk Assess. https://doi.org/10.1007/s00477-020-01963-1

Kim KH, Kabir E, Ara Jahan S (2014) A review of the consequences of global climate change on human health. J Environ Sci Health C 32(3):299–318

Kiss B, Sekulova F, Hörschelmann K, Salk CF, Takahashi W, Wamsler C (2022) Citizen participation in the governance of nature-based solutions. Environ Policy Gov 32(3):247–272

Kolokotsa D, Lilli AA, Lilli MA, Nikolaidis NP (2020) On the impact of nature-based solutions on citizens' health & well being. Energy and Buildings 229:110527

Koutalakis PD, Tzoraki OA, Prazioutis GI, Gkiatas GT, Zaimes GN (2021) Can drones map earth cracks? landslide measurements in north Greece using UAV photogrammetry for nature-based solutions. Sustainability 13(9):4697

Lacoste A, Luccioni A, Schmidt V, Dandres T (2019) Quantifying the carbon emissions of machine learning. arXiv:1910.09700

Lafortezza R, Chen J, Van Den Bosch CK, Randrup TB (2018) Nature-based solutions for resilient landscapes and cities. Environ Res 165:431–441

Lenzen M, Li M, Malik A, Pomponi F, Sun YY, Wiedmann T, Faturay F, Fry J, Gallego B, Geschke A, Gómez-Paredes J (2020) Global socio-economic losses and environmental gains from the Coronavirus pandemic. PLoS ONE 15(7):e0235654

Leuven RS, Poudevigne I (2002) Riverine landscape dynamics and ecological risk assessment. Freshw Biol 47(4):845–865

Li Y, Zhu X, Sun X, Wang F (2010) Landscape effects of environmental impact on bay-area wetlands under rapid urban expansion and development policy: a case study of Lianyungang, China. Lands Urban Plan 94 (3–4):218–227

Li H, Peng J, Jiao Y, Ai S (2022a) Experiencing urban green and blue spaces in urban wetlands as a nature-based solution to promote positive emotions. Forests 13(3):473

Li P, Hua P, Zhang J, Krebs P (2022b) Ecological risk and machine learning based source analyses of trace metals in typical surface water. Sci Total Environ 838:155944

Lin YC, Chen TCT (2022) Type-II fuzzy approach with explainable artificial intelligence for nature-based leisure travel destination selection amid the COVID-19 pandemic. Digital Heal 8:20552076221106320

Liquete C, Udias A, Conte G, Grizzetti B, Masi F (2016) Integrated valuation of a nature-based solution for water pollution control. Highlight Hidden Benefits. Ecosyst Serv 22:392–401

Liu Y, Pang Z, Karlsson M, Gong S (2020) Anomaly detection based on machine learning in IoT-based vertical plant wall for indoor climate control. Build Environ. https://doi.org/10.1016/j.buildenv.2020.107212

Liu J, Wang J, Xiong J et al (2022a) Assessment of flood susceptibility mapping using support vector machine, logistic regression and their ensemble techniques in the Belt and Road region. Geocarto Int. https://doi.org/10.1080/10106049.2022.2025918

Liu Y, Gao Y, Liu L, Song C, Ai D (2022b) Nature-based solutions for urban expansion: integrating ecosystem services into the delineation of growth boundaries. Habitat Int 124:102575

Liu J, Tan J, Zhang S, Zhong C, Lv L, Tara A (2023) Suitability Assessment of Small Dams' Location as Nature-Based Solutions to Reduce Flood Risk in Mataniko Catchment, Honiara, Solomon Islands. Sustainability 15(4):3313

Luz AEO, Negri RG, Massi KG, et al (2022) Mapping fire susceptibility in the Brazilian Amazon forests using multitemporal remote sensing and time-varying unsupervised anomaly detection. Remote Sens 14. https://doi.org/10.3390/rs14102429

Madani A, Hagage M, Elbeih SF (2022) Random forest and logistic regression algorithms for prediction of groundwater contamination using ammonia concentration. Arab J Geosci. https://doi.org/10.1007/s12517-022-10872-2

Mahrad BE, Newton A, Icely JD, Kacimi I, Abalansa S, Snoussi M (2020) Contribution of remote sensing technologies to a holistic coastal and marine environmental management framework: a review. Remote Sens 12(14):2313

McDonald RI, Mansur AV, Ascensão F, Colbert ML, Crossman K, Elmqvist T, Gonzalez A, Güneralp B, Haase D, Hamann M, Hillel O (2020) Research gaps in knowledge of the impact of urban growth on biodiversity. Nature Sustain 3(1):16–24

Mercer J, Kelman I, Alfthan B, Kurvits T (2012) Ecosystem-based adaptation to climate change in Caribbean small island developing states: integrating local and external knowledge. Sustainability 4 (8):1908–1932

Mihalakakou G, Souliotis M, Papadaki M, Menounou P, Dimopoulos P, Kolokotsa D, Paravantis JA, Tsangrassoulis A, Panaras G, Giannakopoulos E, Papaefthimiou S (2023) Green roofs as a nature-based solution for improving urban sustainability: progress and perspectives. Renew Sustain Energy Rev 180:113306

Munawar HS, Mojtahedi M, Hammad AW, Kouzani A, Mahmud MP (2022) Disruptive technologies as a solution for disaster risk management: A review. Sci Total Environ 806:151351

Musse MA, Barona DA, Rodriguez LMS (2018) Urban environmental quality assessment using remote sensing and census data. Int J Appl Earth Obs Geoinf 71:95–108

Mustafa S, Estim A, Tuzan AD, Ann CC, Seng LL, Shaleh SRM (2019) Nature-based and technology-based solutions for sustainable blue growth and climate change mitigation in marine biodiversity hotspots. Environ Biotech 15(1):1–7

Naderpour M, Rizeei HM, Ramezani F (2021) Forest fire risk prediction: a spatial deep neural network-based framework. Remote Sens 13(13):2513

Negi A, Kumar K, Chauhan P (2021) Deep neural network-based multi-class image classification for plant diseases. In: Agricultural informatics: Automation using the IoT and machine learning, pp 117–129

Nishant R, Kennedy M, Corbett J (2020) Artificial intelligence for sustainability: challenges, opportunities, and a research agenda. Int J Inf Manage 53:102104

Palomo I, Locatelli B, Otero I, Colloff M, Crouzat E, Cuni-Sanchez A, Gómez-Baggethun E, González-

García A, Grêt-Regamey A, Jiménez-Aceituno A, Martín-López B (2021) Assessing nature-based solutions for transformative change. One Earth 4(5):730–741
Pontee N, Narayan S, Beck MW, Hosking AH (2016) Nature-based solutions: lessons from around the world. In: Proceedings of the Institution of Civil Engineers-Maritime Engineering, vol 169, No. 1. Thomas Telford Ltd., pp 29–36
Qi C, Wu M, Liu H, Liang Y, Liu X, Lin Z (2023) Machine learning exploration of the mobility and environmental assessment of toxic elements in mining-associated solid wastes. J Clean Prod 401:136771
Quang DN, Linh NK, Tam HS, Viet NT (2021) Remote sensing applications for reservoir water level monitoring, sustainable water surface management, and environmental risks in Quang Nam province, Vietnam. J Water Climate Change 12(7):3045–3063
Ruckelshaus M, Reguero BG, Arkema K, Compean RG, Weekes K, Bailey A, Silver J (2020) Harnessing new data technologies for nature-based solutions in assessing and managing risk in coastal zones. Int J Disaster Risk Reduct 51:101795
Sarabi S, Han Q, de Vries B, Romme AGL, Almassy D (2022) The nature-based solutions case-based system: a hybrid expert system. J Environ Manage 324:116413
Scullion J, Thomas CW, Vogt KA, Perez-Maqueo O, Logsdon MG (2011) Evaluating the environmental impact of payments for ecosystem services in Coatepec (Mexico) using remote sensing and on-site interviews. Environ Conserv 38(4):426–434
Seddon N, Chausson A, Berry P, Girardin CA, Smith A, Turner B (2020) Understanding the value and limits of nature-based solutions to climate change and other global challenges. Philos Trans R Soc B 375 (1794):20190120
Shamsudeen TY (2020) Advances in remote sensing technology, machine learning and deep learning for marine oil spill detection,prediction and vulnerability assessment. Remote Sens 12(20):3416
Slovic P (1999) Trust, emotion, sex, politics, and science: surveying the risk-assessment battlefield. Risk Anal 19 (4):689–701
Somarakis G, Stagakis S, Chrysoulakis N, Mesimäki M, Lehvävirta S (2019) ThinkNature nature-based solutions handbook
Song C (2022) Application of nature-based measures in China's sponge city initiative: current trends and perspectives. Nat-Based Solut 2:100010
Sowińska-Świerkosz B, García J (2022) What are nature-based solutions (NBS)? Setting core ideas for concept clarification. Nat-Based Solut 2:100009
Wu WT, Zhou YX, Tian B (2017) Coastal wetlands facing climate change and anthropogenic activities: a remote sensing analysis and modelling application. Ocean Coast Manag 138:1–10
Xiong J, Li J, Huang J, Bu X, Yang X, Lin C (n.d.) A remote sensing-based approach for spatiotemporal refined assessment of freshwater wetland ecological risk in coastal zones: a case in Jiangsu's coastal zone. Land Degrad Develop
Yekeen ST, Balogun AL (2020) Advances in remote sensing technology, machine learning and deep learning for marine oil spill detection, prediction and vulnerability assessment. Remote Sens 12(20):1–31
Youssef YM, Gemail KS, Sugita M, AlBarqawy M, Teama MA, Koch M, Saada SA (2021) Natural and anthropogenic coastal environmental hazards: an integrated remote sensing, GIS, and geophysical-based approach. Surv Geophys, 1–33
Zhu D, Chen T, Wang Z, Niu R (2021) Detecting ecological spatial-temporal changes by remote sensing ecological index with local adaptability. J Environ Manage 299:113655

Index

S. Talukdar et al. (eds.), *Environmental Risk and Resilience in the Changing World*, GIScience and Geo-environmental Modelling, https://doi.org/10.1007/978-3-031-62442-1

GPSR Compliance

The European Union's (EU) General Product Safety Regulation (GPSR) is a set of rules that requires consumer products to be safe and our obligations to ensure this.

If you have any concerns about our products, you can contact us on ProductSafety@springernature.com

In case Publisher is established outside the EU, the EU authorized representative is:

Springer Nature Customer Service Center GmbH
Europaplatz 3
69115 Heidelberg, Germany

Batch number: 10755764

Printed by Printforce, the Netherlands